CONVERSATIONAL INFORMATICS
AN ENGINEERING APPROACH

Wiley Series in Agent Technology

Series Editor: Michael Wooldridge, *Liverpool University, UK*

The 'Wiley Series in Agent Technology' is a series of comprehensive practical guides and cutting-edge research titles on new developments in agent technologies. The series focuses on all aspects of developing agent-based applications, drawing from the Internet, telecommunications, and Artificial Intelligence communities with a strong applications/technologies focus.

The books will provide timely, accurate and reliable information about the state of the art to researchers and developers in the Telecommunications and Computing sectors.

Titles in the series:
Padgham/Winikoff: Developing Intelligent Agent Systems   0-470-86120-7 (June 2004)
Bellifemine/Caire/Greenwood: Developing Multi-Agent Systems with JADE
978-0-470-05747-6 (February 2007)
Bordini: Programming Multi-Agent Systems in AgentSpeak using *Jason*
978-0-470-02900-8 (September 2007)

CONVERSATIONAL INFORMATICS

AN ENGINEERING APPROACH

Edited by

Toyoaki Nishida
Kyoto University, Japan

John Wiley & Sons, Ltd

Other Wiley Editorial Offices

John Wiley & Sons Inc., 111 River Street, Hoboken, NJ 07030, USA

Jossey-Bass, 989 Market Street, San Francisco, CA 94103-1741, USA

Wiley-VCH Verlag GmbH, Boschstr. 12, D-69469 Weinheim, Germany

John Wiley & Sons Australia Ltd, 42 McDougall Street, Milton, Queensland 4064, Australia

John Wiley & Sons (Asia) Pte Ltd, 2 Clementi Loop #02-01, Jin Xing Distripark, Singapore 129809

John Wiley & Sons Canada Ltd, 6045 Freemont Blvd, Mississauga, ONT L5R 4J3, Canada

Wiley also publishes its books in a variety of electronic formats. Some content that appears in print may
not be available in electronic books.

Library of Congress Cataloguing-in-Publication Data

Nishida, T. (Toyoaki)
    Conversational informatics: an engineering approach / edited by Toyoaki Nishida.
        p.   cm.
    Includes index.
    ISBN 978-0-470-02699-1 (cloth)
    1. Conversation analysis.   2. Conversation analysis–Data processing.   3. Communication models.
    I. Title
    P95.45.N575 2007
    302.3'46—dc22   2007026311

British Library Cataloguing in Publication Data

A catalogue record for this book is available from the British Library

ISBN 978-0-470-02699-1 (HB)

Typeset in 9/11pt Times by Aptara, New Delhi, India.
Printed and bound in Great Britain by Antony Rowe Ltd, Chippenham, England.
This book is printed on acid-free paper responsibly manufactured from sustainable forestry
in which at least two trees are planted for each one used for paper production.

Contents

Preface

Conversation is the most natural and popular means for people to communicate with each other. Conversation is everywhere around us. Even though you may feel that making conversation is relatively effortless, look more closely and you will notice that a tremendous amount of sophisticated interaction is involved in initiating and sustaining conversation.

Conversational Informatics is a field of research that focuses on human conversational behavior as well as on the design of conversational artifacts that can interact with people in a conversational fashion. The field is based on a foundation provided by Artificial Intelligence, Natural Language Processing, Speech and Image Processing, Cognitive Science, and Conversation Analysis. It is aimed at shedding light on meaning, creation and interpretation resulting from the sophisticated mechanisms in verbal/nonverbal interactions during conversation, in search of better methods of computer-mediated communication, human–computer interaction, and support for knowledge creation.

The purpose of this book is to inform a broad audience about the major concerns and results of Conversational Informatics. It begins with an introductory tutorial followed by a collection of surveys describing a list of key questions on various topics in Conversational Informatics and major insights and results obtained in recent research. Because Conversational Informatics covers a wide field of research encompassing linguistics, psychology, and human–computer interaction, interdisciplinary approaches are highly important. Special emphasis is placed on engineering aspects that feature in recent, novel, technical developments such as conversational content acquisition, conversation environment design, and quantitative conversational modeling.

The main body of this book focuses on four subjects. The first of these is *conversational artifacts*. We look at how to build artifacts, such as synthetic characters on the computer screen or intelligent robots that can help the user by making conversation using not only natural language but also eye contact, facial expressions, gestures, or other nonverbal means of communication. The major technological contributions to this area are expressive, knowledgeable embodied conversation agents that can automatically produce emotional and socially acceptable communication behaviors in human–computer interaction, and conversational robots that are empowered by integrating perceptual and cognitive processing to align their behaviors during social interaction with that of the user and other intelligent agents.

The second subject is *conversational content*. We examine the development of a suite of techniques for acquiring, editing, distributing, and utilizing content that is both consumed and produced in a conversation. The major contributions to this subject include a theory called conversation quantization, which relates to a method of content management based on segmenting conversation into small pieces corresponding to plot units; the integration of natural language and computer vision techniques for translating and adapting written documents or videos into conversational content; and a method of video content acquisition and editing that enables intelligent capture and semantic annotation of meaningful scenes from conversations for later use.

The third subject is *conversation environment design*. We look at designing an intelligent environment that can sense conversational behaviors to either help participants become involved in collaboration even though they may be in a separate location, or record conversations together with their atmosphere for later use or review. This area involves the use of ubiquitous sensor rooms to measure and capture conversational behaviors; a technology called real-time human proxy that will produce a uniform virtual classroom by automatically sensing the nonverbal conversational behavior of participants in geographically separate places; and a lecture archiving system that can create a uniform audiovisual stream for a classroom, including both lecture and question-answering activities, by integrating an auditory and visual sensing system based on a theory called environmental media.

The last subject is *conversation measurement, analysis and modeling*. Motivated by scientific interest, we take a data-driven quantitative approach to understanding conversational behaviors by measuring these behaviors using advanced technologies and building detailed, quantitative models of various aspects of conversation. Contributions to this subject include an approach based on building and analyzing conversational corpora, various quantitative models of nonverbal communication and their effects on conversation atmosphere, a theory of verbal communication mechanism of the online communities, and a mutual adaptation model for human–robot conversational interactions.

The book not only describes specific details of new ideas and technologies being generated by Conversational Informatics, it also highlights new goals and challenges that deserve further research. The book will serve as an introduction to the field. The primary audience will be researchers and graduate students in computer science and psychology. It can be read by any person with an undergraduate-level background in computer science or cognitive psychology and should thus be suitable for junior-level graduate students and young researchers who wish to start on new research in artificial intelligence, human–computer interaction, and communication sciences. The secondary audience includes engineers and developers. Engineers in interface technology or knowledge management may also find it useful in designing agent-based interfaces or smart rooms.

I wish to express my sincere gratitude to Kateryna Tarasenko, who carefully read the manuscript. I would like to thank Birgit Gruber, Sarah Hinton, Richard Davies, Brett Wells, Anna Smart, and Rowan January from John Wiley & Sons, and Joanna Tootill for their understanding and continuous support for the production of the book.

Toyoaki Nishida

List of Contributors

Elisabeth André, Multimedia Concepts and Applications, Faculty of Applied Informatics, University of Augsburg, D-86159 Augsburg, Germany. e-mail: andre@informatik.uni-augsburg.de

Daisaku Arita, Laboratory 3, Institute of Systems & Information Technologies/KYUSHU, 2-1-22, Momochihama, Sawara-ku, Fukuoka 814-0001, Japan. e-mail: arita@isit.or.jp

Noboru Babaguchi, Department of Information and Communications Technology, Graduate School of Engineering, Osaka University, 2-1 Yamadaoka Suita, Osaka 565-0871, Japan. E-mail: babaguchi@comm.eng.osaka-u.ac.jp

Christian Becker, Artificial Intelligence Group, Faculty of Technology, University of Bielefeld, D-33594 Bielefeld, Germany. e-mail: cbecker@techfak.uni-bielefeld.de

Justine Cassell, EECS & Communication Studies, Northwestern University, 2240 Campus Drive, 2-148, Evanston, IL 60208. justine@northwestern.edu

Trevor Darrell, Computer Science and Artificial Intelligence Laboratory, Massachusetts Institute of Technology, Cambridge, MA 02139, USA. email: trevor@csail.mit.edu

Yasuharu Den, Dept. of Cognitive and Information Sciences, Faculty of Letters, Chiba University, 1-33 Yayoicho, Inage-ku, Chiba 263-8522, JAPAN. e-mail: den@cogsci.L.chiba-u.ac.jp

Mika Enomoto, Katayanagi Advanced Research Laboratories, Tokyo University of Technology, 1404-1 Katakura, Hachioji, Tokyo 192-0981, JAPAN. e-mail: menomoto@media.teu.ac.jp

Kimberley Ferriman, Department of Psychology and Human Development, Vanderbilt University, Peabody 512, Nashville, TN 37203. email: kim.ferriman@vanderbilt.edu

Nobuhiro Kaji, Institute of Industrial Science, the University of Tokyo, 4-6-1 Komaba, Meguro-ku, Tokyo, 153-8505, Japan. e-mail: kaji@tkl.iis.u-tokyo.ac.jp

Koh Kakusho, Academic Center for Computing and Media Studies, Kyoto University, Yoshida Nihonmatsu-cho, Sakyo-ku, Kyoto 606-8501, Japan. e-mail: kakusho@media.kyoto-u.ac.jp

Daisuke Kawahara, Knowledge Creating Communication Research Center, National Institute of Information and Communications Technology, 3-5 Hikaridai Seika-cho, Soraku-gun, Kyoto, 619-0289, Japan. e-mail: dk@nict.go.jp

Takanori Komatsu, School of Systems Information Science, Future University-Hakodate, 116-2 Kamedanakano, Hakodate 041-8655, Japan. email: komatsu@fun.ac.jp

Masashi Komori, Department of Engineering Informatics, Faculty of Information and Communication Engineering, Osaka Electro-Communication University, 18-8 Hatsucho, Neyagawa, Osaka 572-8530, Japan. e-mail: komori@isc.osakac.ac.jp

Stefan Kopp, Artificial Intelligence Group, Faculty of Technology, University of Bielefeld, D-33594 Bielefeld, Germany. e-mail: skopp@techfak.uni-bielefeld.de

Hidekazu Kubota, Japan Society for the Promotion of Science Research Fellow, Dept. of Intelligence Science and Technology, Graduate School of, Informatics, Kyoto University, Yoshida-Honmachi, Sakyo-ku, Kyoto 606-8501, Japan. e-mail: kubota@ii.ist.i.kyoto-u.ac.jp

Sadao Kurohashi, Dept. of Intelligence Science and Technology, Graduate School of Informatics, Kyoto University, Yoshida-Honmachi, Sakyo-ku, Kyoto 606-8501, Japan. e-mail: kuro@i.kyoto-u.ac.jp

Christopher Lee, Boston Dynamics Inc, 78, Fourth Avenue, Waltham, MA 02451. email: clee@bostondynamics.com

Danilo P. Mandic, Department of Electrical and Electronic Engineering, Imperial College, Exhibition Road, London, SW7 2BT United Kingdom. email: d.mandic@imperial.ac.uk

Kenji Mase, Information Technology Center, Nagoya University, Chikusa, Nagoya, 464-8601, Japan. e-mail: mase@nagoya-u.jp

Naohiro Matsumura, Graduate School of Economics, Osaka University, 1-7 Machikaneyama, Toyonaka, Osaka 560-0043, Japan. e-mail: matumura@econ.osaka-u.ac.jp

Michihiko Minoh, Academic Center for Computing and Media Studies, Kyoto University, Yoshida Nihonmatsu-cho, Sakyo-ku, Kyoto 606-8501, Japan. e-mail: minoh@media.kyoto-u.ac.jp

Louis-Philippe Morency, Computer Science and Artificial Intelligence Laboratory, Massachusetts Institute of Technology, Cambridge, MA 02139, USA. email: lmorency@csail.mit.edu

Chika Nagaoka, Japan Society for the Promotion of Science Research Fellow, Kokoro Research Center, Kyoto University, Yoshida-Honmachi, Sakyo-ku, Kyoto 606-8501, Japan. e-mail: nagaoka@educ.kyoto-u.ac.jp

Yuichi Nakamura, Academic Center for Computing and Media Studies, Kyoto University, Yoshida-Honmachi, Sakyo-ku, Kyoto 606-8501, Japan. email: yuichi@media.kyoto-u.ac.jp

Yukiko I. Nakano, Department of Computer and Information Sciences, Tokyo University of Agriculture and Technology, 2-24-16 Naka-cho, Koganei-shi, Tokyo 184-8588, Japan. e-mail: nakano@cc.tuat.ac.jp

Anton Nijholt, Department of Computer Science, Human Media Interaction, University of Twente, PO Box 217, 7500 AE Enschede, the Netherlands. email: A.Nijholt@ewi.utwente.nl

Toyoaki Nishida, Dept. of Intelligence Science and Technology, Graduate School of Informatics, Kyoto University, Yoshida-Honmachi, Sakyo-ku, Kyoto 606-8501, Japan. e-mail: nishida@i.kyoto-u.ac.jp

Satoshi Nishiguchi, Faculty of Information Science and Technology, Osaka Institute of Technology, 1-79-1 Kitayama, Hirakata-City, Osaka 573-0196, Japan. e-mail: nishigu@is.oit.ac.jp

Igor S. Pandzic, Department of Telecommunications, Faculty of Electrical Engineering and Computing, University of Zagreb, Unska 3, 10000 Zagreb, Croatia. e-mail: igor.pandzic@fer.hr

Matthias Rehm, Multimedia Concepts and Applications, Faculty of Applied Informatics, University of Augsburg, D-86159 Augsburg, Germany. e-mail: rehm@informatik.uni-augsburg.de

Tomasz M. Rutkowski, Brain Science Institute RIKEN, 2-1 Hirosawa, Wako-shi, Saitama 351-0198, Japan. email: tomek@brain.riken.jp

Tomohide Shibata, Dept. of Intelligence Science and Technology, Graduate School of Informatics, Kyoto University, Yoshida-Honmachi, Sakyo-ku, Kyoto 606-8501, Japan. e-mail: shibata@nlp.kuee.kyoto-u.ac.jp

Candace L. Sidner, BAE Systems Advanced Information Technologies, Burlington, MA 01803. e-mail: candy.sidner@baesystems.com

Karlo Smid, Department of Telecommunications, Faculty of Electrical Engineering and Computing, University of Zagreb, Unska 3, 10000 Zagreb, Croatia. e-mail: karlo.smid@fer.hr

Kristina Striegnitz, ArticuLab, Northwestern University, Frances Searle Building Room 2-432, 2240 Campus Drive, Evanston, IL 60208, USA. e-mail: kristina.striegnitz@gmail.com

Yasuyuki Sumi, Dept. of Intelligence Science and Technology, Graduate School of Informatics, Kyoto University, Yoshida-Honmachi, Sakyo-ku, Kyoto 606-8501, Japan. e-mail: sumi@i.kyoto-u.ac.jp

Rin-ichiro Taniguchi, Dept. of Intelligent Systems, Graduate School of Information Science and Electrical Engineering, Kyushu University, 744, Motooka, Nishi-ku, Fukuoka 819-0395 Japan. e-mail: rin@computer.org

Paul A. Tepper, Center for Technology & Social Behavior, Northwestern University, 2240 Campus Drive, Evanston, IL 60208.

Kazuhiro Ueda, Department of General System Studies, Graduate School of Arts and Sciences, The University of Tokyo, 3-8-1 Komaba, Meguro-ku, Tokyo 153-8902, Japan. email: ueda@gregorio.c.u-tokyo.ac.jp

Ipke Wachsmuth, Artificial Intelligence Group, Faculty of Technology, University of Bielefeld, D-33594 Bielefeld, Germany. e-mail: ipke@techfak.uni-bielefeld.de

Sakiko Yoshikawa, Kokoro Research Center, Kyoto University, Yoshida-Honmachi, Sakyo-ku, Kyoto 606-8501, Japan. Email: say@educ.kyoto-u.ac.jp

Goranka Zoric, Department of Telecommunications, Faculty of Electrical Engineering and Computing, University of Zagreb, Unska 3, 10000 Zagreb, Croatia. e-mail: goranka.zoric@fer.hr

1

Introduction

Toyoaki Nishida

1.1 Conversation: the Most Natural Means of Communication

Conversation is the most natural and popular means for people to communicate with each other. Conversation is everywhere around us, and even though you may feel that making conversation is relatively effortless, a closer look shows that a tremendous amount of sophisticated interaction is involved in initiating and sustaining conversation.

Figure 1.1 illustrates a normal conversation scene in our daily research activities. People use a rich set of nonverbal communication means, such as eye contact, facial expressions, gestures, postures and so on, to coordinate their behaviors and/or give additional meaning to their utterances, as shown in Figures 1.1(a) and (b), where participants are passing the initiative in the discussion from one to the other, or keeping it, by quickly exchanging nonverbal signs such as eye contact and facial expression as well as voice. People are skilled in producing and interpreting these subtle signs in ordinary conversations, enabling them to control the flow of a conversation and express their intentions to achieve their goals. Occasionally, when they get deeply involved in a discussion, they may synchronize their behavior in an almost unconscious fashion, exhibiting empathy with each other. For example, in Figures 1.1(c)–(f), a couple of participants are talking about controlling robots using gestures; they start waving their hands in synchrony to refer to the behavioral modality of a particular gesture. After a while, they become convinced that they have established a common understanding. In addition, their behavior enables the third and fourth participants to comprehend the situation and note it as a critical scene during the meeting.

In general, nonverbal means of communication play an important role in forming and maintaining collaborative behaviors in real time, contributing to enhancing engagement in conversation. The process normally takes a short period of time and is carried out almost unconsciously in daily situations.

Nonverbal interaction makes up a significant proportion of human–human interactions (Kendon 2004). Some authors consider that nonverbal interaction reflects intentions at a deeper level and precedes verbal communication in forming and maintaining intentions at the verbal level. McNeill suggests that verbal and nonverbal expressions occur in parallel for some psychological entities called growth points (McNeill 2005).

Conversations consisting of verbal and nonverbal interactions not only provide the most natural means of communication but also facilitate knowledge creation through such mechanisms as heuristic production

Conversational Informatics: An Engineering Approach   Edited by Toyoaki Nishida
© 2007 John Wiley & Sons, Ltd

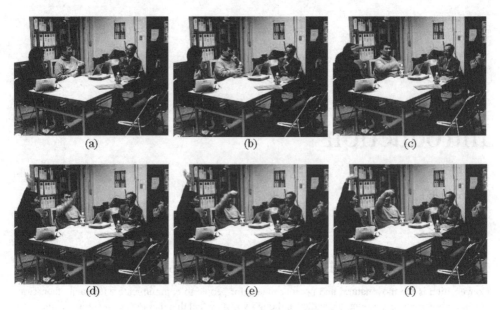

Figure 1.1   Conversation as a knowledge process

of stories from different points of view, tacit–explicit knowledge conversion, and entrainment[1] to the subject.

- *Heuristic production of stories from different points of view.* Conversation can be seen as an improvisational social process that allows each participant to bring together small fragments of stories into larger pieces in a trial-and-error fashion. In a business discussion, for example, participants attempt to find a useful shared story that could benefit themselves and their colleagues. Conversation provides an effective method of negotiating by taking into account the reactions of participants on the fly. The entire process is formulated as a mixture of verbal and nonverbal communication governed by social conventions that reflect a shared cultural background. This shared background constrains choices concerning a common discussion theme, setting up an agenda for the discussion, identifying critical points, raising proposals, arguing for and against proposals, negotiating, establishing a consensus, voting, deriving conclusions, and so on. Effective use of nonverbal signals makes the process of negotiation much more efficient than it would be otherwise.
- *Tacit–explicit knowledge conversion.* Knowledge is explicit or tacit depending on how clearly it is represented by words or numbers. Nonaka pointed out that conversion between tacit and explicit knowledge plays a key role in knowledge creation (Nonaka and Takeuchi 1995). Conversations provide an opportunity and motivation to externalize tacit knowledge. During a discussion, each participant tries to figure out how to express an idea to achieve the intended effect; for example, participants may propose, counter propose, support, challenge, negate, etc., during conversational discourse. In a debate, participants try to be the most expressive speaker in their search for new points that will win the contest. In contrast, conversations often encourage participants to look at knowledge from different perspectives. As a result, they may discover the incompleteness of existing explicit knowledge, which

---

[1] The tendency for two or more objects to behave in a similar rhythm.

may in turn lead to the formation of new knowledge. This might be tacit initially until an appropriate conceptualization is found.

- *Entrainment to the subject.* To gain an in-depth understanding of a given subject, individuals need to possess a sense of reality about the situation and to capture the problem as their own. Conversations help people react subjectively as they get involved in a given situation as a role player. Thus, conversations may remind people of past experiences similar to the given situation that can serve as a source of knowledge creation triggered by new analogies or hypotheses. Subjective understanding is indispensable to formulating mutual understanding in a community.

However, conversations also have shortcomings as a communication medium. Arguments with a logically complicated structure or those loaded with references to multimedia information are not well communicated by spoken language alone. Furthermore, conversations are volatile and not easily extensible beyond space and time. Both the content and context of a conversation may be lost quite easily. Even if conversations are recorded or transcribed, it is difficult to capture the subjective nature of a conversation after it has ended.

Advanced information and communication technologies offer huge potential for extending conversation by compensating for its limitations. For example, there are emerging technologies for capturing a conversation together with the surrounding situation. These technologies enable previous conversations to be recovered in a realistic fashion beyond temporal or spatial constraints. They also offer the possibility of building synthetic characters or intelligent robots that can talk autonomously with people in an interactive fashion on behalf of their owner.

1.2 An Engineering Approach to Conversation

Conversation has been studied from various perspectives. Philosophers have considered the relationship between thought and conversation. Linguists have attempted to model the linguistic structure underlying conversation. Anthropologists have worked on describing and interpreting the conversational behaviors that people exhibit in various situations. Communication scientists have investigated how components of conversation are integrated to make sense in a social context.

We take an engineering approach, using engineering techniques to measure and analyze conversation as a phenomenon (Nishida 2004a,b). Our aim is to exploit state-of-the-art technology to capture the details of conversations from various viewpoints and to automatically index the content. Annotation tools are widely used to analyze quantitative aspects of conversation.

We are also attempting to build artifacts that can participate in conversations. A typical example is the construction of embodied conversational agents (ECAs), which are autonomous synthetic characters that can talk with people (Cassell *et al.* 2000; Prendinger and Ishizuka 2004). This is a challenge because conversation is a sophisticated intellectual process where meaning is associated with complex and dynamic interactions based on collaboration between the speaker and listener.

An even more challenging goal is to build communicative robots capable of becoming involved in the conversational process and producing fluent interactions at an appropriate knowledge level (Figure 1.2). Robots that can merely exchange sentences with humans cannot participate meaningfully in human conversations in the real world because people make extensive use of nonverbal means of communication such as eye contact, facial expressions, gestures, postures, etc., to coordinate their conversational behaviors.

Figure 1.3 suggests how an interaction with a robot could go wrong. Suppose a robot is asked to describe an object. Scenario A shows a normal sequence in which the robot attempts to attract joint attention to the object by pointing and looking at the object. In contrast, in Scenario B, the agent points to the object while making eye contact with the human, which could cause an inappropriate emotional reaction on the part of the human. Thus, even a tiny flaw in nonverbal behavior could result in a significant difference in the outcome.

Figure 1.2   Communicative robots that can be involved in the conversation process

Conversation is a process that is initiated, sustained and ended as the result of collaboration between the speaker and listener. For example, when the speaker looks at an object and starts talking about it, the listener should also look at it to demonstrate that he/she is paying attention to the explanation. If the listener loses the trail of the discourse during the speaker's explanation, he/she may look at the speaker's face and probably murmur to signal the fact (Nakano *et al.* 2003). The speaker should recognize the communication flaw and take appropriate action such as suspending the flow of explanation and supplementing it with more information (Figure 1.4).

Robots need to be able to detect the subtle signs that participants produce, capture the meaning associated with interactions, and coordinate their behavior during the discourse. In other words, robots need to be able to play the role of active and sensible participants in a conversation, rather than simply standing still while listening to the speaker, or continuing to speak without regard for the listener.

In contrast to this view of *conversation-as-interaction*, it is important to think about how meaning emerges from interaction; i.e., the *conversation-as-content* view. Consider a conversation scene where the speaker is telling the listener how to take apart a device by saying "Turn the lever this way to detach the component" (Figure 1.5). Nonverbal behaviors, such as eye contact and gestures made by the speaker, not only control the flow of conversation (e.g., the speaker/listener may look away from the partner to gain time to briefly deliberate) but also become a part of the meaning by illustrating the meaning of a phrase (e.g., a gesture is used to associate the phrase "Turn the lever this way" with more detailed information about the way to turn the lever). It also helps the listener find a referent in the real world by drawing attention to its significant features.

1.3 Towards a Breakthrough

Conversation quantization is a conceptual framework for integrating the *conversation-as-interaction* and *conversation-as-content* views to build artifacts that can create or interpret meaning in conversation by interacting appropriately in real time (Nishida *et al.* 2006). The key idea is to introduce conversation quanta, each of which is a package of interaction and content arising in a quantized segment of conversation.

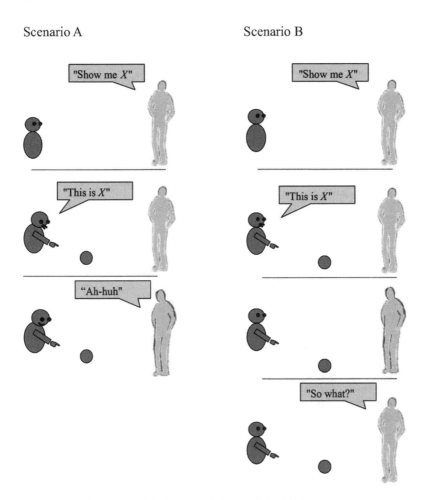

Figure 1.3 Robot is requested to explain object to person

Figure 1.6 illustrates how each conversation quantum represents a fragment of conversation from the situation in Figure 1.5. It contains a description of a visual scene in which the speaker gives an answer to the listener in response to a question that he/she has raised. It also contains a description of an interaction where the speaker is giving an explanation by pointing out a component to the listener, who is listening while paying attention to the object.

Conversation quanta may be acquired and consumed in an augmented environment equipped with sensors and actuators that can sense and affect conversations between people and artifacts. Conversation quanta have a twofold role. Firstly, they may be used to carry information from one conversation situation to another. Secondly, they may serve as a knowledge source that enables conversational artifacts to engage in conversations in which they exchange information with other participants.

Conversation quantization enables conversation quanta to be acquired, accumulated, and reused (Figure 1.7). In addition to the basic cycle of acquiring conversation quanta from a conversational situation and reusing them in a presentation through embodied conversational agents, conversation quanta may be aggregated and visually presented to the user, or may be manipulated, for example by summarization, to

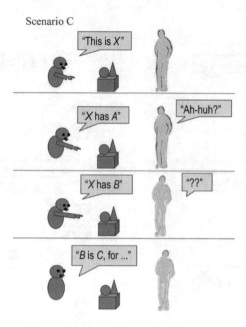

Figure 1.4   Repairing a grounding error

transform one or more conversation quanta into another. They can also be converted to and from various kinds of information archives.

We take a data-intensive approach based on sophisticated measurement and analysis of conversation in order to create a coupling between interaction and content. It will work even if we cannot implement fully intelligent or communicative artifacts from the beginning. In some situations, robots may be regarded as useful even if they can only record and replay conversations. The more intelligent algorithms are introduced, the more autonomous and proficient they will become, which will enable them to provide

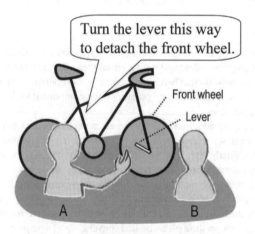

Figure 1.5   Example of a conversation scene

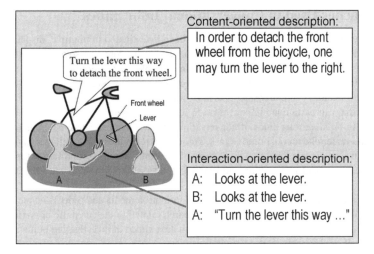

Figure 1.6   Example of a conversation quantum

services in a more interactive fashion. For example, artifacts may be able to follow nonverbal interaction even without an in-depth understanding, if they can mimic conversational behavior on the surface. In fact, a robot will be able to establish joint attention by creating eye contact with an object when the partner is recognized as paying attention to that object. The media equation theory (Reeves and Nass 1996) suggests that superficial similarities might encourage people to behave as if they were real.

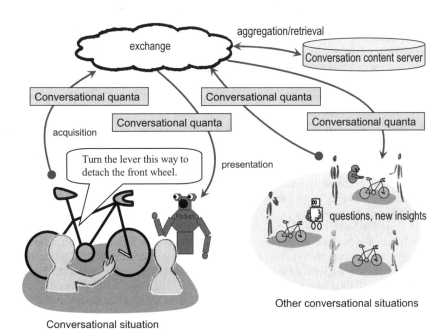

Figure 1.7   Conversation quantization as a framework for circulating conversational content

1.4 Approaches Used in Conversational Informatics

Until recently, various aspects of conversation have been investigated in multiple, disparate fields of research because it is such a complex topic and the theory and technology was so premature that researchers did not consider it feasible to place the entire phenomenon of conversation within a single scope. However, advances in technology have completely changed this situation, enabling us to take an entirely new approach to conversation. Computer scientists have succeeded in developing realistic, embodied conversational agents that can participate in conversation in a realistic setting. Progress in intelligent robotics has enabled us to build various kinds of conversational robots that can share the physical environment with people and undertake a collaborative task. Acoustic stream technology allows us to locate, single out, and track an acoustic stream at a cocktail party, while computer vision and sensor fusion technology is an inexpensive means of recognizing and tracking conversations in the physical environment in real time.

By exploiting these newly developed intelligent information media and processing technologies, conversational informatics brings together fields of research related to the scientific or engineering aspects of conversation. This has given rise to a new research area aimed at investigating human conversational behaviors as well as designing conversational artifacts such as synthetic characters that show up on the computer screen or intelligent housekeeping robots that are expected to interact with people in a conversational fashion.

Conversational informatics covers both the investigation of human behaviors and the design of artifacts that can interact with people in a conversational fashion. It is attempting to establish a new technology consisting of environmental media, embodied conversational agents, and management of conversational content, based on a foundation provided by artificial intelligence, pattern recognition, and cognitive science. The main applications of conversational informatics involve knowledge management and e-learning. Although conversational informatics covers a broad field of research encompassing linguistics, psychology and human–computer interaction, and interdisciplinary approaches are highly important, the emphasis is on engineering aspects, which have been more prominent in recent novel technical developments such as conversational content acquisition, conversation environment design, and quantitative conversational modeling. Current technical developments in conversational informatics center around four subjects (Figure 1.8).

The first of these is conversational artifacts. The role of conversational artifacts is to mediate the flow of conversational content among people. To succeed in this role, conversational artifacts need to be fluent in nonverbal interactions. We address how to build artifacts, such as synthetic characters on a computer screen or intelligent robots that can help the user by making conversation not only using natural language but also using eye contact, facial expressions, gestures, or other nonverbal means of communication.

The second subject is conversational content. Conversational content encapsulates information and knowledge arising in a conversational situation and reuses it depending on a given conversational situation. We examine methods of capturing, accumulating, transforming, and applying conversational content. We address building a suite of techniques for acquiring, editing, distributing, and utilizing content that can be produced and applied in conversation.

The third subject is conversational environment design. The role of a conversation environment is to sense and help actors and artifacts make conversation. We address the design of an intelligent environment that can sense conversational behaviors to either help participants become involved in a collaboration even though they may be in a separate location, or record conversation accompanying the atmosphere of specific conversational behavior for later use or review.

The last subject is conversation measurement, analysis, and modeling. Motivated by scientific interest, we take a data-driven quantitative approach to understanding conversational behaviors by measuring these behaviors using advanced technologies and building detailed quantitative models of various aspects of conversation.

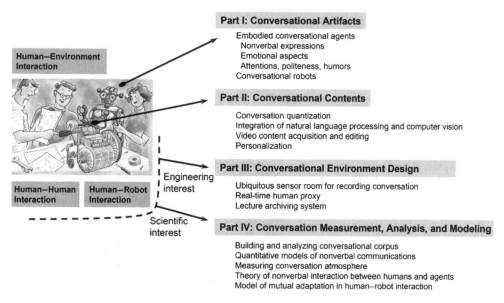

Figure 1.8 Conversational informatics. This figure originally appeared in Toyoaki Nishida: Prospective view of conversational informatics, *Journal of JSAI* (Japanese Society for Artificial Intelligence), Vol. 21, No. 2, pp. 144–149, 2006. (Reproduced by permission of Japanese Society of Artificial Intelligence)

1.5 Conversational Artifacts

Conversational artifacts fall into several categories. The simplest ones are autonomous, text-based dialogue systems, such as ELIZA (Weizenbaum 1996). Even though they do not have physical body, it is known that people sometimes feel as if they really exist due to media equation effects (Reeves and Nash 1996). More complex systems include embodied conversational agents, which are synthetic characters capable of talking with people. Conversational robots are at the high-end. They are physically embodied and share physical spaces with people.

1.5.1 Embodied Conversational Agents

Conversational informatics is mainly concerned with communicating a rich inventory of conversational content with appropriate use of nonverbal means of communication (Figure 1.9).

Examples of major technological contributions include knowledgeable embodied conversation agents that can automatically produce emotional and socially appropriate communication behaviors in human–computer interaction. IPOC (Immersive Public Opinion Channel; see Chapter 5) enables conversation quanta to be expanded in a virtual immersive environment. Users can interact with conversational agents in a story-space in which a panoramic picture background and stories are embedded. The embedded stories are presented on user demand or spontaneously according to the discourse. The stories are used to represent a discourse structure consisting of more than one conversation quantum.

To generate a more complex set of nonverbal behaviors in an immersive environment, theories of nonverbal communication have been extensively studied and incorporated in systems for generating agent behavior. Conversation management includes the ability to take turns, interpret the intentions of participants in the conversation, and update the state of the conversation. Collaboration behavior

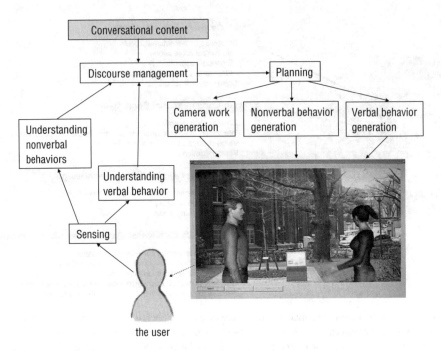

the user

Figure 1.9   Architecture of embodied conversational agents

determines the agent's next action in order to accomplish the goal of the conversation and collaboration with the user. Engagement behaviors consist of initiating a collaborative interaction, maintaining the interaction, and disengaging from the interaction.

The Interaction Control Component (ICC) of the IPOC system interprets inputs from a speech recognizer and a sensor system and generates verbal and nonverbal behaviors performed by conversational agents. The ICC includes a conversation manager (CM), which maintains the history and current state of the conversation, a collaboration behavior generation module (CBG), which selects the next utterance and determines the agents' behaviors in telling a story, and engagement behavior generation (EBG), which determines appropriate engagement behaviors according to the state of the conversation.

In addition to basic communication abilities, it would be much nicer if embodied conversational agents could exhibit humorous behaviors. Nijholt *et al.* (see Chapter 2) consider that humorous acts are the product of an appraisal of the conversational situation. They attempted to have ECAs generate humorous behaviors by making deliberate misunderstandings in a conversation.

To make ECAs believable and sociable, it seems critical to incorporate emotion and personality components. Becker and his colleagues analyzed the notions underlying "emotion" and "personality" and described an emotion simulation system called Max (see Chapter 3). They argue that Max's emotional system increases his acceptance as a coequal conversational partner, and describe an empirical study that yielded evidence that the same emotional system supports the believability and lifelike quality of an agent in a gaming scenario.

Sidner *et al.* (2003) proposed that conversation management, collaboration behavior, and engagement behaviors were communicative capabilities required for collaborative robots. Morency *et al* (see Chapter 7) proposed to combine information from an ECA's dialogue manager and the prediction of head nodding and shaking to predict contextual information for an ongoing dialogue. They point out that a subset of lexical, punctuation, and timing features available in ECA architectures can be used to learn how to predict user feedback.

To make ECAs act politely, Rehm and André (see Chapter 4) applied an empirical study on the impact of computer-based politeness strategies on the user's perception of an interaction. They used a corpus to analyze the co-occurrence of gestures and verbal politeness strategies in the face of threatening situations.

Facial gestures such as various nodding and head movements, blinking, eyebrow movements, and gaze play an important role in conversation. Facial displays are so important as a communication channel that humans use them naturally, often subconsciously, and are therefore very sensitive to them. Zoric *et al.* (see Chapter 9) attempted to provide a systematic and complete survey of facial gestures that could be useful as guideline for implementing such gestures in an ECA.

1.5.2 Robots as Knowledge Media

Another approach is the use of robots as knowledge media, where the role of the robot is to acquire or present information and knowledge contained within the discourse by engaging in appropriate behavior as a result of recognizing the other participants' conversational behaviors.

Nishida *et al.* (2006) focus on establishing robust nonverbal communication that can serve as a basis for associating content with interaction. The proposed communication schema allows two or more participants to repeat observations and reactions at varying speeds to form and maintain joint intentions to coordinate behavior called a "coordination search loop". The proposed architecture consists of layers to deal with interactions at different speeds to achieve this coordination search loop (Figure 1.10).

The lowest layer is responsible for fast interaction. This level is based on affordance (Gibson 1979), which refers to the bundle of cues that the environment gives the actor. People can utilize various kinds of affordances, even though these are subtle. The layer at this level is designed so that a robot can suggest its capabilities to the human, coordinate its behavior with her/him, establish a joint intention, and provide the required service.

The intermediate layer is responsible for interactions at medium speed. An entrainment-based alignment mechanism is introduced so that the robots can coordinate their behaviors with the interaction partner by varying the rhythm of their nonverbal behaviors.

The upper layer is responsible for slow and deliberate interactions, such as those based on social conventions and knowledge, to communicate more complex ideas based on the shared background. Defeasible interaction patterns are employed to describe typical sequences of behaviors that actors are expected to show in conversational situations. A probabilistic description is used to cope with the vagueness of the communication protocol used in human society.

The three-layer model serves as a basis for building listener and presenter robots with the aim of prototyping the idea of robots as embodied knowledge media. The pair of robots serves as a means of

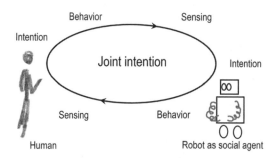

Figure 1.10   Architecture consisting of layers to deal with interactions at different speeds

Listener Robot   Presenter Robot

Figure 1.11   Listener and presenter robots in action

communicating embodied knowledge (Figure 1.11). The listener robot first interacts with the human with knowledge to acquire conversational contents. The presenter robot, equipped with a small display, then interacts with a human to show the appropriate content in situations where this knowledge is considered to be needed. Research is in progress to use conversation quanta as a means for representing knowledge transferred from the listener robot to the presenter robot.

Sidner *et al.* proposed the use of visual processing to extract nonverbal features, in particular those relating to engagement, which would enable robots to talk with humans in a collaborative fashion. As an application, they are working on a robot that collaborates with a human on an equipment demonstration (see Chapter 6).

1.6 Conversational Content

The success of Conversational Informatics depends on the amount of conversational contents circulated in a community. The major technical concern related to conversational content is cost reduction in content production. One approach to this goal is to increase the reusability of content by inventing a method of managing a large amount of conversational content to enable the user to easily create new content by combining and incrementally improving existing content. The other approach is to introduce media processing and artificial intelligence techniques to reduce the cost of content acquisition (Figure 1.12).

- An example of the former approach is the sustainable knowledge globe (SKG), which helps people manage conversational content by using geographical arrangement, topological connection, contextual relation, and a zooming interface. Using SKG, a user can construct content in a virtual landscape and then explore the landscape within a conversational context (see Chapter 10).
- An example of the latter approach is the integration of natural language processing and computer vision techniques. Kurohashi *et al.* (see Chapter 11) succeeded in automatically constructing a case frame dictionary from a large-scale corpus. The case frame dictionary can be used to analyze discourse for content production, including automated production of conversational content from annotated videos or natural language documents.

Nakamura (see Chapter 12) combines an environmental camera module and content-capturing camera modules to capture conversation scenes. In addition, he uses optimization and constraint–satisfaction methods to choose parameters that can be adjusted depending on the purpose of the videos. Using this

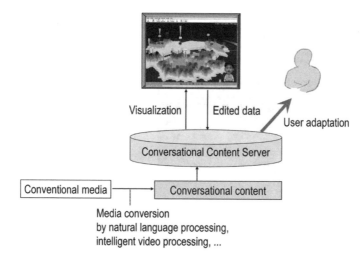

Figure 1.12   Technical issues underlying conversational content

technical platform, he has built a system that can answer and interact with users in a conversational environment.

Content production may also become more effective with the introduction of user adaptation. Babaguchi (see Chapter 13) introduced a technique for acquiring the preferences of the user by observing the user's behavior to implement a personalized video service with tailored video content. The method is applied to personalized summaries of broadcast sports video.

1.7 Conversational Environment Design

One approach to making conversational communication more effective is to embed sensors and actuators into the environment so that conversations can be recorded or measured by sensors and accurate feedback can be given to the participants (Figure 1.13).

Sumi *et al.* (see Chapter 14) built a ubiquitous sensor room for capturing conversations situated in a real space using environment sensors (such as video cameras, trackers, and microphones set up ubiquitously around the room) and wearable sensors (such as video cameras, trackers, microphones, and physiological sensors). To supplement the limited capability of the sensors, LED tags (ID tags with an infrared LED) and IR trackers (infrared signal-tracking devices) are used to annotate the audio/video data with positional information. Significant intervals or moments of activities are defined as interaction primitives called "events". Currently, five event types are recognized: "stay", "coexist", "gaze", "attention", and "facing". Events are captured by the behavior of the IR trackers and LED tags in the room. For example, a temporal interval will be identified as a joint attention event when an LED tag attached to an object is simultaneously captured by IR trackers worn by two users, and the object in focus will be marked as a socially important object during the interval. The accumulated conversation records can be presented to the user in various way, such as using automated video summarization of individual users' interactions by chronologically synthesizing multiple-viewpoint videos, a spatio-temporal collage of videos to build a 3D virtual space for re-experiencing shared events, or an ambient sound display for increasing the level of awareness of the real space by synthesizing spatially embedded conversations.

Nishiguchi *et al.* (see Chapter 16) proposed ambient intelligence based on advanced sensing technology that can record human communication activities and produce feedback without hindering communicative

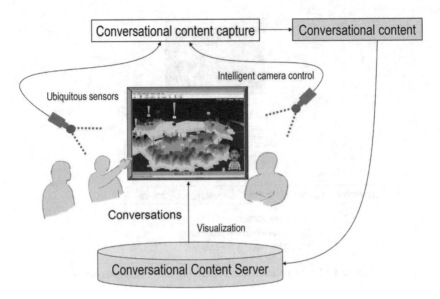

Figure 1.13   Technical issues underlying conversational environment design

activities. The concept, which is called "environmental media", is implemented as an augmented class-room that can automatically highlight and record significant communicative behaviors in a classroom.

Taniguchi and Arita (see Chapter 15) proposed a virtual classroom system using a technique called "real-time human proxy" in which the communicative behaviors of participants at distant sites are encoded using computer vision; a virtual classroom scene in which all the participants are represented by avatars is synthesized in real time.

1.8 Conversation Measurement, Analysis, and Modeling

By exploiting advanced sensor technologies, we are now able to measure various aspects of conversational behaviors, particularly nonverbal behaviors, in greater detail than ever. The insights gained can be applied to the design of conversational artifacts and conversational environments (Figure 1.14).

For example, Xu *et al.* (2006) designed a human–human WOZ (Wizard of Oz) experimental setting to observe mutual adaptation behaviors. The setting enables the researcher to set up an experiment with an appropriate information overlay that may be hidden from subjects. Experiments conducted using the environment, coupled with a powerful annotation and analysis tool such as ANVIL, have yielded interesting findings about mutual adaptation, such as the emergence of pace keeping and timing of matching gestures, symbol-emergent learning, and environmental learning.

Den and Enomoto (see Chapter 17) focused on the responsive actions of listeners including backchannel responses, laughing, head nodding, hand movements, and gazing. They built a multimodal corpus containing records of three-person Japanese conversations annotated with speech transcriptions and the nonverbal communicative behaviors of participants. When they analyzed the distribution of the listeners' responsive actions, they gained some interesting insights; e.g., "there is not only a tendency for the next speaker to gaze at the current speaker toward the end of his turn, but also a tendency for the other listener, who keeps silent during the next turn, to gaze at the next speaker around the end of the current turn and before the next speaker starts speaking".

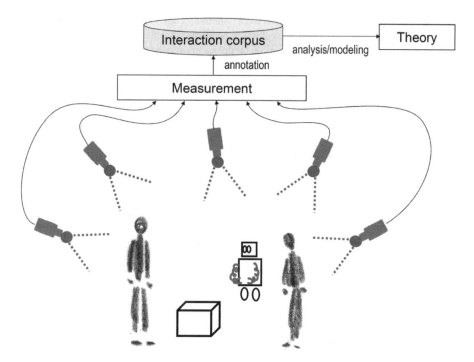

Figure 1.14   Conversation measurement, analysis, and modeling

Kopp *et al.* (see Chapter 8) studied the use of gestures in giving directions. They proposed a framework for analyzing how such gestural images are associated with semantic units (image description features) and morphological features (hand shape, trajectory, etc.).

Nagaoka *et al.* (see Chapter 18) studied the synchrony tendency, a typical phenomenon observed in conversation in which participants' nonverbal behaviors, such as body movement and speech interval, tend to synchronize and become mutually similar. Based on an empirical study in a therapist–client conversation situation, they discuss how various behavioral measures can be used to evaluate the degree of synchrony tendency.

Rutkowski and Mandic (see Chapter 19) proposed three characteristics of a communication atmos phere – based on social features, the mental characteristics of the communicators, and physical features – and discussed how to evaluate the atmosphere by tracking audio-visual signals. Potential applications include semi-automated indexing of multimedia archives and a virtual chat room.

Matsumura (see Chapter 20) proposed a method called an influence diffusion model to identify the roles of individuals in communication (i.e., leader, coordinator, maven, and follower) by analyzing the log of threaded messages. The method is expected to be applied to analyzing conversation records as well as to identifying the influence of various participants.

Interactive agents such as pet robots or adaptive speech interface systems that require forming a mutual adaptation process with users require two competences. One of these is recognizing reward information from the users' expressed paralanguage information, and the other is informing the learning system about the users by means of that reward information. The key issue here is to clarify the specific contents of reward information and the actual mechanism of the learning system by observing how two people could create smooth nonverbal communication, similar to that between owners and their pets.

Ueda and Komatsu (see Chapter 21) conducted a communication experiment to observe how human participants created smooth communication through acquiring meaning from utterances in languages they did not understand. The meaning acquisition model serves as a theory of mutual adaptation that will enable both humans and artifacts to adapt to each other.

1.9 Underlying Methodology

What is common to the above studies? A close look at the details of research in conversational informatics shows that the following methodological features underlie the work described in this book.

1. *Observation, focusing, and speculation.* Conversation is a highly empirical subject. The first thing to do in starting research in this area is to closely observe the conversation process as a phenomenon and focus on a few interesting aspects. Although looking at the entire process of conversation is important, a focus is necessary to derive a useful conclusion. In-depth speculation based on the existing literature is an inevitable part of setting up an innovative research program into highly conceptual issues such as humor or politeness. This approach is effectively used in Chapter 2.
2. *Building and analyzing a corpus.* Intuitions and hypotheses should be verified by data. Building and analyzing an accumulated corpus is effective in both engineering and scientific approaches. In fact, the authors of Chapters 4, 5, and 8 succeeded in deriving useful insights in building embodied conversational agents by analyzing corpora. The authors of Chapter 17 built a novel corpus containing records on multi-party conversations, from which they successfully built a detailed model of eye gaze behaviors in multi-party conversations. The authors of Chapter 14 present a powerful method of using a ubiquitous sensor room to build a conversation corpus.
3. *Sensing and capture.* By exploiting the significant progress made in capturing events taking place in a real-world environment, using a large number of sensing devices embedded in the environment and actors, we can now shed light on conversations from angles that have not been possible before. Conversational environment design, reported in Chapters 14, 15, and 16, has opened up a new research methodology for understanding and utilizing conversations.
4. *Building a computational model and prototype.* A computational model helps us establish a clear understanding of the phenomenon and build a powerful conversation engine. In this book, computational models are used effectively to reproduce the emotional behaviors of embodied conversational agents (in Chapter 3), facial gesture (in Chapter 9), or conversational robots (in Chapters 6 and 7).
5. *Content acquisition and accumulation.* Content is a mandatory constituent of conversation in the sense that we cannot make conversation without content. In addition, conversation can produce high-quality content. Conversational informatics is greatly concerned with capturing and accumulating content that is produced or consumed in a conversation. Chapters 10 and 12 describe new efforts in this direction based on integrating state-of-the-art technologies. Chapter 13 reports the use of personalization to increase the value of content to individual users. These studies are supported by technologies for sensing and capturing.
6. *Media conversion.* Media conversion aims at converting conversational content from/to conventional archives such as documents. Technologies in natural language processing, acoustic stream processing, and image processing are used to achieve these goals, as described in Chapters 11 and 12.
7. *Measurement and quantitative analysis.* Quantitative analysis based on various kinds of measurements provides an effective means of understanding conversation in-depth. Chapter 19 introduces a new measure of evaluating communication atmosphere and Chapter 20 proposes a way of quantifying how much an actor influences other participants in e-mail communications, which may well be applicable to the analysis of face-to-face conversation.

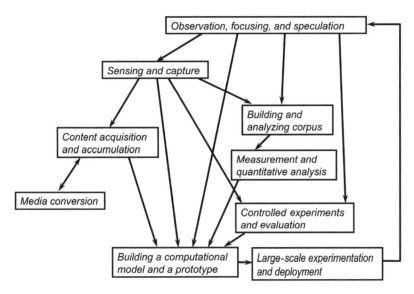

Figure 1.15 Relationship between methodological components: each directed edge represents that the insights or results obtained in the source are used in the destination

8. *Controlled experiments and evaluation.* Designing controlled experiments to verify a working model is a common methodology used in experimental psychology. It can also be applied effectively to experimentally prove novel features of conversation, as shown in Chapters 18 and 21, where models for embodied synchrony and mutual adaptation are validated in psychological experiments.

Figure 1.15 shows the relationship between methodological components. Each arrow shows that the insights or results obtained in the source are used in the destination. One can proceed along the edges as research proceeds. Alternatively, one can focus on one methodological component, assuming the existence of the upstream and evaluating the result in the context of the downstream.

References

Cassell J., Sullivan J., Prevost S. and Churchill E. (eds) (2000) *Embodied Conversational Agents.* MIT Press.

Gibson J.J. (1979) *The Ecological Approach to Visual Perception.* Houghton Mifflin, Boston.

Kendon A. (2004) *Gesture: Visible action as utterance.* Cambridge University Press.

McNeill D. (2005) *Gesture and Thought.* University of Chicago Press.

Nakano Y.I., Reinstein G., Stocky T. and Cassell J. (2003) Towards a model of face-to-face grounding. In *Proceedings of the 41st Annual Meeting of the Association for Computational Linguistics* (ACL 03), pp. 553–561.

Nishida T. (2004a) Towards intelligent media technology for communicative intelligence (Keynote speech). In *Proceedings of International Workshop on Intelligent Media Technology for Communicative Intelligence* (IMTCI), pp. 1–7.

Nishida T. (2004b) Conversational knowledge process for social intelligence design (Invited talk). In A. Aagesen *et al.* (eds): *Intellcomm 2004,* LNCS 3283, Springer, pp. 28–42.

Nishida T., Terada K., Tajima T., Hatakeyama M., Ogasawara Y., Sumi Y., Xu Y., Mohammad Y.F.O., Tarasenko K., Ohya T. and Hiramatsu T. (2006). Towards robots as an embodied knowledge medium (Invited paper). Special Section on Human Communication II, *IEICE Transactions on Information and Systems,* vol. E89-D, no. 6, pp. 1768–1780.

Nonaka I. and Takeuchi H. (1995) *The Knowledge-creating Company: How Japanese companies create the dynamics of innovation*. Oxford University Press.

Prendinger H. and Ishizuka M. (eds) (2004) *Life-Like Characters: Tools, Affective Functions, and Applications*. Springer.

Reeves B. and Nass C. (1996) *The Media Equation: How people treat computers, television, and new media like real people and places*. Cambridge University Press.

Sidner C.L., Lee C. and Lesh N. (2003) Engagement rules for human–robot collaborative interactions. In *Proc. IEEE International Conference on Systems, Man & Cybernetics* (CSMC), vol. 4, pp. 3957–3962.

Weizenbaum J. (1996) "ELIZA": a computer program for the study of natural language communication between man and machine. *Communications of the Association for Computing Machinery* **9**, 36–45.

Xu Y., Ueda K., Komatsu K., Okadome T., Hattori T., Sumi Y. and Nishida T. (2006) WOZ experiments for understanding mutual adaptation. Presented at Social Intelligence Design, 2006.

Part I

Conversational Artifacts

2

Conversational Agents and the Construction of Humorous Acts

Anton Nijholt

2.1 Introduction

Social and intelligent agents have become a leading paradigm for describing and solving problems in human-like ways. In situations where it is useful to design direct communication between agents and their human partners, the display of social and rational intelligence in an embodied human-like agent allows natural interaction between the human and the agent that represents the system the human is communicating with. "Embodied" means that the agent is visualized on the screen as a 2D or 3D cartoon character that shows human behavior through its animations. Research in intelligent agents includes reasoning about beliefs, desires, and intentions. Apart from contextual constraints that guide the agent's reasoning and behavior, other behavioral constraints follow from models that describe emotions. An overview of emotion theories and an implementation in the context of an embodied conversational agent can be found in Chapter 3 of this book. These models often assume that emotions emerge based on appraisals of events taking place in the environment and based on how these events affect the goals that are being pursued by the agents. In current research, it is also not unusual to incorporate personality models in agents to adapt this appraisal process as well as reasoning, behavior, and display of emotions to personality characteristics. So, we can model lots of useful and human-like properties in artificial agents, but, in Roddy Cowie's (2000) words, "If they are going to show emotion, we surely hope that they would show a little humor too." This chapter anticipates the need for such agents by exploring the relevant research questions.

Embodied conversational agents (ECAs) have been introduced to play, among others, the role of conversational partner for the computer user. Rather than addressing the "machine", the user addresses virtual agents that have particular capabilities and can be made responsible for certain tasks. The user may interact with ECAs to engage in an information service dialogue, a transaction dialogue, to solve a problem cooperatively, perform a task, or to engage in a virtual meeting. Other obvious applications can be found in the areas of education (including training and simulation), entertainment, electronic commerce, and teleconferencing. Research projects suggest that in the near feature we might expect that, in addition to being domain and environment experts, ECAs will act as personal assistants, coaches, and buddies.

Conversational Informatics: An Engineering Approach   Edited by Toyoaki Nishida
© 2007 John Wiley & Sons, Ltd

They will accompany their human partners, migrating from displays on handheld devices to displays embedded in ambient-intelligence environments. Natural interaction with these ECAs will require them to display rational and social intelligence and, indeed, also a little humor when appropriate and enjoyable. In this interaction with embodied conversational agents, verbal and nonverbal communication is equally important. Multimodal emotion display and detection are among the research issues in this area of human–computer interaction. And so can be investigations in the role of humor in human–computer interaction.

In previous years researchers have discussed the potential role of humor in the interface. However, when we compare efforts in this area with efforts and experiments that demonstrate the positive role of general emotion modeling in the user interface, then we must conclude that attention is still minimal. As we all know, a computer can be a source of frustration rather than enjoyment. A lot of research is focused on detecting a user's frustration (Klein *et al.* 2002; Picard and Klein 2002) – for example in educational settings – and not on generating enjoyment. Useful observations about the positive role of humor in the interface were made by Binsted (1995) and Stock (1996). Humans use humor to ease communication problems in human–human interaction and in a similar way humor can be used to solve communication problems that arise with human–computer interaction. Binsted emphasizes the role of humor in natural language interfaces. Humor can help to make the imperfections of natural language interfaces more acceptable for the users, and when humor is sparingly and carefully used it can make natural language interfaces much friendlier.

In earlier years the potential role of embodied conversational agents was not at all clear and no attention was paid to their possible role in the interface. In Nijholt (2002) we first discussed the role of humor for embodied conversational agents in the interface. It is a discussion on the possible role of humor support in the context of the design and implementation of embodied conversational agents. This role can be said to follow from the so-called CASA (Computers Are Social Actors) paradigm (Reeves and Nass 1996), assuming that humans contribute human-like properties to embodied agents that can help in obtaining more enjoyable interactions, similar, as such properties, when assigned to a human partner help to make the conversation more enjoyable. More recently, several observations about computational humor appeared in Binsted *et al.* (2006).

In the next section we have a short literature survey of the role of humor in human–human interaction. In section 2.3 we discuss embodied conversational agents and their use in intelligent and social user interfaces. We also make clear why the role of humor in human–human interaction can be translated to a similar role in human–computer interaction, in particular when the interface is inhabited by one or more embodied conversational agents. In section 2.4 we have observations on how to decide to generate a humorous act and on the appropriateness of displaying it. In section 2.5 we look at the incongruity theory in humor research and we consider erroneous ambiguity resolution, in particular erroneous anaphora resolution, as a strategy to generate humorous acts. Some notes on an implementation of the ideas are also provided. An example of humorous act generation is presented. Section 2.6 discusses nonverbal support for humor generation. Possible tools and resources that are needed in future research are discussed in section 2.7. Finally, section 2.8 contains the conclusions of this chapter.

2.2 The Role of Humor in Interpersonal Interaction

In interpersonal interactions humans use humor, humans smile and humans laugh. Humor can be spontaneous, but it can also serve a social role and be used deliberately. A smile can be the effect of appreciating a humorous event, but it can also be used to regulate the conversation. A laugh can be spontaneous but can also mask disagreement or be cynical. Research has shown that laughs are related to topic shifts in a conversation and phases in negotiations or problem solving tasks. In an educational situation humor can be used by the teacher to catch students' attention but also to foster critical thinking. Humor allows criticism to be smoothed, stress can be relieved and students can become more involved in joint classroom activities by the use of humor (Ziv 1988). Humor can also help when it comes to frustration. In an (e-) commerce situation negotiators use humor to induce trust.

Here we discuss the role of humor in human–human interaction. Results from experimental research are surveyed. First we are concerned with general issues, not necessarily connected to a particular domain, but playing a role in human–human interaction in general: humor support in a conversation, interpersonal attraction, and trust. More topics could have been chosen, but the mentioned issues arise naturally when later in this section domains are discussed where in the near future embodied conversational agents will play the roles of one or more of the conversational partners in the current real-life situations. The domains that are chosen are education, information services and commerce, meetings, and negotiations.

2.2.1 General Issues: Support, Attraction, and Trust

It is possible to look at preconceived aims of conversational partners to create humor during a conversation or discussion. However, this chapter rather looks at situations where humor occurs spontaneously during an interaction or where it occurs in a supporting role, for example to hide embarrassment, to dominate the discussion or to change the topic. Some of these roles will get more attention in the next section. The emphasis here is on the role of humor to induce trust and interpersonal attraction and on the appreciation of humor during a conversation.

Humans employ a wide range of humor in conversations. Humor support, or the reaction to humor, is an important aspect of personal interaction and the given support shows the understanding and appreciation of humor. In Hay (2001) it is pointed out that there are many different support strategies. Which strategy can be used in a certain situation is mainly determined by the context of the humorous event. The strategy can include smiles and laughter, the contribution of more humor, echoing the humor, offering sympathy, or contradicting self-deprecating humor. There are also situations in which no support is necessary. In order to give full humor support, humor has to be recognized, understood and appreciated. These factors determine our level of agreement on a humorous event and how we want to support the humor.

Humor support may show our involvement in a discussion, our motivation to continue and how much we enjoy the conversation or interaction. Similarity in appreciation also supports interpersonal attraction, as investigated by Cann *et al.* (1997). This observation is of interest when later we discuss the use of embodied conversational agents in user interfaces. Sense of humor is generally considered a highly valued characteristic of self and others. Nearly everybody claims to have an average to above-average sense of humor. Perceived similarity in humor appreciation can therefore be an important dimension when designing for interpersonal attraction. In Cann's experiments participants had to interact with an unseen stranger. Before the interaction, ratings were made of the attitudes of the participants and they were led to believe that the stranger had similar or dissimilar attitudes. The stranger responded either positively or neutrally to a participant's attempt to humor. The results tell us that similarity in humor appreciation is able to negate the negative effects of dissimilarity for other attitudes when looking at interpersonal attraction. Other studies show how similarity in attitudes is related to the development of a friendship relationship. The development of a friendship relationship requires time, but especially in the initiation phase the kinds of similarities mentioned above can be exploited.

Friendship and intimacy are also closely related. Trust is an essential aspect of intimacy and the hypothesis that there also exists a correlation between humor and trust has been confirmed (Hampes 1999). There are three key factors that help us to understand this relationship. The most important factor is the demonstrated relation between humor and extroversion (Ruch 1994). When we break up extroversion into basic components like warmth, gregariousness, assertiveness, and positive emotions it becomes obvious that extroversion involves trust. Another factor, mentioned above, is the fact that humor is closely related to a high self-esteem. People who are proud of who they are are more likely to trust other persons and to reveal themselves to them. A third factor is that humorous persons are effective in dealing with stress (Fry 1995). They are well qualified to deal with the stress or anxieties involved in interpersonal relationships and therefore more willing to enter relationships.

2.2.2 Conversations and Goal-directed Dialogues

Humor plays a role in daily conversations. People smile and laugh, certainly not necessarily because someone pursues the goal of being funny or tells a joke, but because the conversational partners recognize the possibility to make a funny remark fully deliberately, fully spontaneously, or something in between, taking into account social (display) rules . We will not go deeply into the role of humor in daily conversations, small talk or in entertainment situations. In daily conversations humor very often plays a social role, not only in conversations with friends and relatives (Norrick 1993), but also in the interaction with a real estate agent, a saleswoman, a tourist guide, a receptionist or a bartender. It is difficult to design experiments intended to find the role played by humor in human-to-human interactions, when no specific goals are defined. Even experiments related to rather straightforward business-to-consumer relationships are difficult to find. Rather we have to deal in these situations with regulations protecting a customer from humor by a salesman (never use sarcasm, don't make jokes at the expense of the customer, etc.).

When looking at the more goal-directed situations, teaching seems to be one field where the use of humor has received reasonable attention. Many benefits have been mentioned regarding humor in the teaching or learning process and sometimes made explicit in experiments. Humor contributes to motivation attention, promotion of comprehension and retention of information, a more pleasurable learning experience, a development of affective feelings toward content, fostering of creative thinking, reducing anxiety, etc. The role of humor during instruction, its social and affective functions for teaching, and the implications for classroom practice has been discussed in several papers. However, despite the many experiments, it seems to be hard to generalize from the experiments that are conducted (Ziv 1988).

The role of humor and laughter during negotiations processes is another issue that has received attention. In Adelswärd and Öberg (1998) several tape recordings made during international negotiations have been analyzed. One of the research questions concerned the interactional position of laughter: When do we laugh during interaction? Different phases during negotiation can be distinguished. Laughing events turned out to be related to the phase boundaries and also to discourse boundaries (topic shifts). Hence, laughter serves interactional goals. The distinction between unilateral and joint laughter is also important. Mutual laughter often reflects consensus, unilateral laughter often serves the same function as intonation.

Describing and explaining humor in small task-oriented meetings is the topic of a study conducted by Consalvo (1989). An interesting and unforeseen finding was the patterned occurrence of laughter associated with the different phases of the meeting. The opening phase is characterized by its stiffness and serious tone and the atmosphere of distrust. Humor in this phase is infrequent. This contrasts with the second, transitional phase that lasts only a couple of minutes and the humorous interactions are frequent and for the first time during the meeting all participants laugh. Their laughter conveys the agreement that the problem can be solved and the commitment of the individual participants. The last phase, the problem-solving phase, contains a lot more humorous events than the opening phase, but still less than the transitional phase. In this way humor echoes the progression of a meeting.

2.3 Embodied Conversation Agents

2.3.1 Introduction

Embodied conversational agents (ECAs) have become a well-established research area. Embodied agents are agents that are visible in the interface as animated cartoon characters or animated objects resembling human beings. Sometimes they just consist of an animated talking face, displaying facial expressions and, when using speech synthesis, having lip synchronization. These agents are used to inform and explain

Figure 2.1   Doctor in a simulation environment

or even to demonstrate products or sequences of activities in educational, e-commerce or entertainment settings. Experiments have shown that ECAs can increase the motivation of a student or a user interacting with the system. Lester *et al.* (1997) showed that a display of involvement by an ECA motivates a student in doing (and continuing) his or her learning task. In the following figures we show two examples of ECAs. The virtual human displayed in Figure 2.1 represents a doctor that plays a role in a simulation environment where a US soldier in Iraq has to persuade the doctor to move his medical clinic to a safer area. Such simulation environments are researched at the Institute for Creative Technologies in Marina del Rey.

The second example (Figure 2.2) is the GRETA agent developed by Catherine Pélachaud (Hartmann *et al.* 2005). This ECA has been made available for many applications. In this particular situation it is used

Figure 2.2   GRETA agent

to experiment with different expressivity settings for communication, that is, expressivity in gestures and body postures depending on emotion, personality, culture, role, and gender.

2.3.2 Nonverbal and Affective Interaction for Embodied Agents

An embodied agent has a face. It may have a body, arms, hands, and legs. We can give it rudimentary intelligence and capabilities that allow smooth and natural verbal and nonverbal interaction. Nonverbal signals come from facial expressions, gaze behavior, eyebrow movements, gestures, body posture, and head and body movements. But they are also available in the voice of an ECA. Communicative behavior can be made dependent on the personality that has been modeled in an ECA.

In previous years we have seen the emergence of affective computing. Although many research results on affect are available, it is certainly not the case that a comprehensive theory of affect modeling is available. Reasons to include emotion modeling in intelligent systems are, among others, to enable decision-making in situations where it is difficult, if not impossible, to make rational decisions, to afford recognition of a user's emotions in order to give better and more natural feedback, and to provide display of emotions, again, in order to allow natural interaction. Especially when the interface includes an ECA, it seems rather obvious that the user expects a display of emotions and some recognition of emotions by the embodied agent. On the other hand, in order to improve the interaction performance of embodied agents they should integrate and use multimodal affect information obtained from their human conversational partner. Measurement techniques and technology are becoming available to detect multimodal displayed emotions in human interactants (e.g., cameras, microphones, eye and head trackers, expression glasses, face sensors, movement sensors, pressure sensitive devices, haptic devices, and physiological sensors). In order to recognize and interpret the display of humor emotions by human interactants (and to be perceived by an ECA) we need to look at, among others, computer vision technology and algorithms for the interpretation of perceived body poses, gestures, and facial expressions.

Speech and facial expressions are the primary sources for obtaining information of the affective state of an interactant. Therefore an ECA needs to be able to display emotions through facial expressions and the voice. In speech, emotion changes can be detected by looking at deviations from personal, habitual vocal settings of a speaker because of emotional arousal. Cues come from loudness, pitch, vibrato, precision of articulation, etc. Kappas *et al.* (1991).

To describe emotions and their visible facial actions, facial (movement) coding systems have been introduced (Ekman 1993). In these systems facial units have been selected to make up configurations of muscle groups associated with particular emotions. The timing of facial actions is also described. Using these systems, the relation between emotions and facial movements can be studied and it can be described how emotion representations can be mapped on the contraction levels of muscle configurations. Modalities in the face that show affect also include movements of lips, eyebrows, color changes in the face, eye movement and blinking rate. Cues combine into expressions of anger, into smiles, grimaces or frowns, into yawns, jaw-droop, etc. Happiness, for example, may show in increasing blinking rate. Generally, we need the ability to display intensities and blends of emotions in the face.

2.3.3 Computers and Embodied Conversational Agents as Social Actors

Embodied agents are meant to act as conversational partners for computer users. An obvious question is whether they, despite available verbal and nonverbal communication capabilities, will be accepted as conversational partners.

Can we replace one of the humans in a human-to-human interaction by an embodied conversational agent without being able to observe important changes in the interaction behavior of the remaining human? Can we model human communication characteristics in an embodied conversational agent that guarantee or improve natural interaction between artificial agent and human partner? Obviously, whether something is an improvement or more natural depends very much on the context of the interaction, but being able to model such characteristics allows a designer of an interface employing embodied agents to make decisions about desired interactions.

In the research on the "computers are social actors" (CASA) paradigm (Reeves and Nass 1996) it has been convincingly demonstrated that people interact with computers as if they were social actors. Due to the way we can let a computer interact, people may find the computer polite, dominant, extrovert, introvert, or whatever attitudes or personality (traits) we can think of. Moreover, they react to these attitudes and traits as if a human being displayed them. As an example, consider the situation where a person interacts with the computer in order to perform a certain task. When, after completing the task, the person is asked by the same computer about its (i.e., the computer's) behavior, the user is much more positive than when asked this question while sitting behind another computer.

From the many CASA experiments we may extrapolate that humor, because of its role in human–human interaction, can play an important role in human–computer interactions. This has been confirmed with some specially designed experiments to examine the effects of humor in task-oriented computer-mediated communication and in human–computer interaction (Morkes *et al.* 2000). It was shown that humor could have many positive effects. For example, participants who received jokes during the interaction rated a system as more likable and competent. They smiled and laughed more, they responded in a more sociable manner and reported greater cooperation. This showed especially in the computer-mediated communication situations. Moreover, in their experiments the use of humor in the interface did not distract users from their tasks. According to the authors the study provided strong evidence that humor should be incorporated in computer-mediated communication and human–computer interaction systems.

In the CASA experiments, the cues that were used to elicit the anthropomorphic responses were minimal. Word choice, for instance, elicited personality attribution, voice pitch elicited gender attribution. In ECAs, however, the cues are not even minimal. Gender can be communicated by means of physical appearance and voice, personality can be communicated by means of behavior, word choice and nonverbal communication, much like is the case in human–human interaction. Consequently, the CASA paradigm should be applicable to ECAs at least as well as to computers in general.

In Friedman (1997) research is reported where in the interaction with users the computer is represented as an embodied agent. Again, the starting point is how human facial appearance influences behavior and expectations in face-to-face conversations. For example, a happy, friendly face predicts an enjoyable interaction. In experiments people were asked to talk to different synthetic faces and comparisons were made with a situation where the information was presented by text. General observations were that people behave more socially when communicating with a talking face, they are more attentive, they present themselves in a more positive light, and they attribute personality characteristics to the talking face. It was shown that different faces and facial expressions have impact on the way the user conceives the computer and also on the interaction behavior of the user.

The results of the CASA experiments indicate that users respond to computers as if they were humans. Of course, this doesn't mean that people will interact with computers exactly as they do with humans. Shechtman and Horowitz (2003) conducted experiments to study relationship behavior during keyboard human–computer interaction and (apparently) keyboard mediated human–human interaction. In the latter case participants used much more communion and agency relationship statements, used more words, and spent more time in conversation. Nevertheless, we conclude that it is possible, at least in principle, to design systems and more in particular embodied agents that are perceived as social actors and that can display characteristics that elicit positive feelings about an interaction, even though the interaction is not considered as perfect from the user's point of view. In the next section we study in more detail the possibility of generating humorous acts in order to introduce humor characteristics known to improve interaction in human–human interaction into human–computer interaction.

2.4 Appropriateness of Humorous Acts in Conversations

2.4.1 Introduction

In the previous sections we discussed the role of humor in human–human interaction and a possible role of humor in human–ECA interaction. Obviously, there are many types of humor and it is certainly not the case that every type of humor is suited for any occasion during any type of interaction. Telling a joke among friends may lead to amusement, while the same joke among strangers will yield misunderstanding or be considered as abuse. Therefore, an assessment of the appropriateness of the situation for telling a joke or making a humorous remark is always necessary.

Appropriateness does not mean that every conversational participant has to be in a jokey mood for a humorous remark. Rather, it means that the remark or joke can play a role in the interaction process, whether it is deliberately aimed at achieving this goal, whether there is a mutually agreed moment for relaxing and playing, or whether it is somewhere in between on this continuum. Clearly, it is also the "quality" of the humorous remark that makes it appropriate in a particular situation. Here, "quality" does not refer only to the contents of the remark, which may be based on a clever observation or ingenious wordplay, but in particular on an assessment whether or not to produce the remark at that particular moment.

In what follows we will talk about "humorous acts" (HAs). In telephone conversations a HA is a speech utterance. Apart from the content of what is being said, the speaker can only use intonation and timing in order to generate or support the humorous act. Obviously, we can think of exceptions, where all kinds of non-speech or paralinguistic sounds help to generate – on purpose or by accident – a HA. In face-to-face conversations a humorous act can include, be supported by or even made possible by nonverbal cues. Moreover, references can be made, implicitly or explicitly, to the environment that is perceivable for the partners in the conversation. This situation also occurs when conversational partners know where each of them is looking at or when they are able to look at the same display, display contents (e.g., a web page) or a shared virtual reality environment.

Although mentioned before, we emphasize that participants in a discussion may, more or less deliberately, use humor as a tool to reach certain goals. A goal may be to smooth the interaction and improve mutual understanding. In that case a HA can generate and can be aimed at generating feelings of common attitudes and empathy, creating a bond between speaker and hearer. However, a HA can also be face threatening and be aimed at eliminating an opponent in a (multi-party) discussion. However, whatever the aim is, conversational participants need to be able to compose elements of the context in order to generate a HA and they need to assess the current context (including their aims) in order to determine the appropriateness of generating a HA. This includes a situation where the assumed quality of the HA overrules all conventions concerning cooperation during a goal-oriented dialogue.

Sometimes, conversations have no particular aim, except the aim of providing enjoyment to the participants. The aim of the conversation is to have an enjoyable conversation and humor acts as a social facilitator. In Tannen (1984), for example, an analysis is given of the humorous occurrences in the conversations held at a Thanksgiving dinner. Different styles of humor for each of the dinner guests could be distinguished. All guests had humorous contributions. For some participants more than ten percent of their turns were ironic or humorous. With humor one can make one's presence felt, was one of her conclusions.

Generation (and interpretation) of HAs during a dialogue or conversation has hardly been studied. There is not really a definition, but at least the notion of conversational humor has been introduced in the scientific literature (Attardo 1996), emphasizing characteristics as being improvised and being highly contextual. Conversational humor is in kidding and in teasing and, by acting the improvisation, it is the main ingredient of sitcoms (situation comedies). Many observations on joke telling during conversations are valid as well for HA generation during a conversation. Before going into some more details of generating HAs we repeat some of them here.

Raskin (1985) discussed speaker–hearer couplings for joke telling obtained from the four situations arising from a speaker who intentionally or unintentionally makes a joke and a hearer who expects or does not expect it. Although we won't exclude the situations where a speaker unintentionally performs a HA or where the hearer expects a HA to be generated, we will mostly assume the situation where the speaker intentionally constructs a HA and the hearer is assumed not to expect it.

The next (related) observation concerns the Gricean cooperation principle. Grice's assumption was that conversational partners are cooperative. Jokes are about misleading a conversational partner. However, explicit joke telling is often preceded by some interaction that is meant to obtain agreement about the inclusion of a joke in the interaction. In such a case there is explicit announcement and agreement to make a switch from a bona-fide mode of communication to a non-bona-fide mode of communication. Contrary to a situation where someone starts telling lies, within the constraints of humor the communication can nevertheless be bona-fide since speaker and hearer agree about the switch. When the participants are in the mood for jokes, joke telling occurs naturally and there is some meta-level cooperation. Raskin (1985), Zajdman (1991) and Attardo (1997) all discuss variants and extensions of the Gricean cooperation principle, allowing such "non-cooperative" or non-bona-fide modes of behavior.

Related to this discussion and also useful from the point of view of HA generation and appreciation is the introduction of the double-bond model in Zajdman (1992). In this model the primary bond between speaker and hearer consists of both making the switch from the usual bona-fide mode of communication to the humorous (non-bona-fide) communication mode. It is not necessarily the case that the speaker makes this switch before the hearer does. An unintentional humorous remark may be recognized first by the hearer or it can even be elicited by the hearer in order to create a funny situation. In our HA generation view we assume, however, that the speaker intentionally makes this switch and the hearer is willing to follow. The secondary bond concerns the affirmation of the switch for both parties. When consent is asked and obtained to tell a joke during a conversation, the secondary bond is forged before the primary bond. But it can also occur during or after communicating a humorous message. Some synchronization should be there since otherwise the humorous act will not be successful. Obviously, in face-to-face conversations nonverbal cues can make verbal announcements and confirmations superfluous. And, as mentioned before, when the conversational partners are in the mood for jokes, there is some meta-level cooperation bypassing the requirements for making and acknowledging mode switches.

We emphasize the spontaneous character of HA construction during conversational humor. The opportunity is there and, although the generation is intended, it is also unpredictable and irreproducible. Nevertheless, it can be aimed at entertaining, to show skill in HA construction or to obtain a cooperative atmosphere. HA creation can occur when the opportunity to create a HA and a humorous urge to display the result temporarily overrules Gricean principles concerning truth of the contribution, completeness of the contribution, or relevance of the contribution for the current conversation.

The moment to introduce a canned joke during a conversation can also be related to the situation. It can be triggered by an event (a misunderstanding, the non-availability of information, word choice of a conversational partner, etc.), including a situation where one of the conversational partners does not know what to contribute next to the conversation. However, the contribution – and not the moment – remains canned, although the joke can become merged with the text and adapted to fit a contextual frame (Zajdman 1991). In contrast, conversational HAs are improvised and more woven into the discourse through natural contextual ties. There is no or hardly any signaling of the humorous nature of the act and, in fact, this would reduce or destroy its effect.

2.4.2 Setting the Stage for Humor Generation by Embodied Conversational Agents

When leaving the face-to-face conversations in human–human interaction and entering a situation where one of the humans is replaced by a computer, or more in particular, an embodied conversational agent, we have to reconsider the roles of the conversational partners. To put it simply, one of the partners has to be designed and implemented. While on the one hand we nevertheless need to understand as well as possible the models underlying human communication behavior, it also gives us the freedom to make our own decisions concerning communication behavior of the ECA, taking into account the particular role it is expected to play. We need to be more explicit about what an ECA's human conversational partner can expect from an ECA. From a design point of view, everything is allowed to make an ECA believable for its human partner. Artists have explored the creation of engaging characters, using advanced graphics and animations, in games and movies with tremendous commercial success. They create believable characters, but there is no modeling of (semi-) autonomous behavior of these characters in a context that allows interaction with human partners. Rather than showing intelligence, "appropriately timed and clearly expressed emotion is a central requirement for believable characters" (Bates 1994). In ECA design, rather than adhere to a guideline that says "try to be as realistic as possible", the more important guideline is "try to create an agent that permits the audience's suspension of disbelief".

When looking at embodied conversational agents we need to distinguish four modes of humor interpretation and generation. We mention these modes, but it should be understood that we are far from being able to provide the necessary appropriate models that allow them to display these skills. On the other hand, we don't always need agents that are perfect, as long as they are believable in their application. The first two modes concern the skills of the ECA:

- The ECA should be able to generate HAs. How should it construct and display the HA? When is it appropriate to do so? Apart from the verbal utterance to be used, it should consider intonation, body posture, facial expression, and gaze, all in accordance with the HA. The ECA should have a notion of the effect and the quality of the HA in order to have it accompanied with nonverbal cues. Moreover, when in a subsequent utterance its human partner makes a reference to the HA, it should be able to interpret this reference in order to continue the conversation.
- The ECA should be able to recognize and understand the HAs generated by its human conversational partner. Apart from understanding from a linguistic or artificial intelligence point of view, this also requires showing recognition (e.g., for acknowledgment) and comprehension by generating appropriate feedback, including nonverbal behavior (facial expression, gaze, gestures, and body posture).

These are the two ECA points of view. Symmetrically, we have two modes concerning the skills of the human conversational partner. Generally, we may assume that humans have at least the skills mentioned above for ECAs.

- The human conversational partner should be able to generate HAs and accompanying signals for the ECA. Obviously, the human partner may adapt to the skills and personality of the particular ECA, as will be done when having a conversation with another human.
- The human conversational partner should recognize, acknowledge and understand HA generation by the ECA, including accompanying nonverbal signals. Obviously, the ECA may have different ideas about acts being humorous than its particular conversational partner.

Our aim is to make ECAs more social by investigating the possibility for them to generate humorous acts. This task can certainly not be done completely isolated from the other issues alluded to: the appreciation of the HA by a conversational partner, the continuation of dialogue or conversation, the double-bond issues, and so on. Two observations are in order. Firstly, when we talk about the generation of a HA and corresponding nonverbal communication behavior of an ECA we should take into account an assessment of the appropriateness of generating this particular HA. This includes an assessment of the appreciation of the HA by the human conversational partner, and therefore it includes some modeling of the interpretation of HAs by human conversational partners. That is, a model for generation of HAs requires a model of interpretation and appreciation of HAs. This is not really different from discourse modeling in general. An ECA needs to make predictions of what is going to happen next. Predictions help to interpret a next dialogue act or, more generally, a successor of a humorous act.

A second observation also deals with what is happening after introducing a HA in a conversation. What is its impact on the conversation and the next dialogue acts from a humor point of view? This introduces the issue of humor support; that is, apart from acknowledging, will the conversational partner support and further contribute to the humorous communication mood? Hay (2001) distinguishes several types of humor support strategies: contributing more humor, playing along, using echo, and offering sympathy. Support can also mean the co-construction of a sequence of remarks leading to a hilarious or funny observation starting from a regular discourse situation. Trying to model this requires research on (sequences of) dialogue acts, an issue that is rather distant from current dialogue act research, both from the point of view of non-regular sequences of acts and from the point of view of distinguishing sufficiently many subtleties in dialogue acts that initiate and allow such sequences.

Finally, as a third observation, we need to consider whether HA generation by a computer or by an ECA gives rise to HAs that are essentially different and maybe more easily generated or accepted than human-generated HAs. An ECA may have less background and be less erudite, but it may have encyclopedic knowledge of computers or a particular application. In addition, a computer or an ECA can become easily the focus of humor of a human conversational partner. Being attacked because of imperfect behavior can be anticipated and the use of self-deprecating humor can be elaborated in the design of an ECA. In that case the ECA makes itself the butt of humor, for example, by making references to its poor understanding of a situation, its cartoon-like appearance and facial expressions, or its poor quality of speech recognition and speech synthesis.

All these issues are important, but can only be mentioned here. A start has to be made somewhere and therefore we will mainly discuss an ECA's ability to generate an HA without looking too much at what will happen afterwards. Corresponding nonverbal behavior should be added to the generation of a HA, or better, should be designed in close interaction with the generation of verbal acts (Theune *et al.* 2005). We will return to this in section 2.6.

2.4.3 Appropriateness of Humorous Act Generation

Humor is about breaking rules, such as violating politeness conventions or, more generally, violating Gricean rules of cooperation (see also section 2.4.1). In creating humorous utterances during an interaction people hint, presuppose, understate, overstate, use irony, tautology, ambiguity, etc. (Brown and Levinson 1978); i.e., all kinds of matters that do not follow the cooperative principles as they were formulated in some maxims (e.g., "Avoid ambiguity", "Do no say what you believe to be false", etc.) by the philosopher Grice (1989). Nevertheless, humorous utterances can be constructive – that is, support the dialogue – and there can be a mutual understanding and cooperation during the construction of a HA. The HAs we consider here are, contrary to canned jokes that often lack contextual ties, woven into the discourse. Canned jokes are not completely excluded, since some of them can be adapted to the context, for example by inserting the name of a conversational partner or by mapping words or events of the interaction that takes place on template jokes (Loehr 1996). Nevertheless, depending on contextual clues a decision has to be made to evoke and adapt the joke in order to integrate it in a natural way in the discourse. Such decisions have also to be made when we consider hints, understatements, ambiguities, and other communication acts and properties that aim or can be used to construct a HA.

For HA construction, we need to zoom in on two aspects of constructing humorous remarks:

- recognition of the appropriateness of generating a humorous utterance by having an appraisal of the events that took place in the context of the interaction; dialogue history, goals of the dialogue partners (including the dialogue system), the task domain and particular characteristics of the dialogue partners have to be taken into account; and

- using contextual information, in particular words, concepts and phrases from the dialogue and domain knowledge that is available in networks and databases, to generate an appropriate humorous utterance, i.e., a remark that fits in the context, that is considered to be funny, is able to evoke a smile or a laugh, or that maybe is a starting point to construct a funny sequence of remarks in the dialogue.

It is certainly not the case that we can look at both aspects independently. With some exceptions, we may assume that, as should be clear from human–human interaction, HAs can play a useful and entertaining role at almost every moment during a dialogue or conversation. Obviously, some common ground, some sharing of goals or experiences during the first part of the interaction is useful, but it is also the quality of the generated HA that determines whether the situation is appropriate to generate this act. We cannot simply assess the situation and decide that now is the time for a humorous act. When we talk about the possibility to generate a HA and assume a positive evaluation of the quality of the HA given the context and the state of the dialogue context, then we are also talking about appropriateness.

In order to generate humor in dialogue and conversational interaction we need to continuously integrate and evaluate the elements that make up the interaction (in its context and given the goals and knowledge of the dialogue system and the human conversational partner) in order to decide:

- the appropriateness or non-appropriateness of generating a humorous utterance, and
- the possibility that elements from the dialogue history, the predicted continuation of the dialogue and knowledge available from domain, task and goals of the dialogue partners allow the construction and the generation of a humorous act.

2.5 Humorous Acts and Computational Humor

2.5.1 Computational Humor

Well-known philosophers and psychologists have contributed their viewpoints to the theory of humor. Sigmund Freud saw humor as a release of tension and psychic energy, while Thomas Hobbes saw it as a means to emphasize superiority in human competition. In the writings of Immanuel Kant, Arthur Schopenhauer, and Henri Bergson, we can see the first attempts to characterize humor as dealing with incongruity, that is, recognizing and resolving incongruity. Researchers including Arthur Koestler, Marvin Minsky, and Alan Paulos have tried to clarify these notions, and others have tried to formally describe them.

As might be expected, researchers have taken only modest steps toward a formal theory of humor understanding. General humor understanding and the closely related area of natural-language understanding require an understanding of rational and social intelligence, so we won't be able to solve these problems until we've solved all AI problems. It might nevertheless be beneficial to look at the development of humor theory and possible applications that don't require a general theory of humor; this might be the only way to bring the field forward. That is, we expect progress in application areas – particularly in games and other forms of entertainment that require natural interaction between agents and their human partners, rather than from investigations by a few researchers into full-fledged theories of computational humor.

Incongruity-resolution theory provides some guidelines for computational humor applications. We won't look at the many variants that have been introduced or at details of one particular approach. Generally, we follow Graeme Ritchie's approach (1999). However, since we prefer to look at humorous remarks that are part of the natural interaction between an ECA and its human conversational partner, our starting point isn't joke telling or pun making, as is the case in the work by Ritchie. Rather, we assume a not too large piece of discourse (a text, a paragraph, or a sentence) consisting of two parts. First you read or hear and interpret the first part, but as you read or hear the second part, it turns out that a misunderstanding has occurred that requires a new, probably less obvious interpretation of the previous text. So, we have an obvious interpretation, a conflict, and a second, compatible interpretation that resolves the conflict. Although misunderstandings can be humorous, this is not necessarily the case. Deliberate misunderstanding sometimes occurs to create a humorous remark, and it is also possible to construct a piece of discourse so that it deliberately leads to a humorous misunderstanding.

In both cases, we need additional criteria to decide whether the misunderstanding is humorous. Criteria that have been mentioned by humor researchers deal with a marked contrast between the obvious interpretation and the forced reinterpretation, and with the reinterpretation's common-sense inappropriateness. As an example, consider the following dialogue in a clothing store:

LADY: "May I try on that dress in the window?"
CLERK: [doubtfully] "Don't you think it would be better to use the dressing room?"

The first utterance has an obvious interpretation. The clerk's remark is confusing at first, but looking again at the lady's utterance makes it clear that a second interpretation (requiring a different prepositional attachment) is possible. This interpretation is certainly different, and, most of all, it describes a situation that is considered as inappropriate.

What can we formalize here, and what formalisms are already available? Artificial intelligence researchers have introduced scripts and frames to represent meanings of text fragments. In early humor theory, these knowledge representation formalisms were used to intuitively discuss an obvious and a less obvious (or hidden) meaning of a text. A misunderstanding allows at least two frame or script descriptions of the same piece of text; the two scripts involve overlap. To make it clear that the non-obvious interpretation is humorous, at least some contrast or opposition between the two interpretations should exist. Script overlap and script opposition are reasonably well-understood issues, but until now, although often described more generally, the attempts to formalize this opposition mainly look at word-level oppositions (e.g., antonyms such as hot versus cold). Inappropriateness hasn't been formalized at all.

2.5.2 Generation of Humorous Acts: Anaphora Resolution

It is possible to look at some relatively simple situations that allow us to make humorous remarks. These situations fit in the explanations we gave earlier and they make it possible to zoom in on the main problems of humor understanding: rules to resolve incongruity and criteria that help determine whether a solution is humorous. Below we present an example of constructing a humorous act using linguistic and domain knowledge. The example is meant to be representative for our approach, not for its particular characteristics. It is an example of deliberately misunderstanding, an act that can often be employed in a conversation when some ambiguity in words, phrases or events is present, in order to generate a HA. It can also be considered as a surprise disambiguation.

We can have ambiguities at pragmatic, semantic, and syntactic levels of discourse (text, paragraphs, and sentences). At the sentence level, we can have ambiguities in phrases (e.g., prepositional phrase attachment), words, anaphora, and, in the case of spoken text or dialogue, intonation. As we interpret text that we read or hear, possible misunderstandings will become clear and be resolved, maybe with help from our conversational partner. Earlier, we gave an example of ambiguity that occurred because a prepositional phrase could be attached to a syntactic construct (a verb, a noun phrase) in more than one way.

In this particular example we look at deliberate erroneous anaphora resolution to generate a HA. One problem in natural language processing is anaphoric reference. Anaphorically used words are words that are referring back to something that was earlier mentioned or that is known because of the discourse situation and/or the text as it is read or heard. The anaphorically used word is called "the anaphor", the word or phrase to which it refers "the antecedent". The extra lingual entity they co-refer to is called

Figure 2.3   Attempt at anaphora resolution in a Dilbert cartoon. ©www.bruno.nl Syndicated by Bruno Publications B.V.

"the referent". Anaphora resolution is the process of determining the antecedent of an anaphor. The antecedent can be in the same sentence as the anaphor, or in another sentence. Incorrect resolution of anaphoric references can be used in order to create a humorous remark in a dialogue situation. Consider for example the text used in a Dilbert cartoon (see Figure 2.3) where a new "Strategic Diversification Fund" is explained in a dialogue between the Adviser and Dilbert.

ADVISER: "Our lawyers put your money in little bags, then we have trained dogs bury them around town."

How to continue from this utterance? Obviously, here, in the cartoon, we are dealing with a situation that is meant to create a joke, but all the elements of a non-constructed situation are there too. What are these dogs doing? Burying lawyers or bags? So, a continuation could be:

DILBERT: "Do they bury the bags or the lawyers?"

Surely, this Dilbert question is funny enough, although from a natural-language processing point of view it can be considered as a clarifying question, without any attempt to be funny. There is an ambiguity – that is, the system needs to recognize that generally dogs don't bury lawyers and therefore "them" is more likely to refer to bags than to lawyers. Dogs can bury bags, dogs don't bury lawyers.

We need to be able to design an algorithm that is able to generate this question at this particular moment in the dialogue. However, the system should nevertheless know that certain solutions to this question are not funny at all. It can take the most likely solution, from a common-sense point of view, but certainly this is not enough for our purposes. Here, "them" is an anaphor referring to a previous noun phrase. Its antecedent can be found among the noun phrases in the first sentence. Many algorithms for anaphora resolution are available and generally they come up with a solution that satisfies as many constraints as can be extracted from the sentence (gender, number, recency, emphasis, verb properties, order of words, etc.). We need, however, algorithms for anaphora resolution that decide to take a wrong but humorous solution, that is, a solution that does not necessarily satisfy all the constraints. And, preferably, violating one particular constraint should lead to the determination of an antecedent that, combined with the more obvious antecedent, leads to a question in which both of them appear, as in "Do they bury the bags or the lawyers?"

From a research viewpoint, the advantage of looking at such a simple, straightforward humorous remark is that we can confine ourselves to just one sentence. So, rather than having to look at scripts, frames, and other discourse representations, we can concentrate on the syntactic and semantic analysis of just one sentence. For this analysis, we can use well-known algorithms that transform sentences into feature structure representations and issues such as script overlap and script opposition turn into properties of feature sets.

Contrast and inappropriateness are global terms from (not yet formalized) humor theory. In our approach, determining contrast translates into a heuristic that considers a potentially humorous antecedent and decides to use it because it has many properties in common with the correct antecedent. However, at least one salient property distinguishes the two potential antecedents (a shop window versus a dressing room, a bicycle versus a car, a bag versus a lawyer). A possible approach checks for inappropriateness by looking at constraints associated with the verb's thematic roles in the sentence. For example, these constraints distinguish between animate and inanimate; hence, burying lawyers who are alive is inappropriate. Obviously, more can be said about this. For example, professions or groups for which negative stereotypes exist are often grateful targets of jokes.

2.5.3 Implementation and Experiments

As mentioned earlier, when looking at the fundamental problems in humor research, we must wait until the main problems in artificial intelligence (AI) have been solved and then can apply the results to humor understanding and humor applications. This means, inter alia, that we have to wait until we are able to model "all" common-sense knowledge and the ability to reason with it. Clearly, it is more fruitful to investigate humor itself and see whether solutions that are far from complete and perfect can nevertheless find useful applications. In games and entertainment computing, natural interaction with ECAs for example requires humor modeling. Although many forms of humor don't fit into our framework of humorous misunderstandings, it can be considered a useful approach since it allows us to make the problems rather clear.

Current humor research has many shortcomings, which are also present in the approach discussed in the previous subsection. In particular, the conditions such as contrast and inappropriateness that have been mentioned might be necessary, but they are far from sufficient. Further pinpointing of humor criteria is necessary. Our approach has as starting point well-known theories from computational linguistics rather than trying to put linguistics, psychology and sociology in a comprehensive framework from which to understand humor.

In (Tinholt and Nijholt 2007) a design, an implementation, and experiments are reported that follow the viewpoints and ideas expressed above, in particular the approach to humorous anaphora resolution. This approach has been implemented in a chatbot. One reason to choose a chatbot is that its main task is to get a conversation going. Hence, it might miss opportunities to make humorous remarks, and when an intended humorous remark turns out to be misplaced, this isn't necessarily a problem. An attempt to embed pun-making in the conversational ability of a chatbot has been reported earlier (Loehr 1996). However, unlike our approach, in that case the proposed link between the contents of the pun and the interaction history was very poor.

Implemented algorithms for anaphora resolution are already available. We have chosen a Java implementation (JavaRAP) of the well-known Lappin and Leass' Resolution of Anaphora Procedure (1994). We obtained a more efficient implementation by replacing the embedded natural-language parser with a parser that has been made available to the research community by Stanford University. Experiments were designed to find ways to deal with anaphora-resolution algorithms' low success rate and to consider the introduction of a reliability measure before proceeding with possible antecedents of an anaphor. Other issues we are investigating are the different frequencies and types of anaphora in written and spoken text. In particular, we have been looking at properties from the anaphora viewpoint of conversations with well-known chatbots such as ALICE, Toni, Eugine, and Jabberwocky. Resources that are investigated are publicly available knowledge bases such as WordNet, WordNet Domains, FrameNet, VerbNet, and ConceptNet. For example, in VerbNet, every sense of a verb is mapped to a verb class representing the conceptual meaning of this sense. Every class contains both information on the thematic roles associated

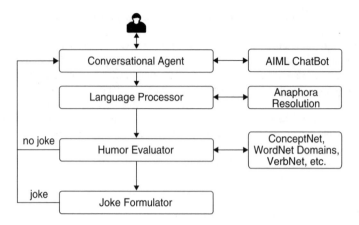

Figure 2.4   Architecture of the humorous anaphora system

with the verb class and frame descriptions that describe how the verb can be used. This makes it possible to check whether a possible humorous antecedent of an anaphor sufficiently opposes the correct antecedent because of constraints that it violates. Unfortunately, because VerbNet only contains about 4500 verbs, many sentences cannot be analyzed.

In Figure 2.4 the four modules that form our system are shown. The "conversational agent" is responsible for receiving input from the user and keeping the conversation with the user going. It forwards all input from the user to the rest of the system. If the system indicates that an anaphora joke can be made, then the conversation agent will make this joke. Otherwise, the conversational agent uses an AIML (Artificial Intelligence Mark-up Language)-based chatbot (Wallace *et al.* 2003) to formulate a response to the user. If the system cannot make an anaphora joke, it uses the reactions of the chatbot. The analysis whether a user utterance contains humorous anaphora ambiguities is done by the "language processor" and the "humor evaluator". In the language processor an adjusted version of JavaRAP is used that locates every pronominal anaphor in the text and returns all antecedents that are not excluded based on eliminating factors like gender and number agreement. If at least two possible antecedents are found, the anaphoric expression is ambiguous and this information is forwarded to the humor evaluator to check whether this ambiguity is humorous.

The humor evaluator implements the check for contrast, described earlier. Roughly this comes down to comparing the real antecedent of the anaphor to the possible other candidate(s). The properties of possible antecedents are retrieved from ConceptNet and they are compared. If there is an acceptable antecedent with a sufficient number of properties in common with the real antecedent and the non-overlapping properties contain at least one pair of antonymous properties, then they are considered to be in contrast. In this implementation we don't perform tests on inappropriateness or taboo. Hence, based on contrast an utterance is generated by the joke formulator, where the (hopefully successful) joke is a simple clarification request that indicates that the anaphoric reference was deliberately misunderstood. In our Dilbert example, the conversational agent is expected to return "The lawyers were buried?"

The system was evaluated by having it analyze a chatbot transcript and a simple story text (Tinholt and Nijholt 2007). It turned out that humorous cross-reference ambiguity was rare. The system was able to make some jokes, but its performance was also very moderate due to the available tools. There is imperfect parser output, the anaphora resolution algorithm is not perfect and has to work with this imperfect output of the parser. In addition, errors are caused by the sparseness of ConceptNet. For these reasons we think that our approach does provide future prospects in generating conversational humor but that for the moment we have to wait for more elaborated versions of ConceptNet or similar resources and for better parsers and anaphora resolution algorithms.

2.5.4 Discussion

Although we have not seen humor research devoted to erroneous anaphora resolution, the approaches in computational humor research in general are not that different from what we saw in the examples presented here. The approaches are part of the incongruity-resolution theory of humor. This theory assumes situations – either deliberately created or spontaneously observed – where there is a conflict between what is expected and what actually occurs. Ambiguity plays a crucial role. Phonological ambiguity, for example in certain riddles, syntactic ambiguity, semantic ambiguity of words, or events that can be given different interpretations by observers. Due to the different interpretations that are possible, resolution of the ambiguity may be unexpected, especially when one is led to assume a "regular" context and only at the last moment it turns out that a second context allowing a different interpretation was present as well. These surprise disambiguations are not necessarily humorous. Developing criteria to generate humorous surprise disambiguations only is one of the challenges of humor theory. Attempts have been made, but they are rather primitive. Pun generation is an example (Binsted and Richie 1997), acronym generation another (Stock and Strapparava 2003). In both cases we have controlled circumstances. These circumstances allow the use of WordNet and WordNet extensions and reasoning over these networks, for example, to obtain a meaning that does not fit the context or is in semantic opposition of what is expected in the context. No well-developed theory is available, but we see a slow increase in the development of tools and resources that make it possible to experiment with reasoning about words and meanings in semantic networks, with syllable and word substitutions that maintain properties of sound, rhyme or rhythm and with some higher-level knowledge concepts that allow higher-level ambiguities.

2.6 Nonverbal Support for Humorous Acts

2.6.1 Introduction

Being allowed to look at computers as social actors also allows us to have more natural interaction and it allows us to influence the interaction on aspects of emotions, trust, personality, attraction, and enjoyment. Interest in these issues grew with the introduction of embodied agents in the interface and the opportunity to add nonverbal cues to support the interaction and to display emotions and individual characteristics in face, body, voice, and gestures. For example, the role of small talk for inducing trust in an embodied real estate agent has been discussed in Cassell *et al.* (2001). The development of long-term relationships with a virtual personal assistant has been discussed in Bickmore and Cassell (2001) and in Bickmore (2003), and in Stronks *et al.* (2002) we presented a preliminary discussion on friendship and attraction in the context of the design and implementation of ECAs.

Facial expressions and speech characteristics have received most of the attention in research on nonverbal communication behavior. Emotion display has become a well-established area, very much stimulated by available theories of emotions on the one hand and theories of human facial expressions and speech intonation on the other hand. It is certainly the case that humor appreciation is associated with positive emotions.

Hardly any research is available on generating accompanying nonverbal behavior in humorous human–human communication. Consequently, there is not yet much research going on in nonverbal issues for embodied agents that interpret or generate humor in the interface. There is, however, a growing interest in translating attainments from social psychology research in face-to-face behavior to the human–ECA situation. Displaying appreciation of humor in face or voice is an issue that, however, has received some attention.

In this section we survey the different approaches in the literature to nonverbal communication by ECAs with, of course, an emphasis on those approaches that seem to be important from a humor generation point of view.

2.6.2 Nonverbal HA Display

In this chapter, the assumption is that it is useful to have ECAs generate humorous remarks. That is, we have ECAs as transmitters that generate HAs and accompanying nonverbal behavior, rather than as hearers or recipients of humorous remarks made by their human conversational partners. Unfortunately, not much can be said about accompanying nonverbal behavior during HA generation. The speaker may enjoy creating and making a humorous remark, but may decide to hide this in order to increase the effect of the act. Spontaneous act generation may be accompanied by displaying an enjoyment emotion in the face and the gestures that are made. Generation may be followed by some nonverbal acknowledgement that a change has been made to a non-bona-fide mode of communication, such as a smile, a particular gesture, a wink or, more likely, a combination of these modalities.

How will an ECA show enjoyment in voice and face? Laughs, smiles or more subtle expressions of enjoyment can be modeled in the expressions an ECA can display in the face and in the voice. See Kappas *et al.* (1991), for example, for a discussion on cues that are related to detecting and generating enjoyment in the voice. From the speech point of view the vocalization of laughter is another interesting issue for ECAs.

Ekman (1985) distinguishes between eighteen different smiles and functions ascribed to them. A smile can be a greeting; it can mean incredulity, affection, embarrassment or discomfort, to mention a few. Smiling does not always accompany positive feelings. The different functions make it important to be able to display the right kinds of smiles at the right time on the face of an ECA. Should it display a felt smile because of a positive emotional experience, should it take the harsh edge of a critical remark, or is it meant to show agreement, understanding or intention to perform?

Frank and Ekman (1993) discuss in some more detail the "enjoyment" smile, the particular type of smile that accompanies happiness, pleasure, or enjoyment. The facial movements that are involved in this smile are involuntary; they originate from other parts in the brain than the voluntary movements and have a different manifestation. Morphological and dynamic markers have been found to distinguish enjoyment smiles from others. The main, best-validated marker is known as the Duchenne marker or Duchenne's smile, the presence of orbicular oculi action (the muscle surrounding the eyes) in conjunction with zygomatic major action (muscles on both sides of the face that pull up the mouth corners). Although some people can produce it consciously, the Duchenne marker is one of the best facial cues for deciding enjoyment[1] and therefore an ECA should show it in the case of sharing humorous events with its human partner. For a survey of hypotheses and empirical findings regarding the involvement of muscles in the laughter facial expression, see Ruch and Ekman (2001). Laughter also involves changes in posture and body movements. Again, we need to distinguish between different types of laughter (spontaneous, social, and suppressed).

2.6.3 Showing Feigned or Felt Support?

In applications using ECAs we have to decide which smiles and laughs to use while interacting with a human conversational partner. When a virtual teacher smiles, should it be a Duchenne smile? Is the

---

[1] The timing of the onset and offset phase are other cues that signal the distinction between a deliberate and a spontaneous smile.

embodied agent "really" amused or does it only display a polite smile because it does not really like the joke made by its human conversational partner? Or should it not laugh or smile at all because of a politically incorrect joke? As mentioned by Cowie (2000):

> People respond negatively to displays of emotion that are perceived as simulated, and that is a real issue for agents that are intended to convey emotion.

Will our attempts to introduce believability not be hampered by the impossibility to convey emotions in a believable way? Maybe we accept poor-quality speech synthesis, maybe we accept poor-quality facial expression (compared with human speech and human facial expressions), but will we accept the same for emotion display, in particular display related to an appreciation of a humorous event conveyed through these channels? Note, that when we talk about a humorous event we include events that appear in a story being told by a virtual agent in interaction with a human conversational partner, events that are interpreted from a sequence of utterances in a dialogue, events that are visualized in a virtual environment, or events that need interpretation of integrated virtual and real-life interactions.

In our view these issues are not different from other observations on believability of embodied agents. In some situations, assuming that quality allows it, a synthesized voice or face may express acted pleasure (or anger), in other situations genuine pleasure (or anger). Whether it sounds or looks sincere depends on being able to suspend disbelief in the human partner of the agent. Interesting in this respect is the work of Marquis and Elliott (1994) who discuss research on embodied poker-playing agents (with a human partner) that can deliberately display false emotions in the face and in the voice.

Humor, when generated by an ECA with appropriate nonverbal display, can help in establishing a sense of well-being by its human partner and this can make it easier to persuade the human partner to make certain decisions or to perform certain actions. This can be for the "good" (persuading someone to continue a fitness program) or for the "bad" (persuading someone to buy goods he or she doesn't really want) of the human partner. Being able to design ECAs that manipulate their human partners requires attention for ethical issues.

2.6.4 Humor, Culture, and Gender

National and cultural differences in humor appreciation have been investigated. Differences were observed, but most authors agree that one should be very careful with interpretation of results because of uncertainty of the samples, low significance in the results, possible differences between humor production and humor appreciation and what has been rated (cartoons, written or spoken jokes), and what humor taxonomies have been used. There is certainly no agreement about definitions of humor classes and there is no agreement in which category to put a certain joke. Some differences that have been mentioned: higher appreciation of sexual humor and lower appreciation of nonsense humor was found for Italian adults when compared with German adults (Ruch and Forabasco 1994); unlike students from Belgium and Hong Kong, US students preferred jokes with sexual and aggressive contents (Castell and Goldstein 1976); Singaporean students supply a greater number of aggressive jokes and fewer jokes with sexual content than American students (Nevo *et al*. 2001); and so on.

There are also gender differences in humor behavior. Again, as we mentioned above, it is not wise to generalize from results obtained in a particular experiment. We observe that in several studies it has been reported that males enjoy self-bragging jokes more than females (Tannen 1994; Chao-chih Liao 2003). It has also been reported that men have a larger repertoire of jokes, enjoy joke telling more than women do, and score higher on humor creation. Women often show more appreciation of jokes than men do. Males enjoy sexual humor more than females, but there are also differences when looking at explicitness (high or low) and the victims of sexual jokes. In the research there are sometimes straightforward conflicting results (Herzog 1999).

Studies also show that there is an anti-feminine bias in humor. See, for example, Lampert and Ervin-Trip (1998) for a review of gender differences in humor behavior, including possible changes. Because of gender roles in society and because of social changes in gender roles we need to be very careful in translating properties in real life to the design of embodied agents that create or appreciate humor.

2.7 Methods, Tools, Corpora, and Future Research

2.7.1 Introduction

When discussing humor research for ECAs and its future development it is useful to distinguish between methods, tools and resources for verbal HA generation and methods and tools that help to have ECAs generate and display HAs using nonverbal communication acts. Nowadays, graphics, animation and speech synthesis technology make it possible to have embodied agents that can display smiles, laughs and other signs of appreciation of the interaction or humorous remarks included in the interaction. Current ECA research, however, hardly pays attention to these issues. Facial expressions are often confined to the display of some basic emotions and subtleties of different smiles are not considered. Especially the display of laughter at appropriate moments will not only meet research challenges about appropriateness, but also challenges in speech synthesis and in facial animation. Current multimodal and affective markup languages need to be extended and detailed in order to include the multimodal presentation of humorous acts in ECA behavior.

2.7.2 Corpora, Annotation Schemes, and Markup Languages

Corpora are needed in order to study the creation of HAs in dialogues and naturally occurring conversations, including conversations that make references to common knowledge, task and domain knowledge, conversation history, and the two- or three-dimensional visualized context of the conversation. By visualized context we mean the ECA and its environment (e.g., a reception desk, a lounge, posters in the environments, a particular training environment, other ECAs, including users and visitors, etc.).

There is a wealth of material available in existing corpora that have been collected, mostly for quite different purposes than for doing humor research. There are corpora that have been collected and annotated for research on dialogue modeling (sometimes obtained from Wizard of Oz experiments) or on dialogue act prediction, there are corpora for specific multimodal tasks that have to be performed, there are corpora of spontaneous conversations, and there are corpora of recorded interactions where the emphasis is on gestures and facial expressions. Transcripts of dialogues and conversations allow investigating when and why humorous acts were created and when there was the opportunity to create a humorous act. That is, which elements of the discourse have been used or could have been used in order to construct a humorous act? Clearly, in goal-directed dialogues we will see less HAs than in naturally occurring conversations that are open-ended, contain many topic shifts, and are also held to maintain a relationship. Hence, it is useful to distinguish between different kinds of domains and tasks when considering corpora that can be used for humor research.

Many corpora have been collected and there are also many initiatives to make them available to researchers. One example is the Linguistic Data Consortium (LDC; http://www.ldc.upenn.edu/), a foundation taking responsibility for making available hundreds of meeting recordings, telephone conversations, and interviews. Many corpora are available where users are asked to plan a trip, rent a car and arrange a hotel. From the so-called Map Task experiments corpora have been collected where two participants engage into a cooperative dialogue to find out how to go from one place to another. Both participants have maps of the environment, but they differ in showing landmarks that can be used for a verbal description of a route. In our own research we collected, using Wizard of Oz experiments, dialogues where the human partner is asking information about theatre performances, performers, dates and locations in natural language. In these examples, most of the interaction is about goal-directed dialogues, rather than on naturally occurring interactions during a conversation. For humor research more corpora are needed containing naturally occurring conversations.

Well known annotation schemes have been introduced to describe and distinguish (verbal) dialogue acts. For example, there is DAMSL (Dialog Act Markup in Several Layers), a scheme that has been developed to annotate two-agent task-oriented dialogues. In the dialogue the agents are supposed to collaborate to solve some problem. In the DAMSL scheme, utterance tags, indicating some particular aspect of the utterance, are assigned to the utterances in the dialogue. The main categories for the tags are the Communicative Status (it records whether the utterance is intelligible and whether it was successfully completed), the Information Level (a characterization of the semantic content of the utterance), the Forward Looking Function (how the current utterance constrains the future beliefs and actions of the participants, and affects the discourse), and the Backward Looking Function (how the current utterance relates to the previous discourse). Although we can emphasize that our interest is in naturally occurring conversations, it is certainly the case that these annotation schemes can be employed to get a grip on models that attempt to include humor appreciation and humor generation acts.

When we turn our attention to corpora that allow us to study nonverbal communication behavior, we first observe that in previous decades it was much easier to collect texts and transcriptions of dialogues and conversations, than to collect easily accessible and manipulatable data on nonverbal communication acts. Rather than digitalized videos of naturally occurring events, researchers had to use transcriptions of videos and pictures of attempts to simulate events. Only in the last decade we see the collecting of recordings of displays of facial expressions, gesture and body posture during experiments. Instead of having to access analogue speech or video data in sequential ways, now we have the possibility to access, select and manipulate data that has been stored digitally. However, unlike what has become usual in speech and language corpora, there are not yet internationally agreed ways of consistently storing such data and agreed ways of making it accessible, for example for research purposes.

Audio-visual data corpora are needed to analyze the role of nonverbal communication acts in HA generation in interaction with verbal communication acts. In corpus research, observations are followed by analysis, analysis tries to confirm ideas, and the results are theories of recognition and interpretation, including learning strategies. Through annotation schemes there is interaction between analysis, restructuring the corpus, and theory building. In this process corpus annotation schemes play an essential role. They connect top-down theory building and bottom-up descriptions of events.

Audio-visual data corpora allow us to analyze captured data about gestures, postures, facial expressions, and speech from a HA generation point of view. We also need tools for model-based annotations, where the models have been designed to include a humor point of view. Those models are not yet available. Nevertheless, we can look at multimodal annotated corpora that include verbal and nonverbal communication acts. Annotation of such corpora allows representation, analysis and synthesis of theory-based characteristics. In this way, annotation tools and annotated corpora have been developed and made available for facial actions and expressions (using the Facial Action Coding System: FACS), visual speech, conversational gestures, deictic gestures, hand shapes, and intonation. Often they build upon tools developed for annotating and retrieving information from spoken language corpora. Video annotation tools (Anvil, FORM) and workbenches (for example, developed in the European NITE project or the German SmartKom project) have been designed for multi-level, cross-level and cross-modality annotation of natural activity. These tools allow annotations from very detailed, on the signal level, to holistic. The former allows, for example, the description of eye aperture and eyelid movements, including duration and intensity aspects; the latter can be found in the SmartKom user-state coding systems, where both the cognitive and the emotional state are labeled (Steininger *et al.* 2002).

From the ECA generation point of view, models that underlie annotation languages can as well be employed in markup languages that define ECA display of natural interaction, such as in the XML-based markup language for affective display APML (Affective Presentation Markup Language) designed by Pélachaud and Poggi (2002), EMOTE (Chi *et al.* 2000), and BEAT (Cassell *et al.* 2001). In APML, for example, we have various face markers (e.g., affective), functional values (e.g., joy as one of the affective markers) and signals (e.g., smile and raised cheek for the functional value joy). A match between annotations and display acts makes it possible, assuming the availability of needed visualization and animation tools, to play back an annotation in order to judge how well phenomena have been captured.

Until now, hardly any experiments can be reported that have been performed from the point of view of embodied humor research and that aimed at building a corpus containing data to be explored from the point of view of humor research, that is, giving attention to analysis, annotation, training of recognition, and generation for humor research and application purposes. However, there have been experiments that, indeed, have been performed for quite different reasons, and that can be used for embodied humor research. Making use of corpora available from emotion research, we can investigate how humor-related emotions (e.g., enjoyment and confusion) are displayed in face, gesture, intonation, and posture. How it is displayed with a choice of dialogue act, utterance, words, and intonation is too far away from current research.

As an example of research performed for quite different reasons than humor research we mention the corpus collected by Hill and Johnston (2001) where they have more than 200 recordings of the facial movements of twelve men and women telling jokes. The aim of their research was to investigate distinctions between male and female facial and head movements. Can a computer make this distinction; that is, in our situation, will an ECA be able to decide from its perception of just these movements that it is communicating with a man or a woman?

The facial movements in the Hill and Johnston corpus were obtained from a two-line question-and-answer joke told to another individual ("Why do cows have bells? Because their horns don't work!"). They used motion-capture with two cameras and many markers on the forehead, temples and nose to recreate the facial movements in an average three-dimensional head of an embodied conversational agent (Figure 2.5). During a conversation or dialogue, having a particular HA or joke schema, an ECA can detect the appropriate moment to generate a particular type of joke or HA and it can use the average three-dimensional head movements to display the joke using verbal and nonverbal humor features.

Average nonverbal communication behavior as described in the previous paragraphs can be adapted by adding personality and emotional characteristics features. See Ball and Breese (2000), linking emotions and personality to nonverbal behavior using Bayesian Networks. In (Allbeck and Badler 2002), the emphasis is on adapting the gestures of ECA to its personality and gestures features.

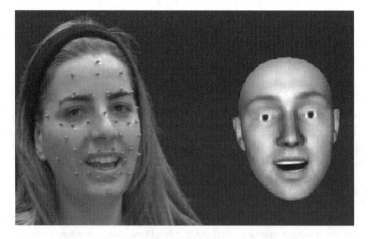

Figure 2.5   Finding the average head in joke telling

2.7.3 Future Research Approaches

In the line of research on autonomous (intelligent and emotional) agents we need an ECA to understand why the events that take place generate enjoyment by its conversational partner and why it should display enjoyment because of its partner's appreciation of a HA. That is, models are needed that allow generation, prediction, detection, and interpretation of humorous events.

What events need to be distinguished, how does the ECA perceive them, and how does it integrate them at a semantic and pragmatic level of understanding of what is going on? There are two approaches to this question when we look at state-of-the-art research. One approach deals with speech and dialogue act prediction. What is going to happen next, given the history and the context of the dialogue? Can an ECA predict the next dialogue act by its conversational partner or can it compute the next dialogue act that is expected by its (i.e., the ECA's) conversational partner? Previous and possibly future dialogue acts are events that need to be "appraised". In earlier research (Keizer *et al.* 2002) we used Bayesian networks in order to predict dialogue acts. While this approach is unconvential from the usual point of view of event appraisal, it is an accepted approach in dialogue modeling research that has been implemented in a number of dialogue systems. Some attempts have been made to introduce multimodal dialogue acts. It seems to be useful to introduce more refined dialogue acts that take into account the willingness of a conversational partner to construct a humorous utterance and that take into account the possibility to give interpretations to (parts of) previous utterances that may lead to humorous acts. Obviously, in order to be able to do so we need corpora of natural conversations that allow us to design, train and test algorithms and strategies. Holistic user-state modeling, as advocated in the previously mentioned SmartKom project, is a possible way to obtain data from which recognition algorithms can be designed.

Clearly, with such an approach we enter the area of emotion research. One of its viewpoints is that of appraisal theory, the evaluation of events and situations followed by categorizing arising affective states. Some of the theories that emerged from this viewpoint have been designed with computation in mind: designing a computational model to elicit and display emotions in a particular situation. A mature theory for calculating cognitive aspects of emotions is the OCC model (Ortony *et al.* 1988), a framework of 22 distinct emotion types. A revised version, presented in the context of believable ECA design was given in Ortony (2001). Can we make a step from event appraisal theories for deciding an appropriate emotion to appraisal theories for deciding the appropriateness of constructing a humorous act? As mentioned, issues that should be taken into account are the ability to construct a HA using elements of the discourse and the appropriateness of generating a HA in the particular context. In human–computer interaction applications some (mostly, stripped-down) versions of the model have been used. Examples are the OZ-project (Reilly and Bates 1992), concerned with the development of a theater world inhabited by emotional agents, and the Carmen project (Marsella *et al.* 2000), where event appraisal is used to recognize feelings of guilt and anger in a setting where an embodied agent talks with a mother of children with leukemia.

It seems also useful to review existing theories and observations concerning the appraisal of (humorous) situations (available as events, in conversations, in verbal descriptions or stories) in terms of possible agent models that include explicit modules for beliefs, desires, intentions, and emotions. Beliefs, desires and intentions (goals) define the cognitive state of an agent. Because of perceptive events, state changes take place. From the humor modeling point of view, agent models of states and state changes need to include reasoning mechanisms about situations where there is the feeling that on the one hand the situation is normal, while at the same time there is a violation of a certain commitment of the agent about how things ought to be. From a humor point of view, relevant cognitive states should allow detection of surprise, incongruity, and reconstruction of incongruity using reasoning mechanisms (Mele 2002). With this view in mind it is useful to look at the violation theory discussed in Veatch (1998), attempts to define degrees of incongruity (Deckers 2001), attempts to define humor in terms of violations of Grice's conversational maxims (Attardo 1993, 1997), and proposals to define and explain humor or laughter in terms of perceptual, affective, and response patterns (Russell 2000).

2.8 Conclusions

This chapter touches upon the state of the art of embodied conversational agents, humor modeling, and affective computing. Using the "computers are social actors" paradigm we made clear that it is useful for natural interaction between human and computer to introduce characteristics of human–human interaction in agent–human interaction, including the generation of humor and the display of appreciation of humor. We introduced the notion of a humorous act in a dialogue or in a conversation. We certainly were not able to give general algorithms for construction of humorous acts or for deciding when to generate a humorous act in a dialogue or conversation. Rather we discussed the issues involved and we presented examples. The first steps towards the implementation of a chatbot constructing humorous acts during a conversation were also presented. Picard and Klein (2002) mentioned that increasingly people spend more time communicating with each other via technology without sufficient affect channels. They argue that this may reduce people's emotional skills and a solution might be to design computers that appropriately support users in handling their emotions. Clearly, a similar argument can be given for a "humor channel" that is lacking in human–computer and computer-mediated interaction. Humor especially appeals to positive emotions and therefore can make human–computer interaction more enjoyable.

Acknowledgments

Part of the research reported in this chapter was supported by the European Future Emerging Technologies (FET) assessment project HAHAcronym (IST-2000-30039), a joint project of ITC-IRST (Trento) and the University of Twente. This chapter is based on two earlier workshop papers (Nijholt *et al.* 2002; Nijholt 2005) and a journal paper (Binsted *et al.* 2006). I am grateful to my student Matthijs Mulder for help in finding and summarizing relevant literature, and my student Hans Wim Tinholt for his work towards implementing our approach to humorous anaphora resolution. Permissions to use the illustrations in this chapter were obtained from Jonathan Gratch of ICT, Marina del Rey (Figure 2.1), Catherine Pélachaud of the Université de Paris (Figure 2.2), United Media, New York (Figure 2.3), and Harold Hill of University College London (Figure 2.4).

References

Adelswärd V. and Öberg B.-M. (1998) The function of laughter and joking in negotiation activities. *Humor: International Journal of Humor Research* **11** (4), 411–429.

Allbeck J. and Badleron N. (2002) Toward representing agent behaviors modified by personality and emotion. Workshop on Embodied conversational agents – let's specify and evaluate them! AAMAS 2002, Bologna, Italy.

Attardo S. (1993) Violations of conversational maxims and cooperation: the case of jokes. *Journal of Pragmatics* **19** (1), 537–558.

Attardo S. (1996) Humor. In J. Verschueren, J.-O. Ostman, J. Blommaert and C. Bulcaen (eds), *Handbook of Pragmatics*. John Benjamins Publishing Co., Amsterdam/Philadelphia, pp. 1–17.

Attardo S. (1997) Competition and cooperation: beyond Gricean pragmatics. *Pragmatics and Cognition* **5** (1), 21–50.

Ball G. and Breese J. (2000) Emotion and personality in a conversational agent. In J. Cassell, J. Sullivan, S. Prevost and E. Churchill (eds), *Embodied Conversational Agents*. MIT Press, Cambridge, MA.

Bates J. (1994) The role of emotion in believable agents. CMU-CS-94-136, Carnegie Mellon University, Pittsburgh; also: *Communications of the ACM*, Special Issue on Agents, July 1994.

Bickmore T. (2003) Relational Agents: Effecting Change through Human-Computer Relationships. MIT Ph.D. Thesis, MIT, Boston, 2003.

Bickmore T. and Cassell J. (2001) Relational agents: a model and implementation of building user trust. In *Proceedings of ACM CHI 2001*, Seattle, Washington, pp. 396–403.

Binsted K. (1995) Using humor to make natural language interfaces more friendly. In Proceedings of AI, ALife, and Entertainment Workshop, International Joint Conference on Artificial Intelligence, 1995.

Binsted K. and Ritchie G. (1997) Computational rules for punning riddles. *Humor: International Journal of Humor Research* **10** (1).

Binsted K., Bergen B., Coulson S., Nijholt A., Stock O., Strapparava C., Ritchie G., Manurung R., Pain H., Waller A. and O'Mara D. (2006) Computational humor. *IEEE Intelligent Systems* **21** (2), 59–69.

Cann A., Calhoun L.G. and Banks J.S. (1997) On the role of humor appreciation in interpersonal attraction: it's no joking matter. *Humor: International Journal of Humor Research* **10** (1), 77–89.

Cassell J. Vilhjalmsson H. and Bickmore T. (2001) Beat: the behavior expression animation toolkit. In *Proceedings of ACM SIGGRAPH*, pp. 477–486.

Castell P.J. and Goldstein J.H. (1976) Social occasions of joking: a cross-cultural study. In A.J. Chapman and H.C. Foot (eds), *It's a Funny Thing, Humour*. Pergamon Press, Oxford, pp. 193–197.

Chao-chih Liao (2003) Humor versus Huaji. *Journal of Language and Linguistics* **2** (1), 25–50.

Chi D., Costa M., Zhao L. and Badler N. (2000) The emote model for effort and shape. In *Proceedings of ACM SIGGRAPH*, New Orleans, LA, pp. 173–182.

Consalvo C.M. (1989) Humor in management: no laughing matter. *Humor: International Journal of Humor Research* **2** (3), 285–297.

Cowie R. (2000) Describing the emotional states expressed in speech. In *Proceedings of the ISCA Workshop on Speech and Emotion*. Belfast, pp. 11–18.

Deckers L. (2001) On the validity of a weight-judging paradigm for the study of humor. *Humor: International Journal of Humor Research* **6** (1), 43–56.

Ekman, P. (1993) Facial expression of emotion, *American Psychologist* **48** (4), 384–392.

Ekman P. (1985) *Telling Lies: Clues to deceit in the market place, politics, and marriages*. New W.W. Norton, New York. Reissued in 2001 with a new chapter.

Frank M.G. and Ekman P. (1993) Not all smiles are created equal: the differences between enjoyment and nonenjoyment smiles. *Humor: International Journal of Humor Research* **6** (1), 9–26.

Friedman B. (ed.) (1997) *Human Values and the Design of Computer Technology*. Cambridge University Press, Cambridge.

Fry P.S. (1995) Perfection, humor, and optimism as moderators of health outcomes and determinants of coping styles of women executives. *Genetic, Social, and General Psychology Monographs* **121**, 211–245.

Grice P. (1989) Logic and conversation. In H. P. Grice (ed.), *Studies in the Way of Words*. Harvard University Press, Cambridge, MA, pp. 22–40.

Hampes W.P. (1999) The relationship between humor and trust. *Humor: International Journal of Humor Research* **12** (3), 253–259.

Hartmann B., Mancini M. and Pélachaud C. (2005) Implementing expressive gesture synthesis for embodied conversational agents. Gesture Workshop, Lecture Notes in Artificial Intelligence, Springer, May 2005.

Hay J. (2001) The pragmatics of humor support. *Humor: International Journal of Humor Research* **14** (1), 55–82.

Herzog T.R. (1999) Gender differences in humor appreciation. *Humor: International Journal of Humor Research* **12** (4), 411–423.

Hill H. and Johnston A. (2001) Categorizing sex and identity from the biological motion of faces. *Current Biology* **11**, 880–885.

Kappas A. Hess U. and Scherer K.R. (1991) Voice and emotion. In R.S. Feldman, *et al.* (eds), *Fundamentals of Nonverbal Behavior*. Cambridge University Press, Cambridge, pp. 200–237.

Keizer S., Akker R. op den, and Nijholt A. (2002) Dialogue act recognition with Bayesian networks for Dutch dialogues. In K. Jokinen and S. McRoy (eds), *Proceedings 3rd SIGdial Workshop on Discourse and Dialogue*, Philadelphia, Pennsylvania, pp. 88–94.

Klein J., Moon Y. and Picard R.W. (2002) This computer responds to user frustration: theory, design, and results. *Interacting with Computers* **14** (2), 119–140.

Lampert M.D. and Ervin-Tripp S.M. (1998) Exploring paradigms: the study of gender and sense of humor near the end of the 20th century. In W. Ruch (ed.), *The Sense of Humor: Explorations of a Personality Characteristic*. Mouton de Gruyter, Berlin, pp. 231–270.

Lappin S. and Leass H. (1994) An algorithm for pronominal anaphora resolution. *Computational Linguistics* **20** (4), 535–561.

Lester J.C., Converse S.A., Kahler S.E., Barlowe S.T., Stone B.A. and Bhogal R. (1997) The persona effect: affective impact of animated pedagogical agents. *Proceedings CHI '97: Human Factors in Computing Systems*, ACM, pp. 359–356.

Loehr D. (1996) An integration of a pun generator with a natural language robot. In J. Hulstein and A. Nijholt (eds), *Computational Humor*. University of Twente, Enschede, pp. 161–171.

Marquis, S. and Elliott, C. (1994) Emotionally responsive poker playing agents. In *Notes for the AAAI-94 Workshop on AI, ALife, and Entertainment*, pp. 11–15.

Marsella S.C., Johnson W.L. and LaBore C. (2000) Interactive pedagogical drama. In *Proceedings 4th International Conference on Autonomous Agents*, ACM, pp. 301–308.

Mele F. (2002) Deliverable on humor research: state of the art. European IST Project HaHacronym on Future Emerging Technologies, 2002.

Morkes J., Kernal H. and Nass C. (2000) Effects of humor in task-oriented human–computer interaction and computer-mediated communication: a direct test of social responses to communication technology theory. *Human–Computer Interaction* **14** (4), 395–435.

Nevo, O., Nevo, B. and Leong Siew Yin, J. (2001) Singaporean humor. A cross-cultural, cross-gender comparison. *Journal of General Psychology* **128** (2), 143–156.

Nijholt A. (2002) Embodied agents: a new impetus to humor research. In O. Stock, C. Strapparava and A. Nijholt (eds), The April Fools Day Workshop on Computational Humor. *Proceedings Twente Workshop on Language Technology* (TWLT 20), ITC-IRST, Trento, Italy, pp. 101–111.

Nijholt A. (2005) Conversational agents, humorous act construction, and social intelligence. In: Proceedings AISB: Social Intelligence and Interaction in Animals, Robots and Agents. Symposium on Conversational Informatics for Supporting Social Intelligence and Interaction: Situational and Environmental Information Enforcing Involvement in Conversation, K. Dautenhahn (ed.). University of Hertfordshire, Hatfield, England, pp. 1–8.

Norrick N.R. (1993) *Conversational Joking: Humor in Everyday Talk*. Indiana University Press, Bloomington, IN.

Ortony A. (2001) On making believable emotional agents believable. In R. Trappl and P. Petta (eds), *Emotions in Humans and Artifacts*. MIT Press, Cambridge, MA.

Ortony A., Clore G.L. and Collins A. (1988) *The Cognitive Structure of Emotions*. Cambridge University Press, Cambridge.

Pélachaud C. and Poggi I. (2002) Subtleties of facial expressions in embodied agents. *Journal of Visualization and Computer Animation* **13** (5), 301–312.

Picard R.W. and Klein J. (2002) Computers that recognize and respond to user emotion: theoretical and practical implications. *Interacting with Computers* **14** (2), 141–169.

Raskin V. (1985) *Semantic Mechanisms of Humor*. (Synthese Language Library 24). D. Reidel, Dordrecht.

Reeves B. and Nass C. (1996) *The Media Equation: How people treat computers, televisions and new media like real people and places*. Cambridge University Press, Cambridge.

Reilly W. and Bates J. (1992) Building emotional agents. Report CMU-CS-92-143, Carnegie Mellon University.

Ritchie G. (1999) Developing the incongruity-resolution theory. *Proceedings of AISB Symposium on Creative Language: Stories and Humour*, Edinburgh, pp. 78–85.

Ruch W. (1994) Extroversion, alcohol, and enjoyment. *Personality and Individual Differences* **16**, 89–102.

Ruch W. and Ekman P. (2001) The expressive pattern of laughter. In A.W. Kaszniak (ed.), *Emotion, Qualia, and Consciousness*. Word Scientific Publisher, Tokyo, pp. 426–443.

Ruch W. and Forabasco G. (1994) Sensation seeking, social attitudes and humor appreciation in Italy. *Personality and Individual Differences* **16**, 515–528.

Russell R.E. (2000) Humor's close relatives. *Humor: International Journal of Humor Research* **13** (2), 219–233.

Shechtman N. and Horowitz L.M. (2003) Media inequality in conversation: how people behave differently when interacting with computers and people. *Proceedings SIGCHI-ACM CHI 2003: New Horizons*. ACM, pp. 281–288.

Steininger S., Schiel F. and Glesner A. (2002) User-state labeling procedures for the multimodal data collection of SmartKom. *Proceedings of the 3rd International Conference on Language Resources and Evaluation*, Las Palmas, Spain.

Stock O. (1996) Password Swordfish: verbal humor in the interface. In J. Hulstein and A. Nijholt (eds), *International Workshop on Computational Humor*, University of Twente, pp. 1–8.

Stock O. and Strapparava C. (2003) An experiment in automated humorous output production. Proceedings of the International Conference on Intelligent User Interfaces (IUI), Miami, Florida.

Stronks B., Nijholt A., Vet P. van der, and Heylen D. (2002) Designing for friendship. In A. Marriott *et al.* (eds), *Proceedings of Embodied Conversational Agents: Let's Specify and Evaluate Them!* Bologna, Italy, pp. 91–97.

Tannen D. (1984) *Conversational Style: Analyzing talk among friends*. Ablex Publishing, Westport.

Tannen D. (1994) *Talking from 9 to 5*. Avon Books New York.

Theune M., Heylen D. and Nijholt A. (2005) Generating embodied information presentations. In O. Stock and M. Zancanaro (eds), *Multimodal intelligent information presentation*. Kluwer Academic Publishers, pp. 47–70.

Tinholt H.W. and Nijholt A. (2007) Computational humour: utilizing cross-reference ambiguity for conversational jokes. In F. Masulli, S. Mitra and G. Pasi (eds), *Proceedings 7th International Workshop on Fuzzy Logic and Applications* (WILF 2007), Camogli, Italy; Lecture Notes in Artificial Intelligence 4578, Springer-Verlag, Berlin, pp. 477–483.

Veatch T. (1998) A theory of humor. *Humor: International Journal of Humor Research* **11** (2), 161–215.

Wallace D.R., Tomabechi D.H. and Aimless D. (2003) Chatterbots go native: considerations for an eco-system fostering the development of artificial life forms in a human world. Manuscript.

Zajdman A. (1991) Contextualization of canned jokes in discourse. *Humor: International Journal of Humor Research* **4** (1), 23–40.

Zajdman A. (1992) Did you mean to be funny? Well, if you say so .... *Humor: International Journal of Humor Research* **5** (4), pp. 357–368.

Ziv A. (1988) Teaching and learning with humor: experiment and replication. *Journal of Experimental Education* **57** (1), 5–15.

3

Why Emotions should be Integrated into Conversational Agents

Christian Becker, Stefan Kopp, and Ipke Wachsmuth

3.1 Introduction and Motivation

SONNY: "I did not murder him."

DETECTIVE SPOONER: "You were emotional. I don't want my vacuum cleaner, or my toaster appearing emotional."

SONNY: [angry] "I did not murder him!"

DETECTIVE SPOONER: "That one's called 'anger'."

From the movie *I, Robot*, Fox Entertainment 2004

Researchers in the field of artificial intelligence (AI) try to gain a deeper understanding of the mechanisms underlying human cognition. Traditionally pure rational reasoning has been the main focus of research in AI, started with the conceptualization of the General Problem Solver (GPS) (Ernst and Newell 1969). However, such heuristic approaches to modeling human problem solving could be applied only to a limited set of well-defined problems, which soon turned out to be insufficient with regard to the wide range of problem spaces a human is naturally confronted with. Consequently, for the next two decades, research focused on so-called expert systems that were used as advice-giving tools for the trained human expert in limited domains such as medical diagnosis (Shortliffe *et al.* 1975). The final diagnosis, still, always remained the responsibility of the human expert, due not only to legal issues but also to the complexity of human physiology. Nowadays rule-based expert systems are used in diverse types of applications and they can help to save money by providing domain-specific advice as soon as a certain amount of expert knowledge has been successfully encoded into their rules and knowledge bases (Giarratano and Riley 2005).

The user interfaces of expert systems are often designed in a dialogue-based fashion. The system consecutively asks for more detailed information to be provided by the human expert, to come up with a set of possible solutions to the initially stated problem. Similarly, dialogue-based routines have been

Conversational Informatics: An Engineering Approach Edited by Toyoaki Nishida
© 2007 John Wiley & Sons, Ltd

used in the famous computer program ELIZA (Weizenbaum 1976) to create the illusion of speaking to a non-directive psychotherapist. Despite the similarities of the dialogue-based interfaces, Weizenbaum did not intend to build an expert system in the domain of psychotherapy. To his own surprise even professional psychotherapists expected his simple program to be of great help in counselling human patients. This might be due to the fact that humans are prone to ascribe meaning to another's responses, even when interacting with machines.

Even if ELIZA successfully created "the most remarkable illusion of having understood in the minds of the many people who conversed with it" (Weizenbaum 1976, p. 189), it did not pass the Turing test (Turing 1950) as proposed by the early pioneer of computer science, Alan Turing (Hodges 2000). The Turing test was an attempt to provide a suitable test for machine intelligence when the question "Can machines think?" became reasonable to ask. After a five-minute conversation – without direct physical contact, such as using a typewriter machine – with both a human and a machine, a human tester has to decide which one of the conversational partners is the machine and which one the human. If in at least 30% of the cases the machine is falsely judged as the human, it has passed the test successfully, which no machine achieved so far. During the Turing-test conversation the human interrogator is free to choose whatever topic comes to her mind. Therefore, Picard (1997) argues for the integration of humor and more general emotions into artificially intelligent systems that are to pass the Turing test (see also Chapter 2). Even recognizing the limitations of the available communication channels during the Turing test, picard concludes: "A machine, even limited to text communication, will communicate more effectively with humans if it can perceive and express emotions."

How can we endow machines with emotions such that they communicate more effectively with humans? One approach to achieve the effectiveness of natural face-to-face communication of humans is followed by researchers in the field of Embodied Conversational Agents (Cassell *et al.* 2000). They are motivated by the idea that computer systems might one day interact naturally with humans, comprehending and using the same communicative means. Consequently, they have started to build anthropomorphic systems, either in the form of virtual characters using advanced 3D computer graphics or in the form of physical humanoid robots. As these agents comprise an increasing number of sensors as well as actuators along with an increase in their expressive abilities, Cassell *et al.* (2000) propose an extended, face-to-face Turing test. In this vein, researchers in the field of Affective Computing (Picard 1997) not only discuss the possibilities to derive human affective states from a variety of intrusive or non-intrusive sensors. They specifically argue for increasing the expressive capabilities of their artificial agents by modeling the influence of simulated emotions on bodily expressions, including facial expression, body posture, and voice inflection. All of these must be modulated in concert to synthesize coherent emotional behavior.

With the beginning of the new millennium the interest in affective computing has increased even more. Also the public has shown a renewed interest in the possible future achievements of AI, up to questions of whether social robots may be integrated members of a future society (recall the movies *I, Robot* or *The Bicentennial Man*). Indeed in recent research in humanoid agents the view that humans are "users" of a certain "tool" is shifting to that of a "partnership" with artificial, autonomous agents (Negrotti 2005). As such agents are to take part in fully fledged social interaction with humans, the integration of psychological concepts like emotions and personality into their architectures has become one major goal. Despite the ongoing debate about the formal definition of such concepts, many computational models have up to now been proposed to simulate emotions for conversational agents.

In the following section we will review the general psychological background that underlies the concepts of emotion and personality. We will distinguish two major approaches toward emotion simulation, emphasizing their similarities and particular advantages. In section 3.3 we discuss motives for the inclusion of emotions and we relate them to our own work on the multimodal conversational agent Max. We describe in detail how Max was made emotional in section 3.4. Considering a scenario where Max is employed as a virtual interactive guide to a public computer museum, we will discuss how Max's emotion system increases his acceptance as a coequal conversational partner. We will further quote an empirical study that yields evidence that his emotion system also supports the believability and lifelikeness of the agent in a non-conversational scenario of gaming.

3.2 How to Conceptualize Emotions

Within the field of psychology two major strands of theories can be distinguished. The *cognitive* emotion theories focus on the cognitive appraisal processes necessary to elicit the full range of emotions in adult humans (Ortony *et al.* 1988). *Dimensional* emotion theories (Gehm and Scherer 1988) are based on the idea of classifying emotions along an arbitrary number of dimensions of connotative meaning (Osgood *et al.* 1957). In the following we will present these strands of theories in more detail.

3.2.1 Cognitive Emotion Theories

The emotion model proposed by Ortony *et al.* (1988) has often been the basis for the integration of emotions into cognitive architectures of embodied characters (e.g., Elliott 1992; Reilly 1996), as it was designed as a computational model of emotions. In subsequent research many extensions have been proposed to circumvent or solve the well-known drawbacks of this cognitive theory (e.g., Bartneck 2002). Later Ortony (2003) has proposed that it might be sufficient to integrate only ten emotion categories, five positive and five negative ones, into an agent's architecture when focusing on believability and consistency of an agent's behavior. When focusing on facial expressions of emotions this number can be further reduced to six "basic emotions" as cross-culturally evaluated by Ekman *et al.* (1980). Ortony (2003) distinguishes between three "emotion response tendencies" by which the effect of an emotion on the agent's cognition can be characterized. Facial display is one indicator of an "expressive response tendency", which also may figure in other involuntary bodily expressions such as "flushing", "clenching one's fist" or "swearing". Another major response tendency type is labeled "information-processing". It is responsible for the diversion of attention that can lead to an "all-consuming obsessive focus" on the emotion eliciting situation and it may explain the observation that humans learn by emotional experience, which can be modeled by updating the beliefs, attitudes and evaluations of an agent with respect to the experienced emotion. The last response tendency is labeled "coping" and deals with the fact that humans are able to cope with their own emotions in a problem-focused or emotion-focused way by either trying to change the situational context (problem-focused) or by trying to reappraise the emotion eliciting situation to manage their own emotions internally (emotion-focused).

Another possibility to conceptualize emotions cognitively was proposed by Damasio (1994). Based on neurophysiological findings Damasio distinguishes "primary" and "secondary" emotions. Primary emotions are elicited as an immediate response to a stimulus, whereas secondary emotions are the product of cognitive processing. Primary emotions are understood as ontogenetically earlier types of emotions and they lead to basic behavioral response tendencies, which are closely related to "flight-or-fight" behaviors. In contrast, secondary emotions like "relief" or "hope" involve higher level cognitive processing based on memory and the ability to evaluate preferences over outcomes and expectations.

3.2.2 Dimensional Emotion Theories

Wundt (1922) claimed that any emotion can be characterized as a continuous progression in a three-dimensional space of connotative meaning. Several researchers (e.g., Gehm and Scherer 1988; Mehrabian 1995) later provided statistical evidence for this assumption. Using principal component analysis they found that three dimensions are sufficient to represent the connotative meaning of emotion categories. These dimensions are commonly labeled Pleasure/Valence (P), representing the overall valence information, Arousal (A), accounting for the degree of activeness of an emotion, and Dominance/Power (D), describing the experienced "control" over the emotion itself or the situational context it originated from. The three-dimensional abstract space spanned by these dimensions is referred to as PAD space.

Recently, the two sub-dimensions Pleasure/Valence and Arousal have been used to drive an agent's facial expression based on statistically derived activation patterns (Grammer and Oberzaucher 2006). The same two dimensions are used for biometrical emotion recognition by Prendinger *et al.* (2004) assuming that skin conductance is positively correlated with arousal and that muscle tension is an indicator of negative valence. After combining sensory values with context information using a decision network, the two-dimensional emotion space is used to classify the derived valence and arousal values according to Lang (1995). The restriction to two dimensions in biometrical emotion recognition is mainly due to the fact that Dominance cannot be derived from such sensor data, as it depends on an actor's subjective feeling of control over a situation. However, the third dimension of Dominance has been proven to be necessary to distinguish between certain emotions such as anger and fear or annoyance and sadness. This dimension can further be seen as a general aspect of an agent's personality trait. Given the same negative Pleasure and medium Arousal, dominant individuals have a much stronger tendency to shout whereas individuals with a submissive personality are more likely to cry in the same situational context. Along these lines, the emotion system of the sociable robots "Kismet" (Breazeal 2002) is based on a similar three-dimensional emotion space.

3.3 Why to Integrate Emotions into Conversational Agents

According to Burghouts *et al.* (2003), researchers in the field of human–computer interaction follow three types of motivations for the integration of emotions into cognitive architectures. When following the *control-engineering motive*, simulated emotional processes are solely used to let an agent generate appropriate and fast reactive responses in a resource-constrained competitive environment, where deliberative reasoning is inappropriate. The *believable-agent motive* is based on the assumption that the simulation of emotions increases the believability of an agent. Researchers who are guided by the *experimental–theoretical motive* are primarily aiming at a deeper understanding of human emotions and human emotional behavior, and they use simulated humanoid agents to verify their theories.

Outside the scope of this chapter, the control-engineering motive is broadly used in the field of Multiagent systems and Socionics; see von Scheve and von Lüde (2005) for a comprehensive introduction. Concerning the integration of emotions into a believable agent, we will first take the more general perspective of modeling social interaction in section 3.3.1. Then, following the experimental–theoretical motive, we will concentrate on the internal effects of emotion modeling by taking the cognitive modeling perspective in section 3.3.2.

3.3.1 Social Interaction Perspective

The empirical research of Reeves and Nass (1998) provides profound evidence for the assumption that humans treat computers as social actors. Their study on flattery, for example, shows that humans have a better opinion about computers that praise them than those that criticize them. This effect remains stable even if the users are made aware of the fact that the computer has no means to evaluate their responses and is simply producing random comments. Concerning the role of emotions in the media the authors first point to neurophysiological evidence supporting the basic assumption that every emotion has an inherent positive or negative quality, a "valence" dimension. In combination with their studies on "arousal" they conclude that people confronted with media content do react with the same variety of emotional responses as in face-to-face interaction between humans. Notably, none of these studies involved any kind of anthropomorphic interface. Rather, the different social and emotional aspects of the computer's responses were encoded only on the textual level; but even this very limited communication channel was efficient enough to support their hypotheses.

With respect to embodied conversational agents such as REA, Bickmore and Cassell (2005) argue for the integration of nonverbal cues, whenever such agents are to take part in social dialogues. In their discussion of the relationship between social dialogue and trust they follow the multi-dimensional model of interpersonal relationship based on Svennevig (1999); see Svennevig and Xie (2002) for a review. This model distinguishes three dimensions, namely *familiarity*, *solidarity* and *affect*, the last of which can be understood as representing "the degree of liking the interactants have for each other" (Bickmore and Cassell 2005, p. 8). In their implementation they couple this dynamic parameter with the social ability of coordination, which in turn is seen as an outcome of fluent and natural smalltalk interaction. Coordination is understood as a means to synchronize short units of talk and nonverbal acknowledgment leading to increased liking and positive affect. Once again it is notable that the additional usage of nonverbal communicative means is sufficient to generate this kind of undifferentiated positive affect in humans. In other words, there is no need to simulate an embodied agent's internal emotional state to affect a human interlocutor positively.

Taking the step from the virtual to the real world, Breazeal (2004b) compares the fields of human–computer interaction (HCI) and human–robot interaction (HRI) and finds similarities and significant differences. In essence she advocates taking the robot's point of view into account when examining and evaluating the relationship of robots and humans in a future society, rather than only taking the human's perspective as is mainly done in HCI. However, the further argumentation seems to neglect the recent endeavors in HCI to follow a cognitive modeling approach in designing virtual humanoid agents instead of strictly focusing on dedicated application scenarios. Nevertheless the integration of socio-emotional intelligence into architectures of robots is advocated, if we want such robots to fulfill an adequate role in human society; see Breazeal (2004a) for details.

3.3.2 Cognitive Modeling Perspective

Cognitive science research is highly interdisciplinary combining philosophy, psychology, artificial intelligence, neuroscience, linguistics, and anthropology. Scientists in this field attempt to build computational models of human cognitive behavior in order to combine and verify the findings of the different disciplines. Emotions are increasingly seen as an important component of cognitive systems of humans as well as humanoids. Arbib and Fellous (2004) present a comprehensive overview of the attempts to improve human–computer interaction through the simulation and expression of emotions. In their discussion of cognitive architectures they give a brief summary of recent attempts to describe the influence of emotions in layered architectures, such as Ortony *et al.* (2005) and Sloman *et al.* (2005).

Gratch and Marsella (2004) present a domain-independent framework for modeling emotions that combines insights from emotion psychology with the methodologies of cognitive science in a promising way. Based on the Belief–Desire–Intention (BDI) approach to model rational behavior by Rao and Georgeff (1991), they describe their efforts to integrate appraisal and coping processes that are central to emotion elicitation and social interaction. Central to their idea are "appraisal frames and variables" by which the emotional values of external and internal processes and events are captured in concrete data structures. By making use of the agent's BDI-based reasoning power based on concepts such as likelihood and desirability, emotions are first generated and then aggregated into a current emotional state and an overall mood. The mood is seen to be beneficial because it has been shown to impact "a range of cognitive, perceptual and behavioral processes, such as memory recall (mood-congruent recall), learning, psychological disorders (depression) and decision-making" (Gratch and Marsella 2004, p. 18). Furthermore this mood value is used as an addend in the calculation of otherwise equally activated emotional states (such as fear and hope at the same time) following the idea of mood-congruent emotions. Remarkably, their framework for modeling emotions is the first fully implemented, domain-independent architecture for emotional conversational agents.

3.3.3 Relation to Our Own Work

The virtual human Max (see Figure 3.1) developed at Bielefeld University's Artificial Intelligence Laboratory has been employed in a number of increasingly challenging scenarios in which Max's conversational capabilities have been (and are being) steadily extended. In a museum application, Max is conducting multimodal smalltalk conversations with visitors to a public computer museum (see section 3.5). In this setting, where he has to follow the basic rules of social dialogue, we have not modeled emotional behavior by devising explicit rules on how to influence the interlocutor's emotional state, as proposed by for example Bickmore and Cassell (2005). Rather, we opted for integration of an emotion simulation module, enabling our agent to "experience emotions of his own", for the following reasons.

The emotion module has an internal dynamics that leads to a greater variety of often unpredictable, yet coherent emotion-colored responses. These responses add to the impression that the agent has a unique personality, and the parameters of the emotion system reflect such aspects of an agent's personality trait. Furthermore, this way of modeling emotions is largely domain-independent, as has been shown by a successful application in a gaming scenario as well as by the combination of an independent simulation of emotion dynamics with a BDI-based reasoning module in order to model the mood-congruent elicitation of secondary emotions. Finally, by integrating a specialized module for the simulation of psychological emotion theories, we can use its high-level parameters to generate and falsify predictions on how to model different personalities following the experimental–theoretical motive.

In the next section we will give a brief overview of our agent's cognitive architecture. It will be followed by a detailed explanation of the emotion dynamics simulation module that provides the so-called awareness likelihoods for primary and secondary emotions. Further, the question of how cognitive processes can give rise to emotions is addressed.

3.4 Making the Virtual Human Max Emotional

Max (Figure 3.1) has been developed to study how the natural conversational behavior of humans can be modeled and made available, for example, in virtual reality face-to-face encounters. Max rests upon an architecture that realizes and tightly integrates the faculties of perception, action, and cognition required

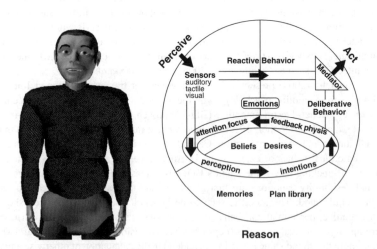

Figure 3.1   Anthropomorphic appearance and outline of the architectural framework of Max (Leßmann *et al.* 2006). (Reproduced by permission of Springer-Verlag)

to engage in such interactions (Leßmann *et al.* 2006). While at large the underlying model employs the classical perceive–reason–act triad, it was conceived such that all processes of perception, reasoning, and action are running concurrently.

Perception and action are directly connected through a reactive connection, affording reflexes and immediate responses to situation events or moves performed by a human interlocutor. Such fast-running stimulus–response loops are utilized to react to internal events. Reactive behaviors include gaze tracking and focusing the current interaction partner in response to prompting signals. In addition, continuous secondary behaviors reside in this layer, which can be triggered or modulated by deliberative processes as well as by the emotional state of the agent. These behaviors make the agent appear more lifelike and include eye blink, breathing, and sway.

Perceptions are also fed into deliberative processes taking place on the reasoning layer. These processes are responsible for interaction management by interpreting input, deciding which action to take next, and composing behaviors to realize it. This is implemented following the BDI approach (Rao and Georgeff 1991) to modeling rational behavior and makes use of an extensible set of self-contained planners. The architecture further comprises a cognitive, inner loop, which feeds internal feedback information upon possible actions to take (e.g., originating from the body model of the agent) back to deliberation.

With regard to the integration of emotions, it is under debate how modular the human emotion system in the brain is. However, with respect to the emotion system of an artificial agent we decided to conceive of it as a single module that is highly interconnected, and runs concurrently, with the agent's other "mental" faculties. This module is based on a dimensional emotion theory as introduced in section 3.2 to realize the internal dynamics and mutual interactions of primary and secondary emotions. With regard to the complex appraisal processes giving rise to secondary emotions it seems justified to extend the architecture by further exploiting the reasoning power of the cognition module described in section 3.4.2.

3.4.1 The Emotion Module

Two major assumptions supported by psychology literature (e.g., Oatley and Johnson-Laird, 1987) underlie our realization of a concurrent emotion dynamics simulation module (Becker *et al.* 2004):

1. An emotion is a short-lived phenomenon and its valence component has a fortifying or alleviating effect on the mood of an individual. A mood, in contrast, is a longer lasting, valenced state. The predisposition to experience emotions changes together with the mood; e.g., humans in a positive mood are more susceptible to positive than negative emotions, and vice versa; see Neumann *et al.* (2001). This "emotion dynamics" is realized in the first component of the emotion module labeled *dynamics/mood* in Figure 3.2.
2. Every primary emotion can be positioned in the PAD emotion space with respect to its inherent degree of Pleasure, Arousal, and Dominance. This assumption underlies the second component of the emotion module labeled *PAD-space* in Figure 3.2. Likewise, parts of the connotative meaning of secondary emotions can be represented as regions in PAD space, as demonstrated in Figure 3.4 by the two regions "relief" and "frustration".

Simulation of Dynamics: Emotions and Moods and their Mutual Interaction

The concept of emotions is linked to the concept of moods using a two-dimensional space defined by an *x*-axis of emotional valence and an orthogonal *y*-axis that represents the valence of moods (Figure 3.3(a)). The system tends to hold both valences in zero because this point constitutes the default state of emotional

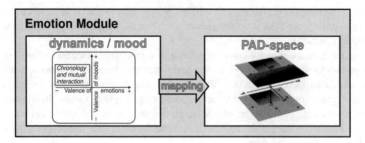

Figure 3.2 The emotion module consists of two components: the dynamics/mood component is used for calculation of the emotion dynamics, and its values are subsequently mapped into PAD space

balance. Therefore two independent spiral springs are simulated, one for each axis, which create two reset forces F_x and F_y whenever the point of reference is displaced from the origin.

The exerted forces are proportional to the value of the corresponding valences x and y just as if the simulated spiral springs were anchored in the origin and attached to the point of reference. The mass–spring model was chosen based on the heuristics that it better mimics the time course of emotions than linear and exponential decreasing models. This assumption is supported by Reisenzein (1994) who showed that in most cases the intensity of emotions in the two-dimensional Pleasure–Arousal theory is not decreasing linearly but more according to a sinoid function.

By adjusting the two spring constants d_x and d_y and the simulated inertial mass m of the point of reference, the dynamics of both concepts can be biased intuitively. These parameters can also be construed as an aspect of the agent's personality trait.

In order to simulate the alleviating and fortifying effects of emotions on moods, the emotional valence is interpreted as a gradient for changing the valence of moods at every simulation step according to the equation $\Delta y / \Delta x = a \cdot x$. The independent parameter a, again, models an aspect of the agent's personality, with smaller values of a resulting in a more "sluggish" agent and greater values of a leading to a more "moody" agent.

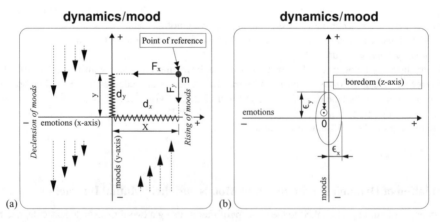

Figure 3.3 Internals of the dynamics/mood component: (a) emotions and moods and their courses over time; (b) the epsilon neighborhood for the simulation of boredom

The Concept of Boredom

In addition to the emotion dynamics described above, a concept of boredom is added to the dynamic component as a third, orthogonal z-axis. Assuming that the absence of stimuli is responsible for the emergence of boredom (as proposed by Mikulas and Vodanovich (1993)), the degree of boredom starts to increase linearly over time if the point of reference lies within an epsilon neighborhood of absolute zero (as given by ϵ_x and ϵ_y, see Figure 3.3(b)). Outside of this neighborhood the value of boredom is reset to zero per default. The co-domain of the boredom parameter is given by the interval $[-1, 0]$, so the agent is most bored if the value of negative one is reached. The linear increase of boredom can be described by the equation $z(t+1) = z(t) - b$, where the parameter b is another aspect of the agent's personality trait.

Categorization: Mapping into PAD Space

The dynamic component provides the following triple at any time step t:

$$D(t) = (x_t, y_t, z_t), \quad \text{with} \quad x_t = [-1, 1], y_t = [-1, 1], z_t = [-1, 0] \tag{3.1}$$

The variable x_t denotes the emotional valence, the variable y_t stands for the actual valence of the mood, and z_t represents the degree of boredom. Given this triple, the mapping into PAD space for categorization (Figure 3.4) is implemented according to the function $PAD(x_t, y_t, z_t)$ as shown in equation (3.2). This mapping results in a triple consisting of the functions $p(x_t, y_t)$ for the calculation of Pleasure, $a(x_t, z_t)$ for Arousal, and $d(t)$ for Dominance.

$$PAD(x_t, y_t, z_t) = (p(x_t, y_t), a(x_t, z_t), d(t)), \quad \text{with}$$
$$p(x_t, y_t) = \frac{1}{2} \cdot (x_t + y_t) \quad and \quad a(x_t, z_t) = |x_t| + z_t \tag{3.2}$$

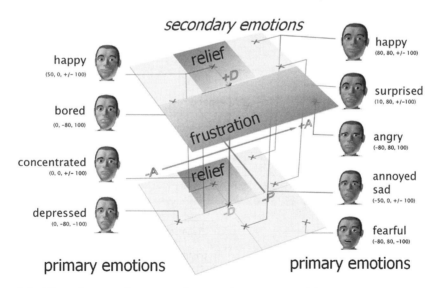

Figure 3.4 Nine primary and two secondary emotions represented in PAD space. Note that some categorized emotions (e.g., happy) may comprise a variety of PAD values; for further explanation see the text

Pleasure is assumed to be the overall valence information in PAD space and therefore calculated as the standardized sum of both the actual emotional valence as represented by x_t and the valence of mood as given by y_t. That way, the agent will feel a maximum of joy when his emotion as well as his mood is most positive and a maximum of reluctance in the contrary case. The agent's arousal ranges from "sleepiness" to a maximum of "mental awareness" and "physiological exertion". As it is assumed that any kind of emotion is characterized by high arousal, only the absolute value of emotional valence is considered in the function $a(x_t, z_t)$. The addition of the (negatively signed) value of boredom reflects its relation to the mental state of inactivity.

The independent parameter of Dominance (or, in the other extreme, Submissiveness) cannot be derived from the dynamic component of the emotion module itself. As explained in section 3.2.2, this parameter describes the agent's "feelings" of control and influence over situations and events versus "feelings" of being controlled and influenced by external circumstances (Mehrabian 1995). By introducing this parameter it is possible to distinguish between anger and fear as well as between sadness and annoyance. Anger and annoyance come along with the feeling of control over the situation whereas fear and sadness are characterized by a feeling of being controlled by external circumstances; Lerner and Keltner (2000) give a theoretical foundation. Therefore, it is in principle not possible to derive such information from the dynamic component. The BDI interpreter within the cognitive module of Max, however, is capable of controlling the state of dominance. In section 3.5.2 we will present one heuristic to control this state of dominance within a card game scenario.

Several primary emotions that have prototypical facial expressions have been anchored in PAD space by defining adequate PAD triples (see Figure 3.4). Some of these emotions are represented more than once because it is assumed unnecessary to distinguish between a dominant and a submissive case for these primary emotions. Furthermore, in case of high pleasure Ekman's set of "basic emotions" contains only one obviously positive emotion, namely happiness (Ekman *et al.* 1980). Thus, in our implementation this primary emotion covers the whole area of positive pleasure regardless of arousal or dominance as it is located in PAD space four times.

In principle, the *awareness likelihood* of a primary emotion increases the closer the point of reference gets to that emotion. If the point of reference gets closer than Φ units to that particular emotion (Figure 3.5), calculation of the awareness likelihood w for that emotion is started according to equation (3.3) until the distance d gets below Δ units.

$$w = 1 - \frac{d - \Delta}{\Phi - \Delta} \tag{3.3}$$

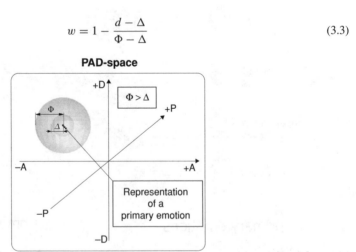

Figure 3.5 Thresholds Φ and Δ for the awareness likelihood of primary emotions. (Reproduced by permission of Springer-Verlag)

The likelihood w is set to 1, if the distance gets below Δ. In equation (3.3), Φ can be interpreted as the *activation threshold* and Δ as the *saturation threshold*, which are both global constants of the emotion system and, thus, valid for every primary emotion.

With respect to the simulation of secondary emotions we propose to represent parts of their connotative meaning in PAD space as well, which readily enables the calculation of their awareness likelihood. But as secondary emotions are understood as resulting from a process of conscious appraisal based on experiences and expectations, it is insufficient for them to be represented in terms of PAD values alone. Furthermore, we believe that one characteristic of secondary emotions is a certain fuzziness when described in terms of these three basic dimensions. Due to this assumption secondary emotions are represented in PAD space as *regions* in contrast to *points* as in the case of primary emotions (see Figure 3.4). For example, we associate the secondary emotion "frustration" with negative pleasure and high dominance and "relief" with positive pleasure and medium arousal.

3.4.2 Connecting Feelings and Thoughts

The emotion module explained above needs so-called valenced emotional impulses together with the actual degree of Dominance as input signals to drive its internal dynamics. In return it provides descriptions of the agent's emotional state on two different levels of abstraction: first, in terms of raw but continuous Pleasure, Arousal and Dominance values; and second, in terms of awareness likelihoods of a number of primary and secondary emotions.

In the following we will first explain how conscious and non-conscious appraisal lead to the elicitation of primary and secondary emotions, respectively. Especially the interplay of conscious reasoning and non-conscious reactive processes together with the emotion dynamics will be explained.

Conscious vs. Non-conscious Appraisal

In the context of emotion simulation it is helpful to divide the cognition module in Figure 3.6 into two layers (based on the ideas of Ortony *et al.* (2005) and Damasio (1994); see also section 3.2.1).

1. Our agent's "conscious", BDI-based deliberation resides in the *reasoning layer*. As the ability to reason about the eliciting factors of one's own emotional state is a mandatory prerequisite for the emergence of secondary emotions, conscious appraisal, taking place on this layer, will lead to secondary emotions. This appraisal process generally includes aspects of the past and the future, making use of different kinds of memories also present on this layer.
2. The *reactive layer* can be understood as resembling ontogenetically earlier processes, which are executed on a more or less "non-conscious", automatic level. These reactive processes include simple evaluations of positive or negative valence and are implemented as hard-wired reactions to basic patterns of incoming sensor information (e.g., fast movement in the visual field). Consequently, non-conscious appraisal leads to primary emotions, which can directly give rise to "non-conscious" reactive behaviors such as approach or avoidance.

As described in section 3.4 every emotion includes certain valence information, which is either positive or negative. This "hedonic" (pleasurable) valence (Gehm and Scherer 1988) is derived from the results of appraisal on both layers and used as the main driving force in the simulation of the agent's emotion dynamics. If, for example, our agent believes that winning the game is desirable (as in the gaming scenario explained in section 3.5.2) and suddenly comes to know that the game is over without him winning, non-conscious appraisal might lead to the emergence of the primary

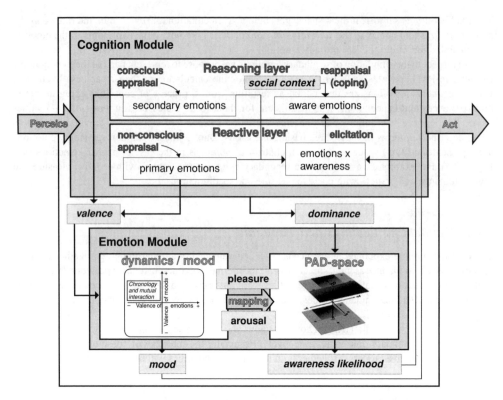

Figure 3.6 Mutual interaction of cognition and emotion. A stimulus is appraised leading to elicitation of both primary and secondary emotions. Emotional valence and dominance values drive the emotion module to continuously update an emotion *awareness likelihood*, which is used to filter the elicited emotions. Finally, resulting aware emotions are reappraised in the social context

emotion "anger" including highly negative valence.[1] However, in our architecture the resulting negative impulse only increases the likelihood of negative emotions such as anger. Thus, our emotional system does not follow a direct perception–action link as present in many purely rule-based, cognitive architectures.

By further representing uncertainty of beliefs as well as memories on the reasoning layer, secondary emotions such as "frustration" and "relief" are derived. For example, if the agent did not achieve a very important goal (such as drawing a desired card from the redraw pile) in several attempts, a general state of anger might turn into "cognitively elaborated" (Ortony *et al.* 2005) frustration. When, however, after a series of unsuccessful attempts the agent finally achieves his long-desired goal, the possible state of undifferentiated happiness might be accompanied or even substituted by the secondary emotion relief (cf. Figure 3.4).

---

[1] One might object that the necessary reasoning capabilities to deduce this kind of "anger" can hardly be conceived as remaining non-conscious. In the current context, however, we only use such a distinction to separate fast reactive emotional appraisal from relatively slower, deliberative (re-)appraisal. Thus we don't feel uncomfortable using symbolic reasoning for the implementation of processes on a so-called non-conscious, reactive level.

Elicitation, Reappraisal, and Coping

After the cognition module has generated "proposals" of cognitively plausible emotions on the basis of conscious and non-conscious appraisal, the inherent valences of these emotions will drive the dynamics/mood part of the emotion module. As described in section 3.4, the values of the dynamics subcomponent will continually be mapped into PAD space for categorization and combined with the actual state of Dominance. This Dominance is provided by the cognition module, which deduces its value from the actual social and situational context. The output of the emotion module in terms of awareness likelihoods for mood-congruent emotions is then fed back to the cognition module. It is combined with the initially generated "proposed emotions" to elicit a set of aware emotions. These aware emotions can be guaranteed to bear a high degree of resemblance in terms of their respective hedonic valences. Finally, reappraisal can take place to implement coping strategies as theoretically motivated in section 3.2.

The conscious appraisal process will be realized following the ideas of Reisenzein (2001) and Scherer (2001). The basic idea is to integrate secondary emotions as "meta-representational states of mind" in our BDI-based approach. If only primary emotions are employed, Max is directly expressing the emotion with the highest awareness likelihood without any kind of reappraisal based on social context. The next section describes how the direct effects of the emotion module on the believability and lifelikeness of our agent can be validated. This involves expressing the agent's internal emotional state as unambiguously as possible.

3.5 Examples and Experiences

To validate a conversational agent's architecture one has to decide on a suitable interaction scenario. Most researchers of the HCI community focus on designated interaction scenarios, which can be divided into competitive and cooperative ones. Virtual anthropomorphic agents in cooperative scenarios are employed as sales assistants in a virtual marketplace (Gebhard *et al.* 2004), pedagogical tutors in learning environments (Gratch and Marsella 2004), story-telling agents (Bernsen *et al.* 2004), conversational guides (Kopp *et al.* 2005), assembly assistants in construction tasks (Kopp *et al.* 2003), or direction-giving agents (Kopp *et al.* 2004). We will first present a smalltalk scenario in which Max is employed as a virtual guide to a public museum. In competitive scenarios (e.g., Rehm and André 2005; Becker *et al.* 2005b), researchers are mainly concentrating on the effect that virtual opponents in computer games have when being presented with a varying degree of naturalness and human-like behavior (see Figure 3.7(b)). In section 3.5.2 we report on the results of an empirical study conducted in the context of a competitive gaming scenario.

3.5.1 The Smalltalk Scenario

In the first scenario our agent Max is employed as a conversational guide in a public computer museum; see Kopp *et al.* (2005). By means of a video camera Max can perceive the presence of museum visitors and engages them in conversations, in which he provides background information about the museum, the exhibition, or other topics (Figure 3.7(a)). As the agent should be able to conduct natural language interactions, constraints on linguistic content (in understanding as well as in producing utterances) should be as weak as possible. Thus, a keyboard was used as input device, avoiding problems that arise from speech recognition in noisy environments.

The continuous stream of visual information provided by the video camera is first analyzed to detect the presence of skin-colored regions. A reactive, gaze-following behavior is triggered whenever a new person enters the visual field of Max. At the same moment a small positive emotional impulse is sent to the emotion module such that Max's mood increases the more people are around. In the absence of interlocutors the emotion module is generating the emotional state of boredom (see section 3.4.1) and

Figure 3.7 Two exemplary interaction scenarios: (a) Max in a smalltalk conversation with a museum visitor; (b) Max playing cards against a human opponent

special secondary behaviors such as leaning back and yawning are triggered. The corresponding physical exertion is modeled to have an arousing effect by automatically setting the boredom value (and thus also the arousal value) to zero. Concerning the Dominance value we decided to let Max never feel submissive in this scenario (although a notion of initiative is accounted for by the dialogue system).

In Figure 3.8 two parts of an example dialogue together with corresponding traces of emotions in Pleasure–Arousal space are presented. In the beginning, Max is in a neutral emotional state labeled *concentrated* until the visitor's greeting is processed by the BDI-based cognition module. In addition to the production of a multimodal utterance, a positive emotional impulse is sent to the emotion module. This impulse drives the internal dynamics of the "dynamics/mood" component as described in section 3.4 and the resulting values are constantly mapped on Pleasure and Arousal values as shown in Figure 3.8(a). The first positive emotional impulse directly leads to the activation of the primary emotion *surprised* at time t_1, modulating Max's facial expression and synthesized voice accordingly (see Figure 3.4). During the next 14 seconds no further impulses affect the emotion module. However, the internal dynamics lead to an increase in the agent's mood together with a decrease of the agent's emotional valence. Hence, the agent's Arousal is decreasing whereas the agent's Pleasure is increasing, such that at time t_2 the reference point in Pleasure–Arousal-space moves to *happy* and this primary emotion gets activated.

After a series of positive emotional impulses due to praising statements by the human dialogue partner, a very intense state of *happiness* is reached at time t_3. The word "pancake" is specially implemented to produce a strong negative impulse (mimicking a very rude insult), which leads to a sudden decrease of arousal and pleasure at time t_4. Notably, the agent does not get *angry* directly but only less *happy*, because he was in a very good mood shortly before. That is, mood-congruent emotions are guaranteed as a result of the internal dynamics of the emotion module.

To the end of the conversation, Max has become *very concentrated* (i.e., non-emotional) again, just before the visitor insults him at time t_1 (see Figure 3.8(b)) resulting in a strongly negative impulse. Within an instant Max is *surprised* at time t_2 and only five seconds later the internal emotion dynamics lets him feel *annoyed* at time t_3. The strongly negative emotional valence causes the mood to become negative within the next five seconds. Thus, when the human insults him again at time t_4, Max gets *angry*, which he becomes aware of himself. This relatively mild form of anger lets Max say that he is starting "to feel unhappy". When he is becoming very angry at time t_5, a kind of situation-focused coping behavior is triggered by leaving the scene. As the visitor asks Max only one emotionally neutral question in the following 15 seconds, Max's emotional state at first slowly shifts from hot to mild *anger* ending in a state of *annoyance* at time t_6. When the visitor is finally apologizing, the resulting positive impulse lets Max feel *concentrated* again at time t_7. In effect, he re-enters the display ready to go on with the conversation.

Human:	Hello Max.
Max:	(t_1, *surprised*) Hello. Let's chat. Use the keyboard to talk to me. What is your name?
Human:	Stefan.
Max:	Hello Stefan! (t_2, *happy*) Nice to meet you. I can explain something to you or we just have small talk.
Human:	How are you?
Max:	I'm fine at the moment and how are you?
	⋮
Human:	You are amazing.
Max:	(t_3, *very happy*) Our conversation gets better and better.
Human:	Pancake. (bad word for Max)
Max:	(t_4, *happy*) Why do you say that?
	⋮
Max:	(t_1, *very concentrated*) Can you do so, too?
Human:	You pig.
Max:	(t_2, *surprised*) Have a look in the mirror (t_3, *annoyed*).
Human:	Pancake.
Max:	(t_4, *angry*) Better think about what you are saying. I start to feel unhappy. (t_5, *very angry*) I'm fed up for now (*leaves the display*).
Human:	(*pause*) Where are you?
Max:	(t_6, *very annoyed*) I will not come back until I calmed down.
Human:	Sorry.
Max:	(t_7, *concentrated*) After all you accept it. Now I calmed down. (*reenters the display*)
Human:	Bye.
Max:	Good bye, Stefan.

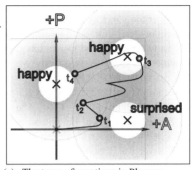

(a) The trace of emotions in Pleasure–Arousal space during the first minute of dialogue: $t_1 = 0{:}10$; $t_2 = 0{:}24$; $t_3 = 1{:}00$; $t_4 = 1{:}05$ (all times in min.)

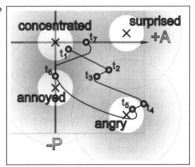

(b) The trace of emotions until the end of the dialogue: $t_1 = 1{:}46$; $t_2 = 1{:}50$; $t_3 = 1{:}55$; $t_4 = 2{:}02$; $t_5 = 2{:}10$; $t_6 = 2{:}25$; $t_7 = 2{:}35$ (all times in min.)

Figure 3.8 A dialogue example from the smalltalk scenario. The subfigures show the corresponding traces of Max's emotions in the Pleasure-Arousal-plane during (a) the first and (b) second part of the dialogue. Dominance is always positive and constant in this scenario

3.5.2 The Gaming Scenario

As a competitive face-to-face interaction scenario between a human and Max we implemented a card game called Skip-Bo (Becker *et al.* 2005b). In this game, both players have the goal of getting rid of their cards following an offensive or defensive strategy (see Figure 3.7(b)). Max, again, always retains control over the game as he corrects the human player in case of a false move (Figure 3.9(b)). Max's emotion module is initialized to reflect this state of high Dominance, but whenever the human player is at least two cards ahead to win the game, this value is changed to reflect a state of Submissiveness (i.e., non-dominance). Consequently, when Max is highly aroused and in a state of negative pleasure, he sometimes shows *fear* (Figure 3.9(a)) instead of *anger*.

In such a playful interaction, the aim was to study the effects that affective and/or empathic agent behavior has on human players. To let Max show positive or negative empathy, physiological indicators of the current emotions of the human players were measured. A galvanic skin response (GSR) sensor was attached to the index finger and the small finger of the non-dominant hand and an electromyography (EMG) sensor was attached to the subject's left (mirror-oriented) cheek to measure the activity of the masseter muscle (see Figure 3.9(a)). By evaluating this data as well as results from a questionnaire we could analyze the observer's emotional experience, solely depending on Max's emotional appraisal and display.

An empirical study was carried out within the gaming scenario. It comprised four conditions designed to assess the effects of simulated emotions and empathic feedback in the context of such a competitive human–computer interaction (cf. (Brave 2003) for a similar set of conditions):

1. *Non-emotional condition*. Max does not display emotional behavior.
2. *Self-centered emotional* condition. Max appraises his own game play only, and displays (e.g., facial) happiness when he is able to move cards.
3. *Negative empathic condition*. Max shows self-centered emotional behavior and responds to those opponent actions that thwart his own goal of winning the game. Consequently, he will, for example, display distress when the human player performs a good move or is detected to be positively aroused. This condition implements a form of "negative" empathy.
4. *Positive empathic condition*. Here, Max is also self-centered emotional, but opponent actions are appraised "positively" such that he is "happy for" the human player's game progress. If the human player is assumed to be distressed, Max will display "sorriness". This condition implements "positive" empathy.

(a)   (b)

Figure 3.9 Two interaction examples from the Skip-Bo gaming scenario: (a) Max is afraid to loose the game; (b) Max corrects an opponent's move

Note that Max follows a competitive playing strategy in all conditions. The study included 14 male and 18 female subjects, who were told in advance that they would get an extra reward if they won against Max. Subjects were randomly assigned to the four conditions (eight in each condition). See Becker *et al.* (2005a) for further details about the experimental setup.

To analyze the recorded physiological data (skin conductance and electromyography), we focused on game situations where emotional reactions in the human player as well as Max were likely to occur. An analysis of the biometrical data revealed that – in this competitive game scenario – the absence of negative empathy in the agent is conceived as stressful (derived from GSR) and irritating, as it might also be experienced when playing against another human. A complementary result is that negatively emphatic agent behavior induces negatively valenced emotions (derived from EMG) in the human player. A detailed discussion of the results can be found in Prendinger *et al.* (2006). Overall, these results suggest that the integration of emotions into an agent's architecture can have an important bearing on the agent's overt behaviors, which in turn can have decisive effects on a human's emotional responses to it. If used in expedient ways, integrating emotions thus has the potential to serve significantly supportive functions in human–machine interaction.

3.6 Conclusions

The initially quoted dialogue from the movie *I, Robot* gave rise to the challenging question "Should conversational agents be able to have emotions?". Most people show a high degree of acceptance concerning the practical outcomes of technological progress in everyday life. When asked how much further technology should or should not advance, few people see a clear borderline of everyday concerns a machine should not interfere with. Thinking of scenarios where artificial agents – be it in the form of virtual characters on a computer screen or in the form of physical humanoid robots – take a role as conversational partners in a future society even raises the issue of *social* acceptance. This issue has been widely recognized by researchers in the field of human–computer interaction, resulting in the development of knowledgable embodied conversation agents rather than unembodied expert systems.

In this line, we have started to simulate human emotion in an embodied agent by combining various research streams in a highly interdisciplinary approach. Our studies and experiences reported in this chapter have shown that believability, lifelikeness, and personality traits of a conversational agent can be increased by the integration of primary and secondary emotions, and that these enhance the acceptance of the conversational agent. Consequently, we understand the simulation of emotions to be one major property of humanoid conversational agents if they are to be more acceptable as co-equal social partners.

But should we also integrate *negative* emotions into conversational agents? Even if we admit the necessity to integrate artificial emotions into an agent's cognitive architecture, the questions remain whether every human emotion has to be present in that architecture, and whether every emotion has to be expressed in an interaction with a human. One might argue, for example, that the occurrence of negative emotions is undesirable in an artificial system built to be of help. However, we are following the experimental–theoretical motive so as to gain a deeper understanding of human emotions and human emotional behavior. As an adequate implementation of a model based on emotion psychology will automatically give rise to internal states similar to, for example, human anger or sadness, it seems inevitable to integrate means of expressing negative emotions. It would likely be irritating to the human interlocutor if the agent was unable to express such an emotional state properly. Additionally, as we are also moving toward simulating empathic behaviors, any limitation on the full spectrum of emotions simulated and expressed by an agent would seem inappropriate. A true human understanding of another's emotional state can only be ascribed to individuals that are believed to "feel" these emotions themselves.

Hence, while at present it is hard to foresee whether it will be appropriate to have agents that "cry", we argue that integrating emotions into conversational agents will make them more believable as social partners.

References

Arbib M.A. and Fellous J.-M. (2004) Emotions: from brain to robot. *TRENDS in Cognitive Sciences* **8** 554–561.

Bartneck C. (2002) Integrating the OCC model of emotions in embodied characters. In *Workshop on Virtual Conversational Characters: Applications, Methods, and Research Challenges*.

Becker C., Kopp S. and Wachsmuth I. (2004) Simulating the emotion dynamics of a multimodal conversational agent. In *Workshop on Affective Dialogue Systems*, pp. 154–165.

Becker C., Nakasone A., Prendinger H., Ishizuka M. and Wachsmuth I. (2005a) Physiologically interactive gaming with the 3D agent Max. In *International Workshop on Conversational Informatics*, Kitakyushu, Japan, pp. 37–42.

Becker C., Prendinger H., Ishizuka M. and Wachsmuth I. (2005b) Evaluating affective feedback of the 3D agent Max in a competitive cards game. In *Affective Computing and Intelligent Interaction*, pp. 466–473.

Bernsen N.O., Charfuelàn M., Corradini A., Dybkjær L., Hansen T., Kiilerich S., Kolodnytsky M., Kupkin D. and Mehta M. (2004) Conversational H.C. Andersen. First prototype description. In *Proceedings of the Tutorial and Research Workshop on Affective Dialogue Systems* (ADS04), pp. 305–308.

Bickmore T. and Cassell J. (2005) Social dialogue with embodied conversational agents. In J. van Kuppevel L. Dybkjaer and N. Bernsen (eds), *Advances in Natural, Multimodal Dialogue Systems*. Kluwer Academic, New York.

Brave S. (2003). User responses to emotion in embodied agents. In N. Suzuki and C. Bartneck (eds), *Proceedings CHI 2003 Workshop on Subtle Expressivity for Characters and Robots*, pp. 25–29.

Breazeal C.L. (2002) *Designing Sociable Robots*. MIT Press, Cambridge, MA.

Breazeal C. (2004a) Function meets style: insights from emotion theory applied to HRI. *IEEE SMC Transactions, Part C* **32**, 187–194.

Breazeal C. (2004b) Social interactions in HRI: the robot view. *IEEE SMC Transactions, Part C* **34**, 81–186.

Burghouts G., op den Akker R., Heylen D., Poel M. and Nijholt A. (2003). An action selection architecture for an emotional agent. In I. Russell and S. Haller (eds), *Recent Advances in Artificial Intelligence*. AAAI Press, pp. 293–297.

Cassell J., Sullivan J., Prevost S. and Churchill E. (eds) (2000) *Embodied Conversational Agents*. MIT Press, Cambridge, MA.

Damasio A.R. (1994) *Descartes' Error, Emotion Reason and the Human Brain*. Grosset/Putnam.

Ekman P., Friesen W. and Ancoli S. (1980) Facial sings of emotional experience. *Journal of Personality and Social Psychology* **29**, 1125–1134.

Elliott C. (1992) *The Affective Reasoner. A process model of emotions in a multi-agent system*. PhD thesis, Northwestern University. Institute for the Learning Sciences.

Ernst G. and Newell A. (1969) *GPS: A Case Study in Generality and Problem Solving*. Academic Press.

Gebhard P., Klesen M. and Rist T. (2004) Coloring multi-character conversations through the expression of emotions. In *Proceedings of the Tutorial and Research Workshop on Affective Dialogue Systems* (ADS'04), pp. 128–141.

Gehm T.L. and Scherer K.R. (1988) Factors determining the dimensions of subjective emotional space. In K.R. Scherer (ed.), *Facets of Emotion Recent Research*. Lawrence Erlbaum Associates, pp. 99–113.

Giarratano J. and Riley G. (2005) *Expert Systems: Principles and programming*. Thomson Course Technology, 4th edition.

Grammer K. and Oberzaucher E. (2006) The reconstruction of facial expressions in embodied systems: new approaches to an old problem. *ZIF Mitteilungen* **2**, 14–31.

Gratch J. and Marsella S. (2004) A domain-independent framework for modelling emotion. *Journal of Cognitive Science Research* **5**, 269–306.

Hodges A. (2000) *Alan Turing: The Enigma*. Walker & Company.

Kopp S., Jung B., Lessmann N. and Wachsmuth I. (2003) Max: a multimodal assistant in virtual reality construction. *KI-Künstliche Intelligenz* **4**(3), 11–17.

Kopp S., Tepper P. and Cassell J. (2004) Towards integrated microplanning of language and iconic gesture for multimodal output. In *Proceedings of the International Conference on Multimodal Interfaces* (ICMI'04), ACM, pp. 97–104.

Kopp S., Gesellensetter L., Krämer N. and Wachsmuth I. (2005) A conversational agent as museum guide: design and evaluation of a real-world application. In *Intelligent Virtual Agents*, pp. 329–343.

Lang P.J. (1995) The emotion probe: studies of motivation and attention. *American Psychologist* **50**(5), 372–385.

Lerner J. and Keltner D. (2000) Beyond valence: toward a model of emotion-specific influences on judgement and choice *Cognition and Emotion* **14**, 473–493.

Leßmann N., Kopp S. and Wachsmuth I. (2006) Situated interaction with a virtual human: perception, action, and cognition. In G. Rickheit and I. Wachsmuth (eds), *Situated Communication*. Mouton de Gruyter, Berlin, pp. 287–323.

Mehrabian A. (1995) Framework for a comprehensive description and measurement of emotional states. *Genetic, Social, and General Psychology Monographs* **121**, 339–361.

Mikulas W.L. and Vodanovich S.J. (1993) The essence of boredom. *The Psychological Record*, **43**, 3–12.

Negrotti M. (2005) Humans and naturoids: from use to partnership. *Yearbook of the Artificial, Cultural Dimensions of the User* **3**, 9–15.

Neumann R., Seibt B. and Strack F. (2001) The influence of mood on the intensity of emotional responses: disentangling feeling and knowing. *Cognition and Emotion* **15**, 725–747.

Oatley K. and Johnson-Laird P.N. (1987) Towards a cognitive theory of emotions. *Cognition and Emotion* **1**, 29–50.

Ortony A. (2003) Emotions in humans and artifacts. In *On Making Believable Emotional Agents Believable*, MIT Press, pp. 189–212.

Ortony A., Clore G.L. and Collins A. (1988) *The Cognitive Structure of Emotions*. Cambridge University Press, Cambridge.

Ortony A., Norman D. and Revelle W. (2005) Affect and proto-affect in effective functioning. In J. Fellous and M. Arbib (eds), *Who Needs Emotions: The brain meets the machine*. Oxford University Press, pp. 173–202.

Osgood C.E., Suci G.J. and Tannenbaum P.H. (1957) *The Measurement of Meaning*. University of Illinois Press.

Picard R.W. (1997) *Affective Computing*. MIT Press, Cambridge, MA.

Prendinger H., Dohi H., Wang H., Mayer S., and Ishizuka M. (2004) Empathic embodied interfaces: addressing users' affective state. In *Workshop on Affective Dialogue Systems*, pp. 53–64.

Prendinger H., Becker C. and Ishizuka M. (2006) A study in users' physiological response to an empathic interface agent. *International Journal of Humanoid Robotics*, 371–391.

Rao A. and Georgeff M. (1991) Modeling Rational Agents within a BDI-architecture. In *Proc. Intl. Conference on Principles of Knowledge Representation and Planning*, pp. 473–484.

Reeves B. and Nass C. (1998) *The Media Equation: How people treat computers, television and new media like real people and places*. CSLI Publications, Center for the Study of Language and Information. Cambridge University Press.

Rehm M. and André E. (2005). Catch me if you can: exploring lying agents in social settings. In *Autonomous Agents and Multiagent Systems*, pp. 937–944.

Reilly W.S.N. (1996) *Believable Social and Emotional Agents*. PhD thesis, Carnegie Mellon University. CMU-CS-96–138.

Reisenzein R. (1994) Pleasure-arousal theory and the intensity of emotions. *Journal of Personality and Social Psychology* **67**, 525–539.

Reisenzein R. (2001) Appraisal processes conceptualized from a schema-theoretic perspective: contributions to a process analysis of emotions. In K.R. Scherer, A. Schorr and T. Johnstone (eds), *Appraisal Processes in Emotion*. Oxford University Press, pp. 187–201.

Scherer K.R. (2001) Appraisal considered as a process of multilevel sequential checking. In K.R. Scherer, A. Schorr and T. Johnstone (eds), *Appraisal Processes in Emotion*. Oxford University Press, pp. 92–120.

Shortliffe E.H., Rhame F.S., Axline S.G., Cohen S.N., Buchanan B.G., Davis R., Scott A.C. Chavez-Pardo R. and van Melle W. J. (1975) Mycin: a computer program providing antimicrobial therapy recommendations. *Clinical Medicine* **34**.

Sloman A., Chrisley R. and Scheutz M. (2005) The architectural basis of affective states and processes. In *Who Needs Emotions?* Oxford University Press.

Svennevig J. (1999) *Getting Acquainted in Conversation*. John Benjamins.

Svennevig J. and Xie C. (2002) Book reviews: Getting acquainted in conversation. a study of initial interactions. *Studies in Language* **26**, 194–201.

Turing A.M. (1950) Computing machinery and intelligence. *Mind* **59**, 433–460.

von Scheve C. and von Lüde R. (2005) Emotion and social structures: towards an interdisciplinary approach. *Journal for the Theory of Social Behaviour* **35**, 303–328.

Weizenbaum J. (1976) *Computer Power and Human Reason: From judgement to calculation*. Freeman.

Wundt W. (1922) *Vorlesung über die Menschen- und Tierseele*. Voss Verlag, Leipzig.

4

More Than Just a Friendly Phrase: Multimodal Aspects of Polite Behavior in Agents

Matthias Rehm and Elisabeth André

4.1 Introduction

During the last decade research groups as well as a number of commercial software developers have started to enhance user interfaces with embodied conversational characters (ECAs) that display facial expressions, gaze pattern, hand and arm gestures in synchrony with their speech (see Figure 4.5 for an example). ECAs offer great promise to more natural interaction because of their potential to emulate verbal and nonverbal human–human interaction. But supplying an interface agent with a body also poses great challenges to the design of appropriate interactions because the user will expect – at least in part – humanlike verbal and nonverbal conversational behaviors of such an agent.

If it is true, as *The Media Equation* by Reeves and Nass (1996) suggests, that people respond to computers as if they were humans, there are good chances that people are also willing to form social relationships with virtual personalities. That is, a virtual character is not just another interface gadget. It may become a companion and even a friend to the user. A prerequisite for this vision to come true is to enrich ECAs with social competencies to render their interactions with the user more natural and entertaining. One pervasive aspect of social interaction is the use of politeness.

4.1.1 Politeness in the Humanities

In their seminal work, Brown and Levinson (1987) analyzed verbal strategies of politeness. Their theory is centered on the notion of face or – to be more precise – keeping one's face. People maintain positive and negative face, which are continuously threatened during interactions, such as by commands or criticism on one's behavior. Positive face describes the positive self-image one has and negative face is the want to autonomously decide on one's actions. Brown and Levinson introduce the notion of a face-threatening act (FTA). Speech acts like commands or criticism that threaten one's face needs are thus called FTAs.

Conversational Informatics: An Engineering Approach   Edited by Toyoaki Nishida
© 2007 John Wiley & Sons, Ltd

Brown and Levinson distinguish four different types of strategies to deliver a FTA. The threat can be delivered directly without any redress, which is the most rude form but sometimes inevitable, such as if it is crucial for the well-being of the addressee: "Your hair is on fire!". Most of the time, speakers try to redress or mitigate such undesirable acts, perhaps by referring to the good looks of the addressee before asking her for a favor. This is a positive politeness strategy taking the self-image of the addressee into account. Negative politeness strategies focus on the addressee's freedom of action and come, for example, in the disguise of apologies or impersonalizations like "Perhaps it would be a good idea to speak louder." The last type are off-record strategies which constitute deliberate violations of the Gricean maxims with the effect that the threat remains very vague and the addressee has to infer the exact meaning of the speaker, leaving him in the position to misunderstand the speaker and thus to not feel offended. An example would be "Well, the volume always differs between people." According to Brown and Levinson, the more severe the threat is, the more abstract a strategy should be; that is, negative politeness is employed for more severe threats than positive politeness.

Previous work has concentrated for the most part on the linguistic aspects of FTAs; i.e., on verbal means to deliver and redress FTAs. But like every interaction, FTAs are multimodal in face-to-face situations like pointing to the projector while saying "You could have done some slides. We have a projector here."

Research on nonverbal communicative behaviors, such as gestures or facial expressions, provides a good impression of the relevance of multimodal aspects of communication (e.g., Allwood 2002; Kendon 1986; Knapp and Hall 1997; Pease 1993), and reveals a bunch of implicit information about the role of gestures and facial expressions in delivering and redressing face threats. However, there is hardly any work that explores the relationship between multimodal means of communication and face threats. An exception is an empirical study by Trees and Manusov (1998) who found that nonverbal behaviors, such as pleasant facial expressions and more direct body orientation, may help to mitigate face threats evoked by criticism. Bavelas et al. (1995) provide a classification of gestures some of which can be directly mapped onto Brown and Levinson's strategies of politeness. Shared information gestures mark material that is part of the interlocutors' common ground. Citing gestures refer to previous contributions of the addressee and aim at conveying the impression that the interlocutors share a common opinion. Elliptical gestures mark incomplete information that the addressee should augment for him- or herself and may take on a similar function as off-record strategies. Seeking agreement gestures directly correspond to Brown and Levinson's approval-oriented strategies. Turn open gestures can be regarded as attempts to satisfy the addressee's desire for autonomy. Linguistic means to deliver FTAs have partly become part of the grammar and Bavelas' classification of gestures suggests that there might be similar principled and standardized connections between nonverbal means of communication and politeness strategies.

4.1.2 Politeness in Conversational Informatics

Walker et al. (1997) have presented one of the first approaches to implement politeness strategies as a means to more flexible dialogue control. They summarize the available strategies into four main categories: (1) direct, (2) approval-oriented, (3) autonomy-oriented, (4) off-record. They describe a selection mechanism that is based on the variables power, social distance, and ranking of speech act, which were introduced by Brown and Levinson. Johnson et al. (2004) describe the value of politeness in a tutoring system. Examining the interactions between a real tutor and his students, they came up with a set of templates to generate appropriate utterances depending on the current situation. The templates have been employed in a planner for tutorial tactics that selects approval and autonomy-oriented strategies based on the type of the expected face threat. To get a politeness rating for each strategy, they conducted the following experiment. Subjects were presented with two hypothetical scenarios in which an embodied agent gives some feedback on a task. In each scenario, the participants were confronted with utterances realizing approval- and autonomy-oriented strategies. In scenario one, participants had to rate these utterances according to the degree of cooperation expressed in the agent's utterance (i.e., relating to positive politeness), and in scenario two according to the degree of freedom of action for the user (i.e., relating to negative politeness). The results showed a significant difference between the different types of the utterances which were in line with rankings proposed by Brown and Levinson, suggesting that the subjects ascribed politeness to the computer tutor as if it were a social actor. In Johnson *et al.* (2005), they investigated how far politeness theory applies equally to tutorial dialogue tactics in English and in German. The results of the cross-cultural experiment revealed that the ratings of German and US students were comparable and highly correlated.

André *et al.* (2004) augmented the model of Brown and Levinson with an emotional layer. The emotion of the addressee as it is observed by the speaker plays a crucial role in determining an appropriate strategy. Bickmore and Cassell (2000) describe how smalltalk is utilized to build up common ground between an embodied conversational agent and the user based on an extension of Brown and Levinson's theory of politeness. Nakano *et al.* (2003) study how people use nonverbal signals, such as eye gaze and head nods, to provide common ground in the context of direction-giving tasks. Even though their work relies on a sophisticated model of gestural communication, they did not investigate how the use of gestures may help to mitigate the face threat for the user. Porayska-Pomsta and Mellish (2004) make use of Brown and Levinson's model in order to motivate linguistic variations of a natural language generator. Prendinger and Ishizuka (2001) consider Brown and Levinson's social variables distance and power in order to control emotional displays of agents. For instance, if the social distance between an agent and its conversational partner is high, the agent would not show anger to the full extent. This behavior can be interpreted as an attempt to reduce the face threat for the conversational partner.

Summing up, it may be said that the implementation of politeness behaviors in an ECA has mainly focused on verbal aspects so far. As a reason we indicate that empirically grounded knowledge regarding multimodal aspects of human politeness tactics is still sparse. Our work provides a first step towards filling this gap by conducting a corpus study to find correlations between gestures and verbal politeness tactics that may be used as a basis for controlling ECA behaviors.

4.2 The Augsburg SEMMEL Corpus

Due to the sparse literature on the use of nonverbal communicative behaviors of politeness, we collected our own corpus based on staged conversations between humans for an empirical grounding of the intended system. To trigger the use of politeness strategies, we had to make sure that the communicative situation was inherently face-threatening for the participants. We therefore decided to record scenarios where an audience had to provide criticism to the speaker. The recorded video material was annotated and analyzed in order to identify frequently occurring combinations of gestures and verbal politeness strategies.

4.2.1 Collecting the SEMMEL Corpus

We devised a scenario that forced the participants to use their (unconscious) knowledge of politeness strategies by confronting them with an inherently face-threatening situation. Criticizing someone is a prototypical example of such a situation. Therefore, we chose seminar talks with subsequent discussion to provide for a more or less "natural" situation for the participants. The focus was on the criticism given by the audience to the speakers on their performance. Students were divided into two groups: audience and speakers. The speakers were asked to give a five-minute talk about one of their hobbies. This topic was chosen to keep the necessary preparatory work for the talk at a minimum and to ensure that the audience had enough knowledge on the topic to easily criticize the speaker.

The initial explanation for this setup that was given to the participants one week before the experiment was our need to collect a corpus of nonverbal communicative behavior. This explanation also accounted

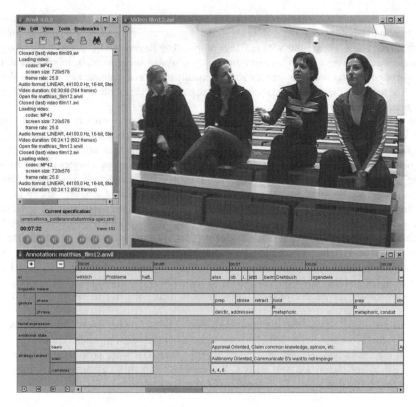

Figure 4.1   Snapshot from the ANVIL annotation system

for the two cameras we were using, one videotaping the speaker, the other one the audience. The initial explanation was detailed on the day of the experiment. The speakers were informed about the real setup to prevent them from reacting in an unwanted way to the critic or the criticism. The audience was told that we were interested in the reaction of the speaker to (potentially unjustified) criticism. In order to ensure that we would collect enough examples of relevant communicative acts, each member of the audience was instructed to criticize the speaker on three different dimensions and received a list of issues that had to be brought up during the discussion: (i) formal aspects, such as too many/too few slides, (ii) content, such as snowboarding is far too dangerous, and (iii) personal, such as the speaker was too nervous. After the experiment, the participants were informed about the actual objective of the data collection.

Twelve students in their first- and second-year courses participated in this data collection, three male, nine female. Four of them (two male, two female) prepared a talk on their hobby and were criticized by four audience members immediately after their presentation (Figure 4.1). The audience for each talk was constituted randomly from the remaining eight students, ensuring that each of them participated two times as an audience member and met one of the other audience members only twice. We held the social variables *distance* and *power* constant and made sure that the speakers and the audience were not from the same year. The resulting SEMMEL corpus (Strategy Extraction for MultiModal Eca controL) contains 66 different acts of criticism, 16.5 on average per talk. An act of criticism covers one of the aspects mentioned above and is always delivered with a mix of strategies and co-occuring gestures. The corpus contains 159 combinations of strategies and gestures.

4.2.2 Annotating the SEMMEL Corpus

The collected material was annotated using ANVIL (Kipp 2004). ANVIL allows a frame-by-frame analysis of the video stream. The annotation features and attributes of the coding scheme can be freely defined to match the goal of the analysis. Figure 4.1 shows a screenshot of the ANVIL system along with annotations of our corpus. Focusing on the interaction of verbal and nonverbal behavior in the use of politeness strategies, the SEMMEL coding scheme features three main layers:

1. *trl*: the transliteration, i.e., the words spoken.
2. *gesture*: the hand gestures of the speaker visible in the video.
3. *strategy*: the politeness strategies employed by the speaker.

The coding of gestures follows Kipp's approach (Kipp 2004) which is based on McNeill's guidelines (McNeill 1992). Accordingly, two different parts of a gesture are distinguished: the gesture phase and the gesture phrase.

- *Track gesture.phase*. This is a primary track, which means that it is directly related to the video. Although gestures are mostly co-verbal (i.e., they accompany speech and add additional meaning to it by visualizing aspects of the mentioned referents), only the stroke of the gesture has verbal–nonverbal synchronization constraints. Thus it does not suffice to bind the gesture only to the transliteration layer but also to the video itself. The most prominent phases of a gesture are preparation, stroke, and retraction. Generally, the hands are brought from a resting position into the gesture space during preparation. The stroke is the phase of the gesture that carries/visualizes its meaning. Afterwards, the hands are brought back to a resting position during the retraction phase.
- *Track gesture.phrase*. The gesture phrase denotes the type of the gesture. It is realized as a secondary track which means it is related to another track of the coding scheme, in this case to the gesture phase. Thus, the gesture phases specify the time dimension of the gesture in regard to the video whereas the gesture phrase gives the interpretation of this specific gesture.

McNeill (1992) distinguishes roughly between adaptor, beat, emblem, deictic, iconic, and metaphoric gestures. Adaptors comprise every hand movement to other parts of the body like scratching one's nose. Beats are rhythmic gestures that may emphasize certain propositions made verbally or that link different parts of an utterance. Emblems are gestures that are meaningful in themselves, without any utterance. An example is the American "OK"-emblem, where the thumb and first finger are in contact at the tips while the other fingers are extended. Deictic gestures identify referents in the gesture space. The referents can be concrete like the addressee or they can be abstract like pointing to the left and the right while uttering the words "the good and the bad". In this case the good and the bad are identified in the gesture space and it becomes possible to refer back to them later on by pointing to the corresponding position. Iconic gestures depict spatial or shape-oriented aspects of a referent, such as using two fingers to indicate someone walking while uttering "he went down the street". Metaphoric gestures are more difficult in that they visualize abstract concepts by the use of metaphors, such as using a box gesture to visualize "a story". This is the conduit metaphor that makes use of the idea of a container in this case a container holding information.

To code politeness strategies, we follow Walker *et al.*'s (1997) categorization of direct, approval-oriented, autonomy-oriented, and off-record strategies. In direct strategies, no redress is used, the speaker just expresses his wishes. Approval-oriented strategies are related to the positive face needs of the addressee, using means to approve of her self-image. Autonomy-oriented strategies on the other hand are related to the negative face wants of the addressee, trying to take care of her want to act autonomously. Off-record strategies are the most vague and indirect form to address someone, demanding an active inference on the side of the addressee to understand the speaker. Depending on variables such as social distance and

Table 4.1 Types of strategies used for coding the SEMMEL corpus and examples
of verbal means to realize these strategies

Strategy	Verbal means
Direct	State the threat directly
Approval-oriented	
Convey interest	Compliments, intensifying adjectives
Claim in-group membership	Address forms, slang, elliptical utterances
Claim common knowledge	White lies, use of "sort of", "in a way", jokes
Indicate knowledge about wants	State to regard addressee's wants
Claim reflexivity	Inclusive "we", give/ask for reasons
Claim reciprocity	State that addressee's owns speaker a favor
Fulfil wants	State sympathy
Autonomy-oriented	
Make minimal assumptions	Hedges "I think", "I guess"
Give option not to act	Subjunctive, use of "perhaps"
Minimize threat	Euphemisms, use of "a little", "just"
Communicate want not to impinge	Avoidance of "you" and "I", state threat as general rule
Indebting	Claim that speaker will owe addressee a favor
Off-record	
Violate relevance maxim	Associations, hints
Violate quantity maxim	Exaggerations like "always"
Violate quality maxim	Irony, rhetorical questions
Violate manner maxim	Ambiguity, elliptical utterances

power, and a culture-specific rating of the speech act, a speaker chooses an appropriate strategy to deliver a face-threatening act (FTA), such as (i) "I really enjoyed your talk but you should be more coherent" vs. (ii) "The talk should be more coherent" vs. (iii) "In a different order ... you know?" In (i) the speaker compliments the addressee on her talk before delivering his criticism, thus employing an approval-oriented strategy. In (ii), an autonomy-oriented strategy is used in impersonalizing the criticism. The speaker refers neither to the addressee nor to himself. In (iii), the speaker opts for an off-record strategy in introducing some ambiguity – the order of what? – and using an elliptical utterance. The coding of strategies uses a simplified version of Brown and Levinson's hierarchy distinguishing between seven different approval-oriented, five different autonomy-oriented, and four different off-record strategies (Table 4.1).

- *Track strategy.basic*. Every strategy that is employed by the speaker is coded and bound to the words in the transliteration track that give rise for this interpretation. For each category of strategies (direct, approval-oriented, autonomy-oriented, off-record), the coder has to decide for a specific type (see Table 4.1).
- *Track strategy.main*. Because a single utterance contains nearly always a mix of strategies, a track is added for coding the main strategy used in a specific utterance. The same elements as in the basic track are used (see Table 4.1), but the elements in this track are not bound to the transliteration but to the basic track.
- *Track.variables*. Brown and Levinson introduce the contextual variables social distance, power relation, and ranking of the imposition to calculate the weight of the face threat that is redressed by the strategy. This track is bound to *strategy.main* assuming that neither of the variables changes during a single utterance.

4.2.3 Analyzing the SEMMEL Corpus

The first part of our analysis concentrated on the distribution of the four basic categories of politeness strategies. Remarkable is the high number of autonomy-oriented strategies. From the 159 strategy/gesture combinations, 63% include autonomy-oriented strategies, 17% off-record, 13% approval-oriented strategies, and 8% direct strategies. By opting for autonomy-oriented strategies, the critics try to leave the choice of action on the side of the addressee. Thus, the criticism is wrapped into some kind of suggestion for the addressee on how to improve the talk. We put this result down to the nature of the power relationship between the speaker and the audience. Since both the speaker and the critics were students, the critics obviously did not feel like being in the position of judging on the performance of their colleagues.

Out of the five autonomy-oriented strategies, only four can be found in the corpus (Table 4.2). Apart from the communicative strategy "Communicate want not to impinge" which relies mainly on the impersonalization of the speech act (reflected by the avoidance of pronouns, such as "you" and "I") and which is used for 34% of the time, the use of the other strategies is equally distributed around 22%. Most communicative acts that correspond to the category "Make minimal assumptions" employ hedging verb phrases, such as "I think", "I guess", or "I suppose". In the case of the strategy "Give option not to act", the subjunctive is widely used along with words such as "perhaps". The strategy "Minimize threat" employs minimizing expressions, such as "a little".

Out of the approval-oriented strategies only the "claim reflexivity" strategy was used regularly (47% of the time). This strategy was realized by giving reasons for the criticism and thus trying to explain to the addressee why the criticism is necessary. Although all off-record strategies identified by Brown and Levinson can be found in the corpus, only one is used regularly: violate manner maxim. To realize this strategy, the critics usually employed elliptical utterances.

Furthermore, we were interested in the distribution of gesture types. Out of the six gesture types that were annotated, only two are exceptional in the frequency of their use: beats and emblems (Figure 4.2). Whereas emblems can be rarely observed (3%), beats are the most frequently used gestures (27%). Emblematic gestures are self-sufficient in that they can be interpreted without any accompanying utterance. Thus, it is not astonishing to find them rarely as co-verbal gestures. Beats are rhythmic gestures that emphasize words in an utterance or relate different parts of an utterance. Thus, the extensive use of beats might be an artifact of the experimental setting because the critics had to "invent" an act of criticism that was not their own on the fly, and thus the beat gesture might be an outward sign of this process indicating that the turn has not yet ended. As noted by McNeill (1992), the number of beats depends among other things on the discourse context. He observed about 25% beats in narrative contexts, which roughly corresponds to our findings, versus 54% beats in extra narrative contexts.

Overall, we did not notice great differences in the distribution of deictic, iconic and metaphoric gestures. However, when analyzing their co-occurrence with politeness strategies, two general tendencies may be observed (Figure 4.3). First, adaptors are used considerably while employing autonomy-oriented strategies (26%). They are used least frequently with off-record strategies (5%). Off-record strategies are the most ambiguous and vague means to deliver a face threat. Given that adaptors often indicate that people are nervous, the more frequent use of adaptors in autonomy-oriented strategies seems plausible because the criticism is delivered more openly, resulting in more stress for the speaker.

Table 4.2   Frequency of autonomy-oriented strategies

Strategy	Frequency
Make minimal assumptions	0.22
Give option not to act	0.21
Minimize threat	0.22
Communicate want not to impinge	0.34

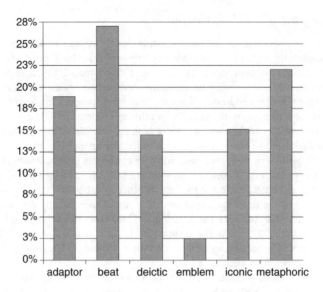

Figure 4.2   Distribution of gestures in the SEMMEL corpus

Second, there is a difference in the use of gestures of the abstract (metaphoric) and gestures of the concrete (iconic and deictic). Nearly all deictic gestures that occurred in our setting referred to the addressee or concrete locations in the space (76.8%). Forty-four percent of all gestures used with the off-record strategies, were metaphoric in nature vs. 15% for iconic and deictic gestures. In contrast to this, 50% of the gestures employed with the direct strategies, and 45% of the gestures employed with the approval-oriented strategies, were iconic and deictic in nature. The same is true to a lesser degree for the autonomy-oriented strategies. In this case, 28% were gestures of the concrete and 20% metaphoric gestures. Thus, the more abstract, vague, and ambiguous the strategies become, the more abstract and vague the primarily employed gesture type becomes (Table 4.3).

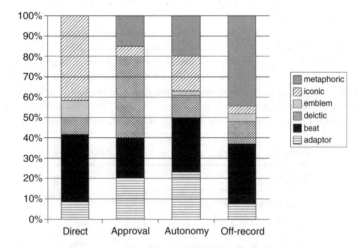

Figure 4.3   Co-occurrence of gestures and politeness strategies in the SEMMEL corpus

Table 4.3 Gestures of the concrete vs. gestures of the abstract

	Direct	Approval	Autonomy	Off-record
Concrete gestures (deictic/iconic)	0.5	0.45	0.28	0.15
Abstract gestures (metaphoric)	0.0	0.15	0.20	0.44

These results confirm the assumptions that not only linguistic regularities can be found in the use of politeness strategies, but that also nonverbal behaviors like gestures play a role in the realization of strategies. Metaphoric gestures relate to abstract concepts and illustrate an aspect of a referent in the utterance by the aid of a metaphor. The best known metaphoric gesture is the conduit metaphor where the hands form a kind of container that symbolizes the concept of a story or narrative. Most of the time, metaphoric gestures contain iconic as well as abstract parts. Why are metaphoric gestures found foremost with off-record strategies? In contrast to direct strategies which do not consider the loss of face of the addressee, and in contrast to approval and autonomy-oriented strategies where the direct criticism is redressed but still visible, off-record strategies just hint at what the speaker intends to deliver as a message, leaving the addressee at a loss to infer the speaker's intention. Being vague and ambiguous does not leave much ground for concrete gestures that refer to aspects of concrete and direct referents. Thus, metaphoric gestures are the first choice for co-verbal gestures while employing off-record strategies. The contrary argument holds for the other types of strategies and the gestures of the concrete. For example, employing a direct strategy, one of the critics said: "... some pictures of the instruments, especially of this cornet [iconic] that you mentioned".[1] The direct referent cornet is iconically visualized by outlining the shape. The left hand is raised like holding the cornet, the index finger of the right hand is extended, and the hand describes a circle. In the off-record case the speaker might try to give only association clues, such as another critic who used an elliptical utterance: "not so clearly to identify ... so of the structure [metaphoric] ... structure you have somehow."[2] Here the verbal information is accompanied by a gesture which comes in the form of the conduit metaphor. The left and right hands indicate holding something like a box.

4.3 Employing the Results for ECA Control

The results from the corpus study may serve as guidelines for the formulation of nonverbal strategies of politeness for an ECA. We illustrate this by means of the GRETA agent system developed by Pelachaud and colleagues (e.g., Pelachaud *et al.* 2002; Hartmann *et al.* 2002). The input to the GRETA agent are APML expressions (Affective Presentation Markup Language) which enable us to express the content and the communicative function of dialogue moves as well as the agent's affective state. In order to enhance GRETA with politeness behaviors, we enrich APML with a new <polite> -tag. Its definition and an example are given in Figure 4.4.

The type-attribute of the <polite> -tag informs the subsequent behavior generation engine which politeness strategy is mainly employed in the utterance. It is not mandatory that the tag ranges over the whole utterance. Furthermore, the <polite> -tag can be used to indicate the different strategies within a single sentence. In the example, the tag covers the whole utterance.

---

[1] Original utterance: "... ein paar Bilder der Instrumente, also gerade dieses Horn[iconic] dass du angesprochen hast".

[2] Original utterance: "nicht so klar erkennen ... so von der Struktur [metaphoric] ... Struktur habt ihr euch irgendwie".

```
<polite type = direct|approval|autonomy|off-record/>

<polite type = autonomy>Perhaps  it would have been better to
present some slides.</polite>
```

Figure 4.4   Definition of the polite tag and an example

Following work done on the BEAT system (Cassell *et al.* 2001), we propose an overgenerate-and-filtering mechanism for enriching the agent's utterances with the appropriate gestural politeness behaviors. This filtering mechanism processes APML statements and enhances them with suitable gestures making use of a word-gesture lexicon (or, in short, gesticon) and rules derived from the statistical analysis of our corpus study. In the first step, a probabilistic process selects a gesture type (iconic, metaphoric, etc.) based on the statistical results of the corpus study. For instance, deictic gestures may be given a higher priority than iconic gestures when suggesting nonverbal behaviors for approval-oriented strategies.

Let us assume a metaphoric gesture shall be generated. In the next step, possible metaphoric gestures for the words and phrases of the utterance are selected from the gesticon (Table 4.4). For each word in the lexicon, a list of suitable gestures is given along with a ranking of their applicability. This value can be modified during the filtering process. For example, if a gesture has been used previously the rating is set to zero to prevent unnecessary repetitions of gestures. Over time the rating will increase to its original value. In our example, the gesture meta01 is chosen for the word slide because it was already decided to generate a metaphorical gesture and this gesture had the highest ranking for this word resulting in an output like

```
Perhaps, it would have been better to present some
<gesture type=meta04>slides.</gesture>
```

The enriched APML structure is passed on to the animation engine. Since nonverbal behaviors are generated independently of each other, the system may end up with a set of incompatible gestures. The set of proposed gestures is therefore reduced to those gestures that are actually realized by the animation module. The findings of our studies may inform not only the generation but also the filtering of gestures. For instance, iconic gestures may be filtered out with a higher probability than metaphoric gestures when realizing off-record strategies.

4.4 Evaluating Multimodal Politeness Behavior

The corpus study revealed a trend to reflect vagueness as a means of politeness by both verbal and nonverbal means. The more vaguely the criticism is formulated, the higher the amount of metaphoric gestures and, vice versa, the more directly the criticism is delivered the higher the amount of iconic

Table 4.4   Structure of the word-gesture lexicon

Concept/word	Gesture	Rating
slide	meta01	0.8
	meta04	0.6
	icon02	0.23
...		

gestures. Thus, if there are preferred categories of gestures accompanying a strategy, will the choice of a less-frequent category alter the interpretation of a FTA when delivered by a computer system?

To shed light on this question, we conducted two experiments: one where subjects had to rank pure textual criticism and another where they had to rank criticism delivered by an embodied conversational agent according to their perceived degree of politeness. The purpose of the first experiment was to investigate whether subjects are able to discriminate the level of politeness in textual criticism and to get a baseline of comparison for the second experiment. In the second experiment, we examined whether the perceived level of politeness changes when an embodied agent is employed. Here, we compared two agents: one that follows the distribution of gestures observed in our corpus and one that behaves differently in that respect.

4.4.1 Subjective Perception of Politeness for Textual Utterances

In the first experiment, the subjects were confronted with a situation where they received textual criticism from a computer system regarding a hypothetical talk they had given before. The subjects had to rank each point of criticism according to the degree to which it expresses (1) respect for the user's choices (negative politeness) and (2) a feeling of working with the user (positive politeness).

The design of the experiments was inspired by an earlier experiment on a cross-cultural evaluation of politeness strategies we conducted together with Johnson *et al.* (2005) (see section 4.1.2). Based on this experiment, we hypothesized that the subjects would ascribe degrees of positive and negative politeness that are in line with the theory of Brown and Levinson. That is, the autonomy-oriented utterances would get a higher rating for the negative politeness tone while the approval-oriented utterances would get a higher rating for positive politeness tone.

Experimental Setting

For each of the two strategies of politeness (autonomy-oriented vs. approval-oriented), four instances were created. Table 4.5 gives an overview of the different versions for the criticism. The points of criticism had been presented in randomized order. Each subject had to rate all eight utterances according to negative and positive politeness.

Eighteen students from the computer science department participated in this experiment, 16 male and 2 female, the ages ranging from 22 to 24. The duration of each session was approximately 10 minutes.

Scales differed according to the face needs addressed by the strategies. The utterances were rated according to the degree of cooperation between agent and user, with 1 meaning "no cooperation at all" and 7 meaning "full cooperation", as well as according to the degree of freedom of action the agent leaves to the user, with 1 meaning "no freedom at all" and 7 meaning "full freedom".

Results and Discussion

Tables 4.6–4.8 summarize the results of the experiment. The results of our earlier experiment could be confirmed. That is, the approval-oriented strategies got higher ratings for cooperation while the autonomy-oriented strategies got higher ratings for freedom of action. Table 4.6, though, reveals a problem with utterance APP2 (Warum haben wir nur kein griffigeres Thema für dich gewählt?). A similar effect was observed in our earlier study and attributed to the fact that the inclusive we (wir) sounds rather patronizing in German. This might be comparable to an example in English when parents tell their children: "Let's wash our faces." A dependent samples *t*-test revealed a highly significant difference between the different types of utterances (approval-oriented vs. autonomy-oriented) on both rating scales. The autonomy-oriented utterances were rated as less appropriate on cooperation than on

Table 4.5 Strategy-specific realization of a given act of criticism

Criticism	Realization[a]
Approval	
Unstructured	Interesting topic. But the talk was a little bit unstructured.[b]
Unfocused	Why haven't we chosen a catchier topic for you?[c]
Vague	Exciting topic! But the talk was somewhat vague.[d]
Not loud enough	Please speak louder and not so fast next time.[e]
Autonomy	
Too slow	How about summarizing the content of the book a little bit faster?[f]
Vague	Perhaps you should meditate a little bit longer on the topic next time.[g]
Unfocused	How about chosing a catchier topic next time?[h]
Not loud enough	Perhaps, you could speak a little bit louder.[i]

[a] Translation of the German sentences by the Authors.
[b] Interessantes Thema! Aber der Vortrag war etwas konfus.
[c] Warum haben wir nur kein griffigeres Thema für dich gewählt?
[d] Spannendes Thema! Aber das Referat war etwas schwammig.
[e] Bitte sprich beim nächsten Mal lauter und nicht so schnell.
[f] Wie wäre es, wenn du den Inhalt des Buches etwas schneller besprichst?
[g] Vielleicht solltest du das nächste Mal ein wenig mehr über das Thema nachdenken.
[h] Wie wäre es, wenn du das nächste Mal ein griffigeres Thema wählst?
[i] Vielleicht könntest Du ein bisschen lauter sprechen.

Table 4.6 Mean ratings for cooperation

Scenario 1: rating cooperation (approval-oriented)							
AUT2	APP2	AUT3	AUT4	AUT1	APP1	APP3	APP4
3.50	3.61	3.72	4.39	4.56	4.83	5.00	5.28

Table 4.7 Mean ratings for freedom of action

Scenario 2: rating freedom of action (autonomy-oriented)							
APP4	APP2	APP1	APP3	AUT3	AUT2	AUT1	AUT4
2.83	3.06	3.17	3.44	4.50	4.78	5.22	5.50

Table 4.8 Mean ratings of utterance types in the two scenarios

	Utterance type					
	Approval-oriented			Autonomy-oriented		
Scenario	1: cooperation	2: freedom of action	t-test	1: cooperation	2: freedom of action	t-test
	4.68	3.13	5.067**	4.04	5.00	−3.637**

$^*p < 0.05$, $^{**}p < 0.01$

freedom of action with $t(71) = 5.067$, $p < 0.01$. On the other hand, the approval-oriented utterances were rated as less appropriate on freedom of action than on cooperation with $t(71) = -3.637$, $p < 0.01$. Thus, the results we obtained for our earlier experiment for instructions could also be confirmed for criticism.

4.4.2 Subjective Perception of Politeness for Gestural Utterances

The purpose of the second experiment was to find out whether a gesturing agent would change the perceived politeness tone compared to that of the textual utterances and whether the subjective rating is influenced by the type of gestures (abstract vs. concrete).

We hypothesize that a distribution of gestures that follows findings from observations of human–human dialogue increases the effect of politeness tactics applied by an embodied agent. An interesting result of the corpus study was the frequent use of concrete gestures in combination with approval-oriented strategies. We therefore assume that approval-oriented strategies that are accompanied by gestures of the concrete will get a more positive rating for cooperation than text-based approval-oriented strategies or approval-oriented strategies that are accompanied by gestures of the abstract. Because the distribution of concrete and abstract gestures does not strongly differ for autonomy-oriented utterances, we assume that there is no such effect for this type of utterance.

Experimental Setting

For the experiment, the same utterances were chosen as in the first experiment. However, for each utterance, two variants were created, one accompanied by a gesture of the concrete, and the other one accompanied by an abstract gesture. Thus, there are 16 different utterances, eight for each strategy, from which four are realized with concrete gestures and the other four with metaphoric ones. The utterances were presented by the GRETA agent in random order, and participants had the opportunity to see each utterance as often as they liked. Figure 4.5 depicts the agent while performing concrete and metaphoric gestures.

Six students from the computer science department participated in the second experiment, five male (age ranging from 24 to 25) and one female (age 28). The duration of each session was approximately 20 minutes.

The same seven-item rating scale was employed for rating positive and negative politeness as in our previous study on politeness. After rating the conveyed tone of politeness, participants were asked to rate the quality of the gestures on a scale from 1 (not at all suitable for this utterance) to 7 (very suitable for this

Figure 4.5　Examples of the GRETA agent performing concrete (left) and abstract (right) gestures

Table 4.9 Mean ratings of utterance types in the two scenarios depending on employed gesture type

		Utterance type					
		Approval-oriented			Autonomy-oriented		
	Scenario	1	2	t-test	1	2	t-test
Gestures	Concrete	4.17	3.21	2.794**	3.63	4.58	−2.673*
	Abstract	4.58	3.17	4.623**	3.71	4.58	−2.482*
	Both	4.38	3.19	5.164**	3.67	4.58	−3.684**

$^*p < 0.05, ^{**}p < 0.01$

utterance) for the eight utterances after the evaluation. They had the opportunity to watch the utterances again and compare the two different gestures for each utterance before writing down their rating.

Results and Discussion

The evaluation with the embodied conversational agent resulted in more data because the additional distinction between the two gesture types had to be taken into account. Dependent sample t-tests revealed that the positive and negative politeness ratings differed significantly for the utterance types (approval-oriented vs. autonomy-oriented) as in the text-only experiment (Table 4.9, last line). The table describes the differences between ratings in scenario one (cooperation) and two (freedom of action) for the two different utterance types and for the different gesture types (concrete and abstract) as well as the combined rating for both gesture types (both). The approval-oriented utterances accompanied by concrete gestures are rated significantly more suitable for co-operaton than for freedom of action with $t(23) = 2.794$, $p < 0.01$. In general, the approval-oriented utterances were perceived as significantly less appropriate for expressing negative politeness, and the autonomy-oriented utterances were perceived as significantly less appropriate for positive politeness. This effect was observed regardless of the employed gesture type, indicating that the gestures have no negative effect on the applicability of the utterances.

In addition, we examined whether the gesture types had any impact on the positive and negative politeness ratings. Table 4.10 gives the results for this analysis. Obviously, there were no differences in the rating of the positive or negative politeness tone for the autonomy-oriented utterances - neither when abstract nor when concrete gestures were used. The same holds true for the approval-oriented utterances. No effect is visible for the employed gesture types. Thus, for approval-oriented utterances our hypothesis could not be confirmed.

Table 4.10 Mean ratings of utterance types for employed gesture type depending on the scenario

		Utterance type					
		Approval-oriented			Autonomy-oriented		
	Gestures	Concrete	Abstract	t-test	Concrete	Abstract	t-test
Scenario	1: cooperation	4.17	4.58	−2.005	3.63	3.71	−0.289
	2: freedom of action	3.21	3.17	0.161	4.58	4.58	0.000
	Both	3.69	3.88	−1.137	4.10	4.15	−0.191

$^*p < 0.05, ^{**}p < 0.01$

If the gestures make no difference, perhaps the animation of the gestures is not convincing enough. To examine this assumption more closely, let us have a look at the quality ratings.

The best rating for a concrete gesture was 4.83 in combination with utterance APP2. The abstract gesture for this utterance was only rated 3.5. The best rating for an abstract gesture was 5.83 in combination with utterance APP3. The corresponding concrete gesture got a very bad rating of 1.67. A comparison of the results obtained for the text-only and the gesture experiment may explain why our hypothesis could not be confirmed.

Above we mentioned the problem with utterance APP2 due to the effect of the inclusive we. A t-test shows that there is no significant difference in positive or negative politeness ratings for this utterance: $t(34) = 0.836$, $p = 0.409$. This result changes for the gesture-based experiment. Here, the t-test shows a highly significant difference between positive and negative politeness ratings with $t(22) = 3.916$, $p < 0.01$. Obviously, using a more suitable gesture in combination with an approval-oriented strategy may have a positive effect on the perception of the agent's willingness to cooperate. The reverse effect was observed for utterance APP3 where the quality of the metaphorical gesture was rated much higher than the quality of the more suitable iconic gesture. Whereas the t-test shows a significant difference for the text-only experiment ($t(34) = 3.39$, $p < 0.01$), this effect is absent in the gesture experiment with $t(22) = 1.089$, $p = 0.288$.

Thus, either the bad quality of the appropriate (concrete) gesture or the good quality of the inappropriate (abstract) gesture has a negative effect on the rating of the utterance. Our analysis reveals that there is indeed an effect of the accompanying gestures on the rating of utterances, but that the quality of the gestures is crucial for the effect on the user. The gestures designed for this study were obviously only partially successful. Nevertheless, it was shown that choosing the right gestures may influence the interpretation of an utterance by an embodied conversational agent.

4.5 Conclusions

We have presented the results of a corpus study we devised to shed light on the question of how face threats are mitigated by multimodal communicative acts. Unlike earlier work on politeness behaviors, we focus on how politeness is expressed by means of gestures. The results indicate that gestures are indeed used to strengthen the effect of verbal acts of politeness. In particular, vagueness as a means of politeness is reflected not only by verbal utterances, but also by gestures. Iconic and deictic gestures were overwhelmingly used in more direct criticism while there was a high frequency of metaphoric gestures in off-record strategies. Obviously, our subjects did not attempt at compensating for the vagueness of their speech by using more concrete gestures. We employed the results in a filtering mechanism for enhancing an agent's utterances with appropriate gestures for politeness strategies.

In order to test if there is an effect on the user's perception of a face threat if the agent's utterances are accompanied by gestures, a user study was conducted. This study was designed along the lines of a previous study testing the effect of different verbal politeness strategies. The results could be replicated in the text-only and in the gesture conditions but we were not able to show a direct effect of the appropriate gesture on the perception of the face threat. A closer look revealed that an effect can be seen depending on the perceived quality of the employed gestures.

This chapter has shown that the observation of human–human conversational behavior is a valuable source of information for the design of agent-based systems. Future work has to show the effectiveness of the introduced filtering mechanism. Questions to be answered include whether the proposed gestures are perceived as a natural choice for the utterances they accompany. Because the believability of the agent's gestures strongly depends on the quality of the gestures, more and qualitatively better gestures have to be realized for the agent. Furthermore, the corpus study may not only inform the design of the agent's nonverbal behavior but it can also be used to evaluate the performance of the agent against the human–human interactions by generating the criticims found in the corpus and comparing the agent's gesture use with the gesture use of the human critics.

Acknowledgments

The work described in this article has been partially supported by the Network of Excellence HUMAINE (Human–Machine Interaction Network on Emotion) IST-2002-2.3.1.6 / Contract no. 507422 (http://emotion-research.net/).

References

Allwood J. (2002) Bodily communication dimensions of expression and content. In B. Granström, D. House and I. Karlsson (eds), *Multimodality in Language and Speech Systems*. Kluwer Academic Publishers, Dordrecht, pp. 7–26.

André E., Rehm M., Minker W. and Bühler D. (2004) Endowing spoken language dialogue systems with social intelligence. In E. André , L. Dybkjaer, W. Minker and P. Heisterkamp (eds), *Affective Dialogue Systems*. Springer, Berlin, pp. 178–187.

Bavelas J.B., Chovil N., Coates L. and Roe L. (1995) Gestures specialized for dialogue. *Personality and Social Psychology Bulletin* **21**, 394–405.

Bickmore T. and Cassell J. (2000) Small talk and conversational storytelling in embodied interface agents. *Proceedings of the AAAI Fall Symposium on Narrative Intelligence*, pp. 87–92.

Brown P. and Levinson S.C. (1987) *Politeness: Some universals in language usage*. Cambridge University Press, Cambridge.

Cassell J., Vilhjalmsson H. and Bickmore T. (2001) BEAT: The behavior expression animation toolkit. *Proceedings of SIGGRAPH '01*, Los Angeles, CA, pp. 477–486.

Hartmann B., Mancini M. and Pelachaud C. (2002) Formational parameters and adaptive prototype instantiation for MPEG-4 compliant gesture synthesis. *Proceedings of Computer Animation 2002*. IEEE Computer Society Press.

Johnson L., Mayer R., André E. and Rehm M. (2005) Cross-cultural evaluation of politeness in tactics for pedagogical agents. *Proceedings of the 12th International Conference on Artificial Intelligence in Education* (AIED).

Johnson W.L., Rizza P., Bosma W., Kole S., Ghijsen M. and van Welbergen H. (2004) Generating socially appropriate tutorial dialog. In E. André, L. Dybkjaer, W. Minker and P. Heisterkamp (eds), *Affective Dialogue Systems*. Springer, Berlin, pp. 254–264.

Kendon A. (1986) Some reasons for studying gestures. *Semiotica* **62**(1–2), 3–28.

Kipp M. (2004) *Gesture Generation by Imitation: From human behavior to computer character animation*. PhD thesis, Universität des Saarlandes, Saarbrücken.

Knapp M.L. and Hall J.A. (1997) *Nonverbal Communication in Human Interaction*, 4th edn. Harcourt Brace College Publishers, Fort Worth.

McNeill D. (1992) *Hand and Mind: What gestures reveal about thought*. University of Chicago Press, Chicago.

Nakano Y.I., Reinstein G., Stocky T. and Cassell J. (2003) Towards a model of face-to-face grounding. *Proceedings of the Association for Computational Linguistics*, Sapporo, Japan.

Pease A. (1993) *Body Language: How to read others' thought by their gestures*, 20th edn. Sheldon Press, London.

Pelachaud C., De Carolis B., de Rosis F. and Poggi I. (2002) Embodied contextual agent in information delivering application. *Proceedings of the First International Joint Conference on Autonomous Agents and Multiagent Systems*, ACM Press, pp. 758–765.

Porayska-Pomsta K. and Mellish C. (2004) Modelling politeness in natural language generation. *Proceedings of INLG*.

Prendinger H. and Ishizuka M. (2001) Social role awareness in animated agents. *Proceedings of Agents '01*, Montreal, Canada, pp. 270–277.

Reeves B. and Nass C. (1996) *The Media Equation: How people treat computers, television, and new Media like real people and places*. Cambridge University Press, Cambridge.

Trees A.R. and Manusov V. (1998) Managing face concerns in criticism: integrating nonverbal behaviors as a dimension of politeness in female friendship dyads. *Human Communication Research* **24** (4), 564–583.

Walker M.A., Cahn J.E. and Whittaker S.J. (1997) Improvising linguistic style: social and affective bases for agent personality. *Proceedings of AAMAS'97*.

5

Attentional Behaviors as Nonverbal Communicative Signals in Situated Interactions with Conversational Agents

Yukiko I. Nakano and Toyoaki Nishida

5.1 Introduction

Linking linguistic messages to the perceived world is one essential aspect of face-to-face communication, and nonverbal behaviors, such as pointing gestures (e.g., "I like THIS design."), play important roles in directing the listener's attention to a specific object, allowing her/him to incorporate the visual information into the conversation (Clark 2003).

In Figure 5.1, for example, two people are working together on a computer. The speaker is looking at the display while pointing at a specific position on the display. The listener is also jointly paying attention to the display. Note that both of the participants are not looking at their partner, but at the shared reference, in this case a specific place in the computer display. Thus, it seems that in such a situation of a shared conversation, the listeners' attention to a shared reference serves as positive feedback to the speaker, and attentional behaviors are indispensable as communicative signals in face-to-face conversation.

Applying these observations to communication between embodied conversational agents (ECAs), which inhabit a virtual world, and human users in the physical world, attentional behaviors towards a shared reference can be classified into four types according to the direction of the attentional behavior and its actor, as shown in Figure 5.2.

(A) Attention towards the physical world:
   (A1) From an ECA to a physical object
   (A2) From a user to a physical object
(B) Attention towards the virtual world:
   (B1) From an ECA to a virtual object
   (B2) From a user to a virtual object

Conversational Informatics: An Engineering Approach   Edited by Toyoaki Nishida
© 2007 John Wiley & Sons, Ltd

Figure 5.1   Situated conversation. Reproduced by permission of ACL, Nakano *et al.*, 2003

Attention type (A) is directed towards the physical world: (A1) is displayed via character animations generated by ECA systems; (A2), on the other hand, which represents the user's attention, is recognized by ECA systems. Likewise, attention type (B) is directed towards the virtual world: (B1) is generated by ECA systems; and type (B2), which represents the user's attention to objects in the virtual world, is displayed by the user and recognized by the system.

Because ECAs and users inhabit different worlds, sharing a reference between them is difficult. To resolve this problem, it is indispensable to seamlessly integrate these two different worlds and broaden the communication environment in human–agent interaction. For this purpose, this chapter proposes ECAs that can integrate the physical world and the virtual world by using these four types of attentional behaviors.

For example, by checking the user's attention (A2) or (B2), the agent can judge whether the user is paying proper attention to the object focused in the current conversation. When the user does not pay attention to the focused object, the agent is able to redirect the user's attention using the agent's own attentional behaviors (A1) or (B1). In human communication competence, if the agent looks at an object, the user would follow the agent's gaze direction to establish a joint attention. It is expected that exchanging these kinds of attentional behaviors between the user and the agents will improve the

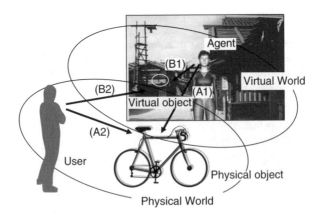

Figure 5.2   Attentional behaviors in an immersive conversational environment

naturalness of human–agent communication, and would also have the power of involving the user in a conversation with the agent.

In the following sections, we will show how these four types of attentional behaviors are implemented in ECAs. The next section describes related work. Section 5.3 will present an ECA system MACK, which can recognize and generate attentional behaviors towards a physical object. Then, in section 5.4, another ECA system IPOC, which can exploit attentional behaviors towards objects in the virtual world, is introduced. These two sections will show how these ECA systems recognize and generate the four types of attentaional behaviors illustrated in 5.2. Finally, future work will be discussed in section 5.5.

5.2 Related Work

5.2.1 Functions of Attentional Behaviors in Human Conversation

Attentional behaviors are used as communication signals at different levels and in different aspects of conversation. A number of studies of face-to-face communication have discussed how listeners return nonverbal feedback, by using such signals as head nods and eye gaze, as visual evidence indicating whether a conversation is on the right track (Argyle and Cook 1976; Clark and Schaefer 1989; Clark 1996; Duncan 1972, 1974; Kendon 1967). In his detailed study of gaze, Kendon (1967) observed that speakers look up at grammatical pauses to obtain feedback on how utterances are being received by the listener, and to see if the listener is willing to carry on the conversation. Goodwin (1981) reported how speakers pause and restart until they obtain the listener's gaze. By contrast, Novick *et al.* (1996) reported that during conversational difficulties, mutual gaze was held longer at turn boundaries. These results suggest that speakers distinguish different types of listener's gazes. In fact, Argyle *et al.* (1973) claimed that gaze is used in order to send positive feedback accompanied by nods, and smiles etc., as well as to collect information from a partner.

Other results evoke another argument for attentional behaviors. Argyle and Graham (1977) reported that in comparing a situation without any shared object between the conversational participants and one in which complex objects (e.g., a map) are shared during the conversation, the percentage of time spent gazing at another drops from 76.6% to 6.4%. Anderson *et al.* (1997) found similar results in which mutual gaze fell below 5% in such a situation. These results suggest that not only looking at a partner, but also looking at a shared reference, is essential in face-to-face conversation.

Interestingly, these results support discussions in video-mediated communication (VMC) research, which aims at proposing technologies for supporting mediated communication with visual information. Whittaker and O'Conaill (1993) and Whittaker (2003) reported that when tasks require complex reference to, and joint manipulations of, physical objects, both speaker and listener observe the changes in the objects. They claimed that such shared objects serve as implicit conversational common ground, to which the conversants do not have to refer explicitly. Moreover, Whittaker and O'Conaill (1993) compared communication effectiveness with and without shared workspace. The shared workspace enabled people to share visual material such as documents or designs as well as to type, draw, and write. They reported that participants with the shared workspace took fewer turns to identify pieces because they were able to refer to pieces deictically. Similar results were also reported in comparing speech, speech/video, and face-to-face communication (Kraut *et al.* 1996; Olson *et al.* 1995).

5.2.2 Attentional Behaviors in Conversational Agents

The significant advances in computer graphics over the last decade enable the implementation of animated characters capable of interacting with human users. First, as an extension of multimodal presentation systems, presentation agents were proposed to emulate presentation styles common in human–human communication (Andre 1999; Noma and Badler 1997; Wahlster *et al.* 2001). For instance, these agents are able to point at objects on the screen using deictic gestures or a pointer so as to demonstrate comprehensible explanation about multimedia contents.

Pedagogical agents are autonomous animated characters that cohabit learning environments with students to create rich face-to-face learning interactions (Johnson *et al.* 2000). In pedagogical agents, nonverbal behaviors are primarily used to enhance the quality of advice to learners. The agents demonstrate the task to the users by manipulating objects in the virtual environment, and navigate a user's attention using pointing and gaze (Rickel and Johnson 1999; Lester *et al.* 1999). They also nonverbally express the evaluation of a learner's performance, using nods and shakes, and puzzled, pleasant, or more exaggerated facial expressions (Stone and Lester 1996; Lester *et al.* 1999; Rickel and Johnson 1999; Shaw *et al.* 1999). More recently, in their Mission Rehearsal Experience (MRE) project (Swartout *et al.* 2001), Traum and Rickel (2002) proposed a multimodal conversation model that controls a multiparty conversation in an immersive virtual world.

Obviously, virtual agents on a screen cannot change the physical world whereas communication robots can. Although these two types of communication artifact inhabit different worlds, common research issues between these two areas do exist. As a study employing an empirical method, based on the analyses of face-to-face conversations, Sidner *et al.* (2003) proposed a humanoid robot designed to mimic human conversational gaze behavior in collaboration conversation. They analyzed use of attentional behaviors in human face-to-face conversation, and proposed engagement rules to maintain a conversation with a user.

In sum, attentional behaviors have been studied in both human communication and conversational agents, and results in previous studies suggest that attentional behavior is one of the most effective devices used to maintain and control face-to-face conversations. Moreover, previous studies also suggest that attention towards a shared reference has different functions from that towards the partner, and implementing such attentional functions into conversational agents would be effective in improving the communication capability of communication artifacts.

5.3 Nonverbal Grounding using Attentional Behaviors Towards the Physical World

This section focuses on attentional behaviors towards the physical world (attention type (A) in Figure 5.2), and describes a conversational agent, MACK (Media Lab Autonomous Conversational Kiosk), an interactive public information ECA kiosk guiding a user in the Media Lab building by sharing a physical floor map with the user.

Figure 5.3(a) shows a snapshot of MACK. MACK is capable of multimodal interaction with the user. As shown in the picture, the shared paper map is on a table with an embedded Wacom tablet, by which the user's pen gestures are recognized. In addition, the user's eye gaze and head nods are recognized via a head-pose tracker. MACK also allows the user speech interaction using the IBM ViaVoice speech recognizer. As for the output side, MACK's pointing gestures are expressed as highlighting on the paper map by an LCD projector (Figure 5.3(b)). The Microsoft Whistler Text-to-Speech (TTS) API is used as Mack's voice.

5.3.1 Attention Towards a Shared Workspace Serving as Evidence of Understanding

Conversation can be seen as a collaborative activity to accomplish information-sharing and to pursue joint goals and tasks. Under this view, what the interlocutors understand to be mutually shared is called common ground, and adding what has been said and what is meant to the common ground is called grounding (Clark and Schaefer 1989). According to Clark (1996) and Clark and Schaefer (1989), eye gaze is the most basic form of the positive evidence of understanding displayed by the listener. Head nods have a similar function to verbal acknowledgements such as "uh huh", "I see".

Clark and Brennan (1991) also claimed that the way of displaying positive evidence of understanding is different depending on communication modality. Extending this discussion, Brennan (2000) provided experimental evidence showing how communication modality affects the cost for accomplishing common

 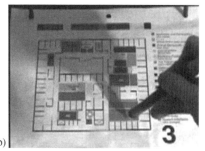

(a)   (b)

Figure 5.3   Snapshot of MACK on a display (a) and highlighting on a map (b)

ground. Her study revealed that the grounding process became shorter when more direct evidence, such as listener's displaying correct task manipulation, was available.

As a similar experimental study for multimodal communication, Dillenbourg (1996) found that grounding is often performed across different modes. The subjects are in the virtual environment where they can use three modes of communication; verbal communication: action commands to change the virtual environment, and whiteboard drawing. The results show that information presented verbally was grounded by an action in the virtual environment, and vice versa.

5.3.2 Empirical basis for MACK attentional functions

Based on the discussion about nonverbal evidence of understanding, as MACK's attentional functions, we focus on grounding mechanisms under a shared workspace situation. To accomplish this goal, first we collected and analyzed 10 face-to-face direction giving conversations where conversants share a map.[1]

Data Coding

As a unit of verbal behavior, we tokenized a turn into utterance units (UU) (Nakatani and Traum 1999), corresponding to a single intonational phrase (Pierrehumbert 1980). Each UU was categorized using the DAMSL coding scheme (Allen and Core 1997). In the statistical analysis, we concentrated on the following four categories with regular occurrence in our data: Acknowledgment, Answer, Information request (Info-req), and Assertion. For nonverbal behaviors, the following four types of nonverbal behaviors were coded:

- Gaze at partner (gP): Looking at the partner's eyes, eye region, or face
- Gaze at map (gM): Looking at the map
- Gaze elsewhere (gE): Looking away elsewhere
- Head nod (Nod): Head moves up and down in a single continuous movement on a vertical axis, but eyes do not go above the horizontal axis.

By combining the basic categories for eye gaze and head nod, six complex categories (gP with nod, gP without nod, etc.) are generated,[2] and 16 combinations of these categories are defined as a dyad's nonverbal status (NV status), as shown in Table 5.1. For example, gP/gM stands for the combination of the speaker's gaze at a partner and the listener's gaze at a map.

---

[1] A detailed description of our empirical study is reported in Nakano *et al.* (2003).
[2] Only categories with more than 10 instances were used in our statistical analysis.

Table 5.1 Variations of NV status. Reproduced by permission of ACL, Nakano *et al.*, 2003

		Listener's behavior			
Combinations of NVs		gP	gM	gMwN	gE
Speaker's	gP	gP/gP	gP/gM	gP/gMwN	gP/gE
	gM	gM/gP	gM/gM	gM/gMwN	gM/gE
behavior	gMwN	gMwN/gP	gMwN/gM	gMwN/gMwN	gMwN/gE
	gE	gE/gP	gE/gM	gE/gMwN	gE/gE

Analysis of Correlation Between Verbal and Nonverbal Behaviors

We analyzed NV status shifts with respect to type of verbal communicative action. Table 5.2 shows the most frequent target NV status (shift to these statuses from other statuses) for each speech act type. "Probability" indicates the proportion to the total number of transitions.

As for "Acknowledgement", within an UU, the dyad's NV status most frequently shifts to gMwN/gM (e.g., speaker utters "OK" while looking at the map and nodding, and listener looks at the map). At pauses between UUs, a shift to gM/gM (both the speaker and the listener look at the map) is the most frequent.

The most frequent shift within an "Answer" UU is to gP/gP (the speaker and the listener look at each other). This suggests that speakers and listeners rely on mutual gaze to ensure an answer is grounded. In addition, we found that speakers frequently look away at the beginning of an answer, as they plan their reply (Argyle and Cook 1976).

In "Info-req", the most frequent shift within a UU is to gP/gM (the speaker looks at the partner and the listener looks at the map). At pauses between UUs, shift to gP/gP is the most frequent. This suggests that speakers obtain mutual gaze after asking a question to ensure that the question is clear, before the turn is transferred to the listener for reply.

Finally, in "Assertion", listeners look at the map most of the time, and sometimes nod. By contrast, speakers frequently look at the listener. Therefore, a shift to gP/gM is the most frequent within an UU. This suggests that speakers check whether the listener is paying attention to the shared referent mentioned in the Assertion. This implies that not only the listener's gazing at the speaker, but also paying attention to a referent, works as positive evidence of understanding. A typical example of Assertion is shown in Figure 5.4. At [690], the speaker (the direction giver) glances at the receiver while speaking, so that the

Table 5.2 Salient transitions

	Within UU			Pause		
	Speaker's gaze shift to	Listener's gaze shift to	probability	Speaker's gaze shift to	Listener's gaze shift to	probability
Acknowledgment	map while nodding	map	0.495	map	map	0.888
Answer	partner	partner	0.436	map	map	0.667
Info-req	partner	map	0.38	partner	partner	0.5
Assertion	partner	map	0.317	map	map	0.418

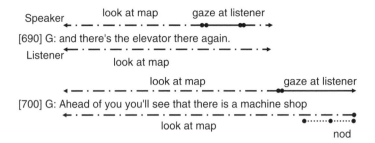

Figure 5.4   Example of nonverbal acts for Assertion

NV status of the dyad shifts to gP/gM. Then, at a pause after the UU, the speaker's gaze direction moves back to the map, and the NV status of the dyad shifts to gM/gM. These are, again, the typical NV status shifts in Assertion, as explored by statistical analysis.

In sum, it is already known that the eye gaze can signal a turn-taking request (Duncan 1974), but turn-taking cannot account for all our results. Gaze direction changes within as well as between UUs, and the use of these nonverbal behaviors differs depending on the type of conversational action. Note that subjects rarely demonstrated communication failures, implying that these nonverbal behaviors represent positive evidence of understanding.

Analysis of Correlation Between Speaker and Listener Behavior

Thus far we have demonstrated the difference in distribution of nonverbal behaviors with respect to the speaker's conversational action. But, to uncover the function of these nonverbal signals, we must examine how the listener's nonverbal behavior affects the speaker's next action. Thus, we looked at two consecutive Assertion UUs by a direction-giver, and analyzed the relationship between the NV status of the first UU and the direction-giving strategy in the second UU. The giver's second UU is classified as *go-ahead* if it gives the next leg of the directions, or as *elaboration* if it gives additional information about the first UU, as in the following example:

> [U1] S: And then, you'll go down this little corridor.
> [U2-a] S: It's not very long. *(elaboration)*
> [U2-b] S: Then, take a right. *(go-ahead)*

Suppose that the first UU is [U1]. If the second UU is [U2-a], this is an *elaboration* of [U1]. If the second UU is [U2-b], this is categorized as *go-ahead*.

Results are shown in Table 5.3. When the listener begins to gaze at the speaker somewhere within an Assertion UU, and maintains gaze until the pause after the UU, the speaker's next UU is an elaboration of the previous UU for 73% of the time. On the other hand, when the listener keeps looking at the map during an UU, only 30% of the next UU is an elaboration ($z = 3.678$, $p < 0.01$). Moreover, when a listener keeps looking at the speaker, the speaker's next UU is *go-ahead* only 27% of the time. In contrast, when a listener keeps looking at the map, the speaker's next UU is *go-ahead* 52%[3] of the time ($z = -2.049$, $p < 0.05$). These results suggest that speakers interpret listeners' continuous gaze as

---

[3] The percentage for map does not sum to 100% because some of the UUs are cue phrases or tag questions which are part of the next leg of the direction, but do not convey content.

Table 5.3 Relationship between direction-receiver's NV and direction-giver's next verbal behavior

	Receiver's NV	Evidence of	Giver's next UU
Assertion	within: look at map pause: look at map & nod	Understanding (positive evidence)	go-ahead: 0.7
			elaboration: 0.3
	within: look at partner within: look at partner	Not-understanding (negative evidence)	go-ahead: 0.27
			elaboration: 0.73
Answer	within: look at partner pause: look at map	Understanding (positive evidence)	go-ahead: 0.83
			elaboration: 0.17
	pause: look at partner	Not-understanding (negative evidence)	go-ahead: 0.22
			elaboration: 0.78

evidence of not-understanding, and they therefore add more information about the previous UU. Similar findings were reported for a map task by Boyle *et al.* (1994) who suggested that, at times of communicative difficulty, interlocutors are more likely to utilize all the channels available to them. In terms of floor management, gazing at the partner is a signal of giving up a turn, and here this indicates that listeners are trying to elicit more information from the speaker.

Analyzing spoken dialogues, Traum and Heeman (1996) reported that grounding behavior is more likely to occur at an intonational boundary, which we use to identify UUs. This implies that multiple grounding behaviors can occur within a turn if it consists of multiple UUs. However, in previous models, information is grounded only when a listener returns verbal feedback, and acknowledgment marks the smallest scope of grounding. If we apply this model to the example in Figure 5.4, none of the UU has been grounded because the listener has not returned any spoken grounding clues.

Our results suggest that considering the role of nonverbal behavior, especially eye gaze, allows a more fine-grained model of grounding, employing the UU as a unit of grounding. Moreover, we found that speakers are actively monitoring positive evidence of understanding as well as the absence of negative evidence of understanding (that is, signs of miscommunication). When listeners continue to gaze at the task, speakers continue on to the next leg of directions.

5.3.3 MACK System Architecture

On the basis of empirical results reported in the previous section, MACK is implemented by employing the system architecture shown in Figure 5.5. The Understanding Module (UM) handles three types of inputs: speech, head movement, and pen gesture. It interprets the multiple modalities both individually and in combination by operating these inputs as parallel threads, and passes the interpretation results to the Dialogue Manager (DM). The DM consists of two primary sub-modules, the Response Planner, which determines MACK's next action(s) and adds a sequence of utterance units (UUs) to Agenda, and the Grounding Module (GrM), which updates the Discourse Model. The GrM also decides when the Response Planner's next UU should be passed on to the Generation module (GM). Finally, the GM converts the UU into speech, gesture, and projector output, sending these synchronized modalities to the TTS engine, the Animation Module (AM), and the Projector Module.

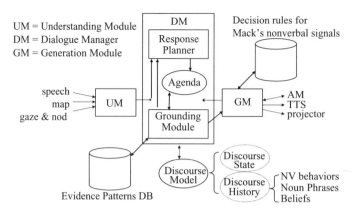

Figure 5.5   MACK system arhitecture

5.3.4 Recognizing the User's Attention to a Physical Shared Reference

Eye gaze and head nod inputs are recognized via a stereo-camera-based 6-degree-of-freedom head-pose tracker, which calculates rotations and translations in three dimensions based on visual and depth information taken from two cameras (Morency *et al.* 2003). The calculated head pose is translated into "look at MACK", "look at map", or "look elsewhere". The rotation of the head is translated into head nods, using a modified version of Kapoor and Picard's (2001) detector. These head nod and eye gaze events are time-stamped and logged within the nonverbal component of the Discourse History.

5.3.5 Grounding Judgment by User's Nonverbal Signals

When MACK finishes uttering a UU, to update the Discourse Model, the Grounding Module (GrM) judges whether or not the UU is grounded, based on the user's verbal and nonverbal behaviors during and after the UU. In the current implementation, verbal evidence is considered stronger than nonverbal evidence in the judgment of grounding. If the user returns an acknowledgement (e.g., "OK") as verbal evidence, the GrM judges the UU grounded. If the user explicitly reports failure in perceiving MACK's speech (e.g., "what?"), or not-understanding (e.g., "I don't understand"), the UU remains ungrounded.

On the other hand, if no explicit verbal evidence is observed, then the nonverbal grounding judgment process starts. Because of the incremental nature of grounding, we implement a nonverbal grounding functionality based on a process model that describes steps for a system to judge whether a user understands system contributions.

- *Step 1: Preparing for the next UU*. By referring to the Agenda, the DM can identify the speech act type of the next UU, which is the key to specifying the positive/negative evidence in grounding judgment later.
- *Step 2: Monitoring*. The GrM sends the next UU to GM, and the GM begins to process the UU. At this time, the GM logs the start time in the Discourse Model. When it finishes processing (as it sends the final command to the animation module), it logs the end time. The GrM waits for this speech and animation to end (by polling the Discourse Model until the end time is available), at which point it retrieves the timing data for the UU, in the form of time-stamps for the UU start and finish. This timing data is used in the following Judging step.

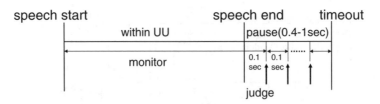

Figure 5.6   Process of grounding judgment

- *Step 3: Judging*. When the GrM receives an end signal from the GM, it starts the judgment of grounding using nonverbal behaviors logged in the Discourse History. Accessing the Evidence Patterns DB and looking up the Grounding Model shown in Table 5.3 by the type of the UU, the GrM identifies the positive/negative evidence in grounding judgment. Note that these combinations of positive/negative evidence shown in the model were found in our empirical study reported in section 5.3.2.

The grounding decision process is illustrated in Figure 5.6. As a first trial, the nonverbal behaviors within the UU and for the first tenth of a second of a pause are used for the judgment. If these two behaviors ("within" and "pause") match the positive evidence pattern for a given speech act, then the GrM judges the UU grounded. Suppose that the current UU type is "Assertion". As shown in Table 5.3, if the user's nonverbal behavior within the UU is "look at map (user looks at the map)" and the user's behavior during a pause after the UU is also "look at map" or "nod", then the GrM judges the UU grounded. If they match a pattern for negative evidence, the UU is not grounded. If no pattern is matched during the first tenth of a second of a pause, then MACK monitors the user's nonverbal behaviors for another one-tenth of a second, and judges again. The GrM continues looping in this manner until the UU is either grounded or ungrounded explicitly, or the timeout. According to previous studies, the length of a pause between UUs is between 0.4 and 1 second (Traum and Heeman 1996). Thus, we set the timeout at one second into a pause. If the GrM does not have evidence until then, the UU remains ungrounded.

In summary, attention to the workspace over an interval following an utterance unit triggers grounding. Gaze at the speaker in the interval means that the contribution stays provisional, and triggers an obligation to elaborate. Likewise, if the system times-out without recognizing any user feedback, the segment remains ungrounded. This process allows the system to keep talking across multiple utterance units without getting verbal feedback from the user. From the user's perspective, explicit acknowledgment is not necessary, and minimal cost is involved in eliciting elaboration.

5.3.6 Updating the Discourse State

After judging grounding, the GrM updates the Discourse Model. The Discourse State maintained in the Discourse Model is similar to the frame-based information state in TRINDI kit (Matheson *et al.* 2000), except that we store nonverbal information. There are three key fields: GROUNDED, UNGROUNDED, and CURRENT. The current UU being processed is saved in the CURRENT field. If the current UU is judged grounded, its belief is added to the GROUNDED field, which stores a list of grounded UUs. If ungrounded, the UU is added to the UNGROUNDED field, which stores a list of pending (ungrounded) UUs. If a UU has subsequent contributions such as elaboration, these are stored in a single discourse unit, and grounded together when the last UU is grounded.

5.3.7 Determining the Next Action after Updating the Discourse State

After judging the UU's grounding, the GrM decides what MACK does next according to the result of the judgment. Probabilities shown in "Giver's next UU" in Table 5.3 are used as the decision rules. For example, if the type of the previous UU is Assertion and it is successfully grounded, MACK goes on to the next leg of the explanation for 70% of the time. On the contrary, if the previous assertive UU has not been grounded yet, MACK elaborates the previous UU for 73% of the time by describing the most recent landmark in more detail. These rules, again, are based on our empirical results.

MACK's attentional behaviors are also decided according to the selection rules based on our empirical data. For example, when MACK generates a direction segment (an Assertion), he keeps looking at the map for 66% of the time. When elaborating a previous UU, MACK gazes at the user for 47% of the time. Commands for these nonverbal acts are packed with UU (verbal content), and are sent to the GM, where these verbal and nonverbal contents are generated in a synchronized way.

5.3.8 Example

Figure 5.7 shows an example of a user's interaction with MACK. The user asks MACK for directions, and MACK replies using speech and pointing (using a projector) to the shared map.

When the GrM sends the first segment in the Agenda to the GM, the starting time of the UU is noted and it is sent to the AM to be spoken and animated. During this time, the user's nonverbal signals are logged in the Discourse Model. When the UU has finished, the GrM evaluates the log of the UU at the very beginning of the pause (by waiting a tenth of a second and then checking the nonverbal history). In this case, MACK noted that the user looked at the map during the UU[2], and continued to do so just afterwards. This pattern matches the positive evidence for Assertion. The UU is judged as grounded, and the grounded belief is added to the Discourse Model.

MACK then utters the second segment, UU[3], as before, but this time the GrM finds that the user was looking up at MACK during most of the UU as well as after it, which signals that the UU is not grounded. Therefore, the Response Planner generates an elaboration in UU[4]. This UU is judged to be grounded both because the user continues looking at the map, and because the user nods, and so the final stage of the directions is spoken. This is also grounded, leaving MACK ready for a new inquiry.

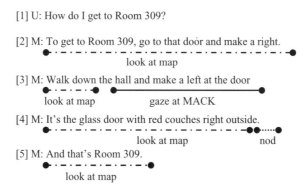

Figure 5.7 Example of user (U) interacting with MACK (M). Reproduced by permission of ACL, Nakano *et al.*, 2003

Figure 5.8   Snapshot of a user interacting with IPOC. Reproduced by permission of University of Tokyo Press

5.4 Dialogue Management using Attentional Behaviors Towards the Virtual World

This section focuses on the second type of attentional behavior, attention towards the virtual world (attention type (B) in Figure 5.2), and presents how this type of behavior is useful in dialogue management in human–agent interaction. For this research purpose, we developed an immersive conversational environment, IPOC, where a conversational agent is aware of a user's attentional behaviors towards the virtual world and is capable of displaying such behaviors in managing a conversation with the user. Figure 5.8 shows a snapshot of a user interacting with the IPOC. In the IPOC, a panoramic picture is used as a virtual world that a conversational agent inhabits, and the agent acts as a guide or an instructor who tells stories about objects and events shown in the picture.

Interaction with the IPOC agent is managed by the Interaction Control Component (ICC). The ICC architecture is shown in Figure 5.9. First, the Understanding Module (UM) interprets the user's inputs via the Julius speech recognizer (Lee *et al.* 2001) and a head tracking system using two CCD cameras. Next, the Conversation Manager (CM) updates the state of the conversation according to the user's verbal and nonverbal input, and selects an appropriate story according to the conversational and perceptual situation. The selected story text is sent to the Agent Behavior Planning Module (ABPM), where the agent's verbal and nonverbal behaviors are decided, and a time schedule for agent animations is calculated by accessing a text-to-speech (TTS) engine. Finally, according to the time schedule, agent animations and speech sounds are produced in a synchronized way. Agent animations are produced by the Haptek animation system and Japanese speech sounds are synthesized by Hitachi HitVoice.

5.4.1 Recognizing a User's Attention to a Virtual Shared Reference

We employed a head-pose tracking system developed by (Sato *et al.* 2004). This system, similar to that installed in MACK, does not require the user to wear a special cap or glasses, so that it does not harm the natural interaction between the user and a conversational agent. Moreover, this system is suitable for recognizing the user's gaze direction because (1) by controlling the increase in the number of hypotheses that should be maintained in the motion model, the system can achieve high speed and accuracy in estimating the user's head pose when a user is gazing at one point, and (2) the system also can keep

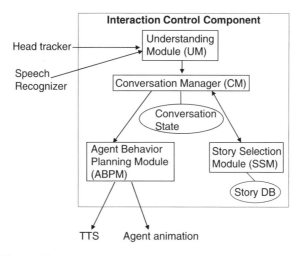

Figure 5.9 Architecture of the Interaction Control Component

track of the user's head even if the user suddenly moves her/his head. Using this method, this head-pose tracking system can recognize head movements caused by gaze-direction shifting.

This head-pose tracking system calculates the user's head position and rotation every 1/30th of a second, and the average of data in a 0.3-second window is used as a recognition result. Using the head-pose recognition results, the UM estimates the user's gaze direction on a 70-inch large display. In the current implementation, the whole display is split into six areas, and the system judges which of these areas the user's gaze falls into. Although this method does not recognize accurate gaze direction, it is still very useful for perceiving the user's attention without the user having to wear a device.

5.4.2 Dialogue Management using Attentional Behaviors

The Conversation Manager (CM) maintains the history and the current state of the conversation based on the Information State approach (Matheson *et al.* 2000), which is also employed in MACK. The conversation history is a list of stories that have been already told. The conversation state is represented as a set of values including the current story ID, topic focus, utterance ID and its semantic representation, start and end time of the utterance, as well as the log of the user's gaze direction change with a time-stamp. In addition, the CM calculates the major gaze area for a given utterance (i.e., the area that the user was looking at for the longest period of time during the utterance). The CM updates the conversational state when the system receives the user's verbal and/or nonverbal input (e.g., change of a user's focus of attention), or produces agent actions such as telling a story and confirming the user's intention. According to the updated conversational state, the CA determines the next agent behavior.

In IPOC, the user's gaze direction is used in judging whether the user is engaged in the conversation, and in identifying the user's interest in the virtual environment. Basically, if the user is properly paying attention to the focused object for the current topic and establishes a joint attention with the agent, the system presumes that the conversation is going well. In the current implementation, the system classifies the user's eye gaze into four types: attention (a) to the agent, (b) to a joint reference, (c) to other objects in the background, and (d) to a different direction from the display. When the user's attentional state is in (a)–(c), the system interprets the communicative function of the gaze behaviors according to the conversational context. Simplified rules for determining the agent's next actions are shown as follows.

(a)   If look-at-agent
          If first-look-at-agent
<1>   Then start-conversation
          ElseIf mid-of-story
<3>   Then continue-current-story
<2-2> Else select-next-story-randomly
ElseIf look-at-object-x
          If mid-of-story
              (b) If object-x is focused
<3>   Then continue-current-story
<4>   (c) Else askif-shift-to-topic-x
<2-1> Else suggest-next-story-x

- **<1>** *Initiating a conversation*. When the user looks at the agent for the first time, the system interprets that the user's gaze is signaling the user's request to start a conversation with the agent. In this case, the CA generates verbal and nonverbal greeting behaviors to start the conversation.
- **<2>** *Selecting the next topic*. **<2-1>** After finishing one story, the user has a chance to choose a new topic by looking at an object in the background picture. If the user is gazing at an object different from one that was focused on in the previous story, the CM assumes that the user is interested in a new object, and then suggests a new story about the object, by saying something like: "Would you like to listen to a story about this building?" At that moment, by pointing and gazing at the object, the agent establishes joint attention with the user. **<2-2>** Otherwise, if the user keeps looking at the agent even after finishing a story, the system assumes that the user expects to receive information from the agent. In this case, the system chooses a new topic.
- **<3>** *Continuing the current topic*. If the user pays attention to the agent or the focused object during a story, the system judges that the user is engaged in the agent's narrative and continues on with the current story. To maintain this relationship, the CM generates the agent's glance at the user, and demonstrates that the agent checks the user's attentional status.
- **<4>** *Changing a topic*. If the user's eye gaze is redirected to an object "x" different from the current focus in the middle of a story, the CA interprets it as a nonverbal interruption by the user. In such a case, the system asks the user whether s/he would like to listen to a new story about object "x" by quitting the current story. For example, the system asks: "Although it is in the middle of the story, would you like to move to a new topic concerned with this building?" At this moment, the agent demonstrates the awareness of the user's attention by looking at the object to which the user is paying attention.

5.4.3 Presenting a Story with Nonverbal Behaviors

To generate a story related to a user's interests that are suggested via the user's utterance or the object to which the user is paying attention, the CA sends the keywords in the utterance or the object name to the Story Selection Module (SSM). Then, the SSM retrieves short paragraphs from the story DB by using the keywords.

Once the next story is selected, the Agent Behavior Planning Module (ABPM) starts planning agent animations for presenting the story. Agent gestures emphasizing important concepts in a story are automatically determined and scheduled using an agent behavior generation system CAST (Conversational Agent System for neTwork applications) (Nakano *et al.* 2004). Taking Japanese text as input, CAST automatically selects agent gestures and other nonverbal behaviors, calculates an animation schedule, and produces synthesized voice output for the agent. As shown in Figure 5.10, first, in the Language Tagging Module (LTM), the received text is parsed and analyzed using a Japanese part-of-speech tagger and a syntactic analyzer (Kurohashi and Nagao 1994). According to the linguistic information calculated by the

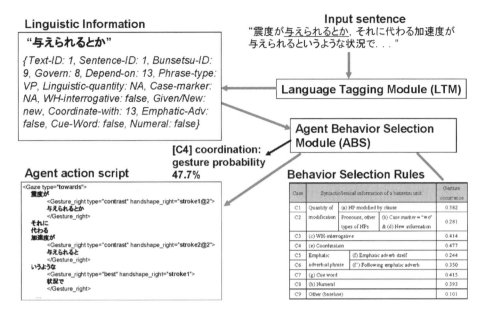

Figure 5.10   Agent behavior decision process in CAST

LTM, the Agent Behavior Selection Module (ABS) selects appropriate gestures and facial expressions by looking up behavior selection rules. The output from the ABS is an agent behavior script in XML format, where the agent's nonverbal behaviors are assigned to Japanese phrases. Finally, by obtaining phoneme timings from the text-to-speech engine, CAST calculates a time schedule for the XML. As the result, CAST produces synthesized speech and a set of animation instructions that can be interpreted and executed by the agent animation system.

Finally, the ABPM plays synthesized speech and sends commands to the animation system to generate the speech and the animation in a synchronized way.

5.4.4   Example

Figure 5.11 shows an example interaction between the user and the agent in the IPOC. White circles in the pictures indicate the areas where the user is paying attention. In [1:A], the system recognizes the user's first look at the agent, and the agent initiates a conversation. This is a case for <1> in section 5.4.2. Next, the user looks at the agent or the focused object (palanquin house). Therefore, the system assumes that the user is engaged in the conversation, and continues on with the story. This is a case for <3>.

After ending the first story, in [11:A], the agent asks the user what s/he would like to know next, while observing the user's attention. Since the system recognizes the user's attention to an accessory shop at the lower left on the display, the agent suggests a new topic while pointing at the shop in [12:A]. This is a case for <2-1> because the next topic is determined according to the user's atttentional behavior. After confirming the user's interest in the accessory shop in [13:U], the agent starts telling a story about the accessory shop in [14:A].

In the middle of this story, while uttering [21:A], the agent perceives the user's attention shifting towards a fire watchtower at the upper left on the display. Then, in [22:A], the agent asks the user whether s/he would like to change the topic. After getting the user's verbal acknowledgment in [23:U], the agent starts a new story about the fire watchtower in [24:A]. This sequence is an example for case <4>.

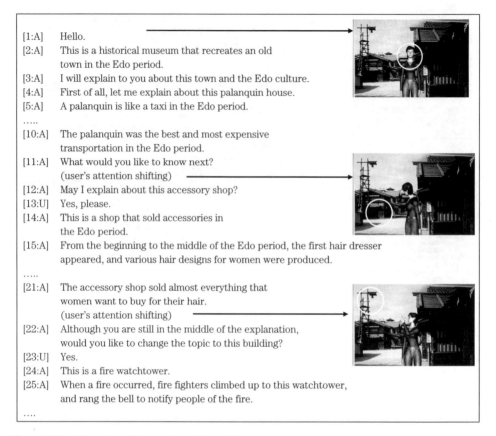

[1:A]	Hello.
[2:A]	This is a historical museum that recreates an old town in the Edo period.
[3:A]	I will explain to you about this town and the Edo culture.
[4:A]	First of all, let me explain about this palanquin house.
[5:A]	A palanquin is like a taxi in the Edo period.
.....	
[10:A]	The palanquin was the best and most expensive transportation in the Edo period.
[11:A]	What would you like to know next? (user's attention shifting)
[12:A]	May I explain about this accessory shop?
[13:U]	Yes, please.
[14:A]	This is a shop that sold accessories in the Edo period.
[15:A]	From the beginning to the middle of the Edo period, the first hair dresser appeared, and various hair designs for women were produced.
.....	
[21:A]	The accessory shop sold almost everything that women want to buy for their hair. (user's attention shifting)
[22:A]	Although you are still in the middle of the explanation, would you like to change the topic to this building?
[23:U]	Yes.
[24:A]	This is a fire watchtower.
[25:A]	When a fire occurred, fire fighters climbed up to this watchtower, and rang the bell to notify people of the fire.
....	

Figure 5.11 Example of interaction with IPOC. Reproduced by permission of Japanese Society of Artificial Intelligence

5.5 Conclusions

To leverage human communication protocols in human–computer interaction, research in ECAs has not only focused on human verbal and nonverbal communication behaviors, but also has proposed how to implement the communication functions into ECA systems. From this perspective, this chapter has focused on attentional behaviors as one of the most effective nonverbal communication behaviors, and has presented how these behaviors are effective in improving the naturalness of human–agent interaction.

The first system, MACK, is aware of a user's attention to a shared physical map, and uses attentional information in judging whether linguistic messages are grounded or not. The second system, IPOC, can perceive where on the display the user is paying attention, and use this information to estimate the user's interest and engagement in the conversation. In a preliminary evaluation experiment for MACK, we found that nonverbal behaviors exchanged between MACK and the user are strikingly similar to those in our empirical study of human-to-human communication. This result suggests that ECAs with attention capability have the potential for interacting with a human user using human–human conversational protocols.

The next step is to integrate these two systems, so that conversational agents have attentional capability to both the virtual world and the physical world. Such an ideal system is also expected to contribute to broadening a communication channel and improving the robustness of communication in human–computer interaction.

Acknowledgments
Thanks to Professor Justine Cassell and Dr. Candy Sidner for supervising the study in nonverbal grounding, and Gabe Reinstein and Tom Stocky for implementing MACK. Thanks to Professor Sato and Dr. Oka, who kindly helped us use their head tracking technology for IPOC. This study uses OKAO vision technology provided by OMRON Corporation.

References

Allen J. and Core M. (1997) *Draft of DMSL: Dialogue act markup in several layers*. Available from: http://www.cs.rochester.edu/research/speech/damsl/RevisedManual/

Anderson A.H. *et al.* (1997) The effects of face-to-face communication on the intelligibility of speech. *Perception and Psychophysics* **59**, pp. 580–592.

André E., Rist T. and Muller J. (1999) Employing AI methods to control the behavior of animated interface agents. *Applied Artificial Intelligence*, **13** (4–5): pp. 415-448, Taylor & Francis, Philadelphia, PA.

Argyle M. *et al.* (1973) The different functions of gaze. *Semiotica* **7**, pp. 19–32.

Argyle M. and Cook M (1976) *Gaze and Mutual Gaze*. Cambridge University Press.

Argyle M. and Graham J. (1977) The Central Europe experiment: looking at persons and looking at things. *Journal of Environmental Psychology and Nonverbal Behaviour* **1**, 6–16.

Boyle E., Anderson A. and Newlands A. (1994) The effects of visibility on dialogue and performance in a cooperative problem solving task. *Language and Speech* **37** (1), 1–20.

Brennan S. (2000) Processes that shape conversation and their implications for computational linguistics. In *Proceedings of 38th Annual Meeting of the ACL*, pp. 1–11.

Clark H.H. (1996) *Using Language*. Cambridge University Press, Cambridge.

Clark H.H. (2003) Pointing and placing. In S. Kita (ed.), *Pointing: Where language, culture, and cognition meet.*, Erlbaum, Hillsdale, NJ, pp. 243–268.

Clark H.H. and Brennan S.E. (1991) Grounding in communication. In L.B. Resnick, J.M. Levine and S.D. Teasley, (eds), *Perspectives on Socially Shared Cognition*. American Psychological Association, Washington, DC, pp. 127–149.

Clark H.H. and Schaefer E.F. (1989) Contributing to discourse. *Cognitive Science* **13**, 259–294.

Dillenbourg P. (1996) Some Technical Implications of Distributed Cognition on the Design of Interactive Learning Environments. *Journal of Artificial Intelligence in Education* **7** (2), 161–179.

Duncan S. (1972) Some signals and rules for taking speaking turns in conversations. *Journal of Personality and Social Psychology* **23** (2), 283–292.

Duncan S. (1974) On the structure of speaker-auditor interaction during speaking turns. *Language in Society* **3**, 161–180.

Goodwin C. (1981) Achieving mutual orientation at turn beginning, In *Conversational Organization: Interaction between speakers and hearers*. Academic Press, New York, pp. 55–94.

Johnson W.L., Rickel J.W. and Lester J.C. (2000) Animated pedagogical agents: face-to-face interaction in interactive learning environments. *International Journal of Artificial Intelligence in Education* **11**, 47–78.

Kapoor A. and Picard R.W. (2001) A real-time head nod and shake detector. In *Proceedings of Workshop on Perceptive User Interfaces*.

Kendon A. (1967) Some functions of gaze direction in social interaction. *Acta Psychologica* **26**, pp. 22–63.

Kraut R., Miller M. and Siegel J. (1996) Communication in performance of physical tasks: effects on outcomes and performance. In *Proceedings of Conference on Computer Supported Cooperative Work*, ACM New York, pp. 57–66.

Kurohashi S. and Nagao M. (1994) A syntactic analysis method of long Japanese sentences based on the detection of conjunctive structures. *Computational Linguistics* **20** (4), 507–534.

Lee A., Kawahara T. and Shikano K. (2001) Julius: an open source real-time large vocabulary recognition engine. In *Proceedings of European Conference on Speech Communication and Technology* (EUROSPEECH), pp. 169–1694.

Lester J.C. *et al.* (1999) Deictic believability: coordinated gesture, locomotion, and speech in lifelike pedagogical agents. *Applied Artificial Intelligence* **13** (4–5), 383–414.

Matheson C., Poesio M. and Traum D. (2000) Modelling grounding and discourse obligations using update rules. In *Proceedings of 1st Annual Meeting of the North American Chapter of the Association for Computational Linguistics* (NAACL2000).

Morency L.P., Rahimi A. and Darrell T. (2003) A view-based appearance model for 6 DOF tracking. In *Proceedings of IEEE Conference on Computer Vision and Pattern Recognition*.

Nakano Y.I. *et al.* (2003) Towards a Model of Face-to-Face Grounding. In *Proceedings of 41st Annual Meeting of the Association for Computational Linguistics* (ACL03), pp. 533–561.

Nakano Y.I. *et al.* (2004). Converting text into agent animations: assigning gestures to text. In *Proceedings of Human Language Technology Conference of the North American Chapter of the Association for Computational Linguistics* (HLT-NAACL 2004), Companian Volume, pp. 153–156.

Nakatani C. and Traum D. (1999) *Coding Discourse Structure in Dialogue (version 1.0)*. University of Maryland, Institute for Advanced Computer Studies Technical Report UMIACS-TR-99–03.

Noma T. and Badler N. (1997) A virtual human presenter. In *Proceedings IJCAI Workshop on Animated Interface Agents: Making Them Intelligent*, pp. 45–51.

Novick D.G., Hansen B. and Ward K. (1996) Coordinating turn-taking with gaze. In *Proceedings of ICSLP-96*, pp. 1888–1891.

Olson J.S., Olson G.M. and Meader D.K. (1995). What mix of video and audio is useful for remote real-time work? In *Proceedings of CHI95*, ACM, pp. 362–368.

Pierrehumbert J.B. (1980) *The Phonology and Phonetics of English Intonation*. Ph.D. thesis Massachusetts Institute of Technology.

Rickel J. and Johnson W.L. (1999) Animated agents for procedural training in virtual reality: perception, cognition and motor control. *Applied Artificial Intelligence* **13** (4–5), 343–382.

Sato Y. *et al.* (2004) Video-based tracking of user's motion and its use for augmented desk interface. In *Proceedings of IEEE Conference Automatic Face and Gesture Recognition* (FG 2004), pp. 805–809.

Shaw E., Johnson W.L. and Ganeshan R. (1999) Pedagogical agents on the web. In *Proceedings of Third International Conference on Autonomous Agents*, pp. 283–290.

Sidner C.L. and Lee C. (2003) Engagement rules for human–robot collaborative interactions. In *Proceedings of IEEE International Conference on Systems, Man & Cybernetics* (CSMC), vol. 4, pp. 3957–3962.

Stone B. and Lester J. (1996) Dynamically sequencing an animated pedagogical agent. In *Proceedings of Thirteenth National Conference on Artificial Intelligence*, pp. 424–431.

Swartout W. *et al.* (2001) Toward the holodeck: integrating graphics, sound, character and story. In *Proceedings of 5th International Conference on Autonomous Agents*, 2001, pp. 409–416.

Traum D. and Heeman P. (1996) Utterance units and grounding in spoken dialogue. In *Proceedings of ICSLP*, pp. 1884–1887.

Traum D. and Rickel J. (2002) Embodied agents for multi-party dialogue in immersive virtual worlds. In *Proceedings of The First International Joint Conference on Autonomous Agents and Multi-agent Systems* (AAMAS 2002), pp. 766–773.

Wahlster W., Reithinger N. and Blocher A. (2001) SmartKom: multimodal communication with a life-like character. In *Proceedings of Eurospeech 2001*, vol. 3, pp. 1547–1550.

Whittaker S. (2003) Theories and Methods in Mediated Communication. In Graesser A., Gernsbacher M. and Goldman S. (eds.), *Handbook of Discourse Processes*. Erlbaum, NJ, pp. 243–286.

Whittaker S. and O'Conaill B. (1993) Evaluating videoconferencing. In *Companion Proceedings of CHI93: Human Factors in Computing Systems*. ACM Press, pp. 135–136.

6

Attentional Gestures in Dialogues Between People and Robots

Candace L. Sidner and Christopher Lee

6.1 Introduction

Gestures are fundamental to human interaction. When people are face-to-face at near or even far distance, they gesture to one another as a means of communicating their beliefs, intentions, and desires. When too far apart or in too noisy an environment to converse, gestures can suffice, but in most human face-to-face encounters, conversation and gesture co-occur. According to the claims of McNeill (1992), they are tightly intertwined in human cognition. However cognitively entangled, people gesture freely, and the purposes of those gestures are fundamental to our work on human-robot interaction.

Some gestures, especially those using the hands, provide semantic content; these gestures supplement the content of utterances in which the gestures occur (Cassell 2000). However, some gestures indicate how the conversation is proceeding and how engaged the participants are in it. Attentional gestures are those that involve looking with head movements and eye gaze, and those involving body stance and position. Hand gestures by speakers also can be used to direct attention toward the gesture itself (Goodwin 1986). All these gestures convey what the participants are or should be paying attention to. The function of these gestures is distinguished by their role in the ongoing interaction. Nodding gestures also can indicate that the participants are paying attention to one another. Nodding *grounds* (Clark 1996) previous comments from the other speaker, while looking conveys attention for what is coming.

We call these attentional gestures "engagement behaviors"; that is, those behaviors by which interlocutors start, maintain and end their perceived connection to one another. The process by which conversational participants undertake to achieve this connection is the process of *engagement*. These behaviors indicate how each participant is undertaking their perceived connection to one another. They also convey whether the participants intend to continue their perceived connection. Of course, linguistic gestures (that is, talking) are a significant signal of engagement. However, linguistic and non-linguistic gestures of engagement must be coordinated during face-to-face interactions. For example, a speaker's ongoing talk while he or she looks away from the hearer(s) into blank space for a long time conveys contradictory information about his or her connection to the hearer; she's talking but doesn't appear interested. While other researchers, notably Clark and his students, have been investigating how conversational participants

Conversational Informatics: An Engineering Approach   Edited by Toyoaki Nishida
© 2007 John Wiley & Sons, Ltd

come to share common ground, and have considered some aspects of nonverbal gesture in that light, the overall problem of how participants come to the attention of one another, how they maintain that attention, how they choose to bring their attention to a close, and how they deal with conflicts in these matters is an open matter of investigation. In this article, all of our observations concern gestural phenomena that are typical of North Americans interacting. While we believe that engagement is natural in human–human interactions in all cultures, the details of how this is signaled will vary and will require further study.

It is essential to understand the process by which engagement transpires. It is part and parcel of our everyday face-to-face interactions, and it may well be relevant even in non face-to-face circumstances such as phone conversations. Furthermore a clear picture of the engagement process is necessary to reproduce natural communication behavior for artificial agents, either those in embodied on-screen characters that interact with computer users or those in the form of humanoid robots.

6.2 Background and Related Research

Head nods are a well-studied behavior in human conversation (Duncan 1973; Yngve 1970). They are used for establishing common ground, that is, shared knowledge between conversational participants (Clark 1996), and generally are accompanied by phrases such as "uh-huh, ok" or "yes". They also occur in positive responses to yes/no questions and as assents to questions or agreements with affirmative statements. McClave (2000) observed that head nods serve as backchannels and as encouragement for the listener. Bickmore (2002) observed that people talking to people in an experiment with a hand-held device nodded 1.53 times per turn either as acknowledgments, agreements or, in the case of the speaker, for emphasis.

Argyle and Cook (1976) documented the function of gaze as an overall social signal and noted that failure to attend to another person via gaze is evidence of lack of interest and attention. Other researchers have offered evidence of the role of gaze in coordinating talk between speakers and hearers, in particular, how gestures direct gaze to the face and why gestures might direct it away from the face (Kendon 1967; Duncan 1973; Goodwin 1986). Nakano *et al.*'s (2003) work on grounding reported on the use of the hearer's gaze and the lack of negative feedback to determine whether the hearer has grounded the speaker's turn. Our own studies of conversational tracking of the looks (using head directions only, not eye gaze) between a human speaker and human hearer observed that the hearer tracked the speaker about 55% of the time. The non-tracking times were characterized by attention to other objects, both ones that were relevant to their conversation and ones that were evidence of multitasking on the part of the hearer. However, sometimes for very short looks away on the part of the speaker, the hearer simply did not give evidence of attending to these looks (Sidner *et al.* 2005).

While we have investigated people talking to people in order to understand human behavior, we are also interested in how these behaviors translate to situations where people and robots converse. Robots that converse but have no gestural knowledge (either in production or in interpretation) may miscue a human and miss important signals from the human about engagement. Nodding on the part of humans is so unconscious, that people nod in conversations with robots, even though the robot has no ability to recognize the nods, as we will show later in this article. A portion of the time, they also nod without accompanying verbal language, so the multimodal aspects of gesture and language cannot be overlooked in human–robot interaction or in conversations with embodied conversational characters.

Previous work in human–robot interaction has largely explored gaze and basic interaction behavior. Breazeal's work on infantoid robots explored how the robot gazed at a person and responded to the person's gaze and prosodic contours in what might be called pre-conversational interactions (Breazeal and Aryananda 2002). Other work on infantoid robot gaze and attention can be found in Kozima *et al.* (2003). Minato *et al.* (2002) explored human eye gaze during question answering with an android robot; gaze behavior differed from that found in human–human interaction. More recent work (Breazel *et al.* 2004) explores conversation with a robot learning tasks collaboratively, but the robot cannot interpret nods during conversation. Ishiguro *et al.* (2003) report on development of Robovie with reactive transitions between its behavior and a human's reaction to it; they created a series of episodic rules and modules to control the robot's reaction to the human. However, no behaviors were created to interpret human head nods. Sakamoto *et al.* (2005) have experimented with cooperative behaviors on the part of the robot (including robot nodding but not human nodding) in direction-giving.

Thus we are interested in imbuing the robot with the ability to engage gesturally by looking, standing and nodding appropriately in interactions with humans and by properly interpreting such behaviors from people. This chapter reports on our progress in such efforts.

6.3 A Conversational Robot

The robot we have developed interacts with a single user in a collaboration that involves: spoken language (both understanding and generation) using a mouth that opens and closes in coordination with the robot's utterances, gestures with its appendages, head gestures to track the user and to turn to look at objects of interest in the interaction, recognition of user head gestures in looking at objects, and recognition of user head nods. The robot also initiates interactions with users, and performs typical preclosings and goodbyes to end the conversation. All these capabilities increase the means by which the robot can engage the user in an interaction.

The robot, called Mel and embodied as a penguin, has the following hardware:

- Seven DOF in the body (one in the beak/mouth, two in the head, two in each of the two wings)
- body mounted on a Pioneer II mobile robot platform for floor navigation and body positioning
- stereo camera
- two far-distance microphones (one for speech recognition, one for speech detection)
- two onboard laptop computers and an onboard PC-104 computer for all software.

The robot is depicted in Figure 6.1, and the architecture for the robot is displayed in Figure 6.2.

The architecture of this robot is divided into a conversational subsystem and a sensorimotor subsystem. The conversational subsystem is based on the Collagen conversation and collaboration manager (Rich *et al.* 2001). The sensorimotor subsystem controls the robot's physical hardware and performs sensor fusion. Together, these subsystems maintain a model of the dialogue and of the user's engagement, and use this model to generate appropriate verbal and gestural behaviors.

The sensorimotor subsystem is based on a custom, task-based blackboard robot architecture. It detects and localizes human speech, and uses vision algorithms of Viola and Jones (2001) to detect and recognize faces and of Morency *et al.* (2002) to track the 6-DOF position and orientation of one face. The subsystem fuses the sound and vision information to locate the robot's conversational partner, and roughly track that person's gaze from their head direction. In addition, it interprets certain of the human's head movements as head nods (Morency *et al.* 2005; see also Chapter 7). The sensorimotor subsystem receives information from the conversational subsystem about the dialogue state, commands for head and arm gestures, and commands to alter body stance (via the mobile base), and gaze.

Dialogue state information from the conversation subsystem is vital to the sensorimotor subsystem, both for proper generation of robot motions and for interpretation of sensor data. For example, the robot's gaze typically tracks the face of the human conversational partner, but the robot glances away briefly when it takes the floor and starts to speak. Wing gestures of the penguin signal new information in its utterances (so called *beat* gestures; see Cassell *et al.* (2001)) and provide pointing gestures to objects in the environment. The robot turns away from the conversational partner when it must point to an object in the demo, and it monitors the orientation of the person's head to ensure that the person follows the pointing motion. Proper tracking of the robot's pointing by the human is interpreted semantically, based

Figure 6.1 Mel, the robot penguin. Reproduced by permission of the ACM

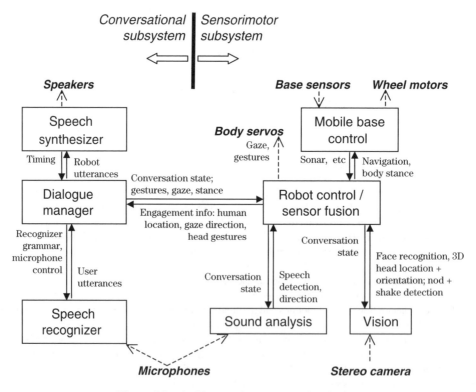

Figure 6.2 Architecture for a conversational robot

on dialogue context, by the conversational subsystem ("user looks at cup"). Failure to track within a reasonable amount of time may result in further prompting and gesturing ("The cup is here to my right").

Before the robot finishes speaking an utterance, it prepares for the human's response by activating speech recognition grammars tailored to the expected range of verbal responses. It also may adjust its priors for recognizing head motions as nods with respect to the content and timing of the robot's utterance (Morency *et al.* 2005). When the robot finishes speaking, it activates the microphones for speech detection and recognition. If the robot detects a head nod at this time, it checks to see if it also detects speech. If no speech is detected, the nod is immediately interpreted as "yes" or an acknowledgment ("OK"). Otherwise, the robot waits a short time for a result from speech recognition, and only interprets the nod if no recognition result is generated or the speech was not understood.

The robot participates in two different interactions with human participants. The first of these is a demonstration of IGlassware, a technology developed at Mitsubishi Electric Research Laboratories. IGlassware is a specially designed cup that senses fluid level using capacitance and communicates the level to a special table by inductance (Dietz *et al.* 2002). Mel was programmed to demonstrate this invention to a human visitor. A typical interaction is shown in Figure 6.3; the robot's gestures with its head towards the table are noted in the text. However, many different conversations are possible, including one that is not about the demo at all (if the human participant refuses to be shown a demo).

The robot can also demonstrate its own capabilities in an interactive conversation. A portion of one such conversation can be found in Figure 6.4. During this conversation, the robot gestures at its own body parts when relevant, and in the case of its mobile base, moves around to demonstrate its locomotive abilities. We have observed that people interacting with the robot during this conversation maintain their gaze on the robot through most of the conversation with relatively few looks away (in contrast to the IGlassware conversations in which users direct much of their attention to the IGlassware cup and table). We surmise that gaze maintenance is due to the fact that there is nothing else that is relevant to the conversation, and due to the fact that the robot is pointing out parts of itself in the conversation.

When initiating both interactions, Mel uses his vision system to find a conversational partner. Thereafter Mel tracks the conversational partner's face and adjusts his "gaze" towards the partner, even when the

M: Hi, I'm Mel a robotic penguin.
U: Hi.
M: What's your name?
U: Sam.
M: Sam, I'd like to show you a demo. OK?
U: No.
M: But it's really interesting. Come on. Try it!
U: Okay.
M: It would be really nice to know what kind of progress people are making in their dining.
M: IGlassware stands for Instrumented glassware!
U: Okay.
M: Sam, right there <robot turns head to table> is the IGlassware cup <robot turns to user>.
U: Okay.
M: All right, now we'll see how to use the cup.
U: All right.

M: Please pour enough water from the pitcher into the glass to fill the cup. Then put it on the table.
U: Okay.
M: I'm waiting for a reading from the table for the glass.
M: Good. Take a look at the bar on the readout.
U: Yes.
M: It shows that the cup is full, right?
U: Yes.
M: Would you like me to explain how this works?
U: Sure.
M: You see the copper in the glass? <robot turns head to glass>....
    [continued conversation not illustrated]
M: Well, Sam, that's about all.
    See Candy Sidner or Chris Lee for more about IGlassware. So long.
U: Good-bye.

Figure 6.3 Excerpts from the IGlassware Demo with Mel. Reproduced by permission of the ACM

M: Hello, I'm Mel, a robotic penguin.
U: Hello.
M: What's your name?
U: Mike.
M: Mike, I'd like to tell you about myself. Okay?
U: Ok.
M: First I'll explain my body.
U: Ok.
M: Well. I have 7 degrees of freedom in my body parts plus my mobile base can turn wherever I want it to. <robot rotates base during conversation>
U: Ok. [continuing conversation not illustrated]
M: How about we discuss my vision and speech systems?
U: Ok.
M: Shall I explain my vision system, or my speech abilities first?
U: Your vision system.

M: Well, above my head is a black camera with two lenses. Do you see it? <robot looks up above head during this exchange>
U: Yes.
M: I see with my two lens eyes.... [continuing conversation not illustrated]
M: Well, that's all my parts. Are you curious about what I can do with all that?
U: Yes.
M: Mostly I can talk about things. And I give demos. And I'd like to learn how to tell jokes well.
U: Ok.
M: So you wanna hear a joke?
U: No.
M: Well, Mike that's about all.
U: Ok.
M: See Candy Sidner or Chris Lee, for more about me! So long!
U: Good-bye.
M: Good-bye.

Figure 6.4   Excerpts from a Mel self-demo. Reproduced by permission of the ACM

partner moves about. Mel has eyes in his head, but they do not see, and his camera is above his head, so his gaze serves only to communicate his focus of attention to the partner. Mel does not look at the human partner at all times. During the demos, he turns to look at the table and its contents or to parts of himself. Mel also prompts a partner who fails to look at the table to notice the objects there. After the demo and explanation conclude, Mel wishes the partner goodbye, waves, and drops his head to his chest to indicate that he is no longer available.

6.4 Looking Behaviors for the Robot

Mel the robot is able to follow the looking gestures of a human partner using the vision technology discussed above. The robot also displays looking ability when he turns to relevant objects in the room, points at them, and then turns back to look at his human interlocutor. Does this robot's nonverbal gestural behavior have an impact on the human partner? The answer is a qualified yes. We conducted an experiment with 37 human participants who were given an IGlassware demo by the robot. Twenty of these participants interacted with the robot with full gestural behaviors and with tracking of the human head movements (the robot indicated the tracking by its head movement), while 17 participants interacted with a robot that only moved his beak but produced no other gestures (Sidner *et al.* 2005). This "wooden" robot looked straight at the position where it first saw the human's head and did not change its head position thereafter. Just as the moving robot did, the wooden robot also moved its beak with his utterances and noticed when the human looked at objects in the scene.

Our "wooden" robot allowed us to study the effects of just having a linguistic conversation without any looking behavior. In comparison, the moving robot performed looking gestures by tracking the person's face, looking at relevant objects (for a brief period of time) and then returning to look at the user. These behaviors permitted us to study the effects of looking in concert with conversation. Informally we observed that Mel, by talking to users and tracking their faces, gave users a sense of being paid attention to. However, whether this "sense" had any effect on human robot interaction led us to the study with people interacting with the two versions of our robot.

Both "wooden" and moving robots assessed human engagement. Humans who did not respond to their turn in the conversation were judged as possibly wanting to disengage, so in both conditions, the robot would ask "Do you want to continue the demo?" when humans failed to take a turn. Human looks away were not judged by themselves as expressing disengagement. In fact most looks away were to the demo objects, or in the case of IGlassware, to the area under the table to ascertain how the table worked. However, when humans did not perform shared attention to the demo objects when the robot introduced them into the discussion, the robot prompted the human with an additional utterance and looking gesture to encourage them to view the object. Thus our robot both makes decisions about where to look itself and assesses aspects of joint attention with its human conversational partner. As Nakano and Nishida demonstrate in Chapter 5, shared attention to objects in the environment is a critical attentional behavior in human–agent interactions.

In our experiment, each person had a conversation with the robot (before which the person was told that they would see a demonstration with a robot) and after filled out a questionnaire about their experience. From the questionnaires, we learned that all participants liked the robot. However, those who talked to the moving robot rated it as having more appropriate gestures than did participants with the wooden robot.

While questionnaires tell us something about people's response to a robot, videotapes of their interactions revealed more. We studied videotapes of each conversation between person and robot. From these studies, we discovered that in both conditions, about 70% of the time, the human participants who were looking away from the robot looked back to the robot when it was their turn to speak in the conversation. It seems that conversation alone is a powerful mechanism for keeping people engaged. In particular, in human–human interactions, people look at their conversational partners when they have the conversational turn. People behaved similarly when their conversational partner was a robot, even one whose looking gestures were nonexistent.

However, two other aspects of human looking produced different effects in the "wooden" versus moving condition. First, the human participants looked back at the moving robot significantly more often even when it was not their turn than their counterparts did with the "wooden" robot (single-factor ANOVA, $F[1, 36] = 15.00$, $p < 0.001$). They determined that the robot was to be paid attention to and looked at it even when they were also trying to look at the demo objects.

Second, there was a weak effect of the robot's looking by turning its head towards objects relevant to the demo. We observed that human participants tracked very closely the turn of the robot's head by themselves turning to look at the IGlassware table or glass. This behavior was more common than for participants who only had utterances (such as "Right here is the IGlassware cup") to direct them to objects. Looking gestures of this type provide a strong signal of where to look and when to start looking, and our participants responded to those signals.

In sum, the moving robot's looking gestures affected how people participated in conversation. They paid closer attention to the robot, both in terms of what it said and in terms of where it looked at objects. All these behaviors are indicative of engagement in the interaction. Without overt awareness of what they were doing, people's gestures indicated that they were engaged with the moving robot.

6.5 Nodding at the Robot

Head nodding is one of the many means by which participants indicate that they are engaging in the conversation. In the experiments discussed in the previous section, we observed that the participants nodded to the robot to acknowledge his utterances, to answer "yes" to questions, and to assent to some statements. However, the robot had no ability to interpret such nods. Thus we were determined to see if this behavior would continue under other circumstances. Furthermore, because nodding is a natural behavior in normal human conversation, it behooved us to consider it as a behavior that a robot could interpret.

In our experiments (Sidner *et al.* 2006), human participants held one of two conversations with the robot, shown in the previous section, to demonstrate either its own abilities or the IGlassware equipment. During these conversations people nodded at the robot, either because it was their means for taking a turn after the robot spoke (along with phrases such as "OK" or "yes" or "uh-huh"), or because they were answering in the affirmative a yes/no question and accompanied their linguistic "yes" or "OK" with a nod. Participants also shook their heads to answer negatively to yes/no questions, but we did not study this behavior because too few instances of "no" answers and headshakes occurred in our data. Sometimes a nod was the only response on the part of the participant.

The participants were divided into three groups, called the MelNodsBack group, MelOnlyRecognizesNods group, and the NoMelNods group. The NoMelNods group with 20 participants held conversations with a robot that had no head nod recognition capabilities. This group served as the control for the other two groups.

The MelNodsBack group with 15 participants, who were told that the robot understood some nods during conversation, participated in a conversation in which the robot nodded back to the person every time it recognized a head nod. It should be noted that nodding back in this way is not something that people generally do in conversation. People nod to give feedback on what another has said, but having done so, their conversational partners only rarely nod in response. When they do, they are generally indicating some kind of mutual agreement. Nonetheless, by nodding back, the robot gives feedback to the user on their behavior. Due to mis-recognition of nods, this protocol meant that the robot sometimes nodded when the person did not nod.

The MelOnlyRecognizesNods group with 14 participants held conversations without the robot ever nodding back, and without the knowledge that the robot could understand head nods although the nod recognition algorithms were operational during the conversation. We hypothesized that participants might be affected by the robot's nodding ability because (1) when participants responded only with a nod, the robot took another turn without waiting further, and (2) without either a verbal response or a nod, the robot waited a full second before choosing to go on. Hence participant nods caused the robot to continue talking. Therefore, we hypothesized that, over the whole conversation, the participants might have gotten enough feedback to rely on robot recognition of nods. This group provided an opportunity to determine whether the participants were so affected. Note also that, as with the Mel NodsBack group, nod mis-recognition occurred although the participants got no gestural feedback about it.

Every conversation with the robot in our total of 49 participants varied in the number of exchanges held. Hence every participant had a varying number of opportunities to give feedback with a nod depending on when a turn was taken or what question was asked. This variation was due to: different paths through the conversation (when participants had a choice about what they wanted to learn), the differences in the demonstrations of IGlassware and of the robot itself, speech recognition (in which case the robot would ask for re-statements), robot variations in pausing as a result of hearing the user say "OK," and instances where the robot perceived that the participant was disengaging from the conversation and would ask the participant if they wished to continue.

In order to normalize for these differences in conversational feedback, we coded each of the individual 49 conversations for feedback opportunities in the conversation. Opportunities were defined as the end of an exchange where the robot paused long enough to await a response from the participant before continuing, or exchange ends where it waited only briefly but the participant chose to interject a verbal response in that brief time.

So for each participant, our analysis used a "nod rate" as a ratio of total nods to feedback opportunities, rather than the raw number of nods in an individual conversation. Furthermore, the analysis made three distinctions: nod rates overall, nod rates where the participant also uttered a verbal response (nod rates with speech), and nod rates where no verbal response was uttered (nod-only rates).

Our study used a between-subjects design with Feedback Group as our independent variable, and Overall Nod Rate, Nod with Speech Rate, and Nod Only Rate as our three dependent variables.

A one-way ANOVA indicates that there is a significant difference among the three feedback groups in terms of Overall Nod Rate ($F[2,46] = 5.52$, $p < 0.01$). The mean Overall Nod Rates were 42.3%, 29.4%, and 20.8% for MelNodsBack, MelOnlyRecognizesNods, and NoMelNods groups respectively. A post-hoc LSD pairwise comparison between all possible pairs shows a significant difference between the MelNodsBack and the NoMelNods groups ($p = 0.002$). No other pairings were significantly different. The mean Overall Nod Rates for the three feedback groups are shown in Figure 6.5.

A one-way ANOVA indicates that there is also a significant difference among the three feedback groups in terms of Nod with Speech Rate ($F[2,46] = 4.60$, $p = 0.02$). The mean Nod with Speech Rates was 32.6%, 23.5%, and 15.8% for the MelNodsBack, MelOnlyRecognizesNods, and NoMelNods groups respectively. Again, a LSD post-hoc pairwise comparison between all possible pairs of feedback groups shows a significant difference between the MelNodsBack and NoMelNods groups ($p = 0.004$). Again, no other pairs were found to be significantly different. The mean Nod with Speech Rates for the three feedback groups are shown in Figure 6.6.

Finally, a one-way ANOVA found no significant differences among the three feedback conditions in terms of Nod Only Rate ($F[2,46] = 1.08$, $p = 0.35$). The mean Nod Only Rates were more similar to one another than the other nod measurements, with means of 8.6%, 5.6%, and 5.0% for the MelNodsBack, MelOnlyRecognizesNods, and NoMelNods groups respectively. The mean Nod Only Rates for the three feedback groups are shown in Figure 6.7.

These results indicate that, under a variety of conditions, people will nod at a robot as a conversationally appropriate behavior. Furthermore, even subjects who get no feedback about nodding do not hesitate to nod in a conversation with the robot. We conclude that conversation alone is an important feedback effect for producing human nods, regardless of the robot's ability to interpret it.

It is worthwhile noting that the conversations our participants had with robots were more than a few exchanges. While they did not involve the human participants having extended turns in terms of verbal contributions, they did involve their active participation in the purposes of the conversation. Future researchers exploring this area should bear in mind that the conversations in this study were extensive, and that ones with just a few exchanges might not see the effects reported here.

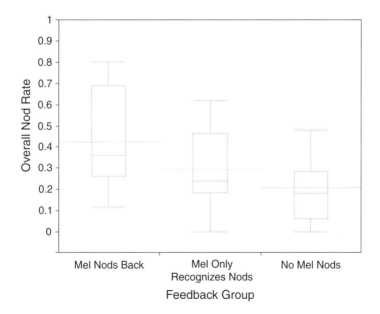

Figure 6.5 Overall nod rates by feedback group. Reproduced by permission of the ACM

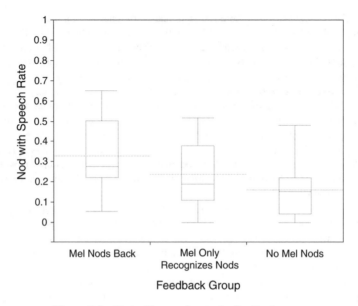

Figure 6.6   Nod with speech rates by feedback group

The two statistically significant effects for nods overall and nods with speech that were found between the NoMelNods group and the MelNodsBack group indicate that providing information to participants about the robot's ability to recognize nods and giving them feedback about it makes a difference in the rate at which they produce nods. This result demonstrates that adding perceptual abilities to a humanoid robot that the human is aware of and gets feedback from provides a way to affect the outcome of the human and robot's interaction.

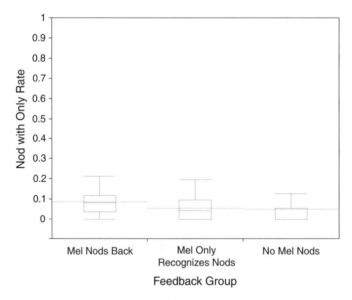

Figure 6.7   Nod only rates by feedback group

The lack of statistical significance across the groups for nod rates without any verbal response (nod-only rates) did not surprise us. The behavior of only nodding (without accompanying speech) in human conversation is a typical behavior, although there are no statistics that we are aware specifying rates of such nods. It is certainly not as common as nodding and making a verbal response as well. Again, it is notable that this behavior occurs in human–robot interaction and under varying conditions. By count of participants, in the NoMelNods group, 9 of 20 participants nodded at least once without speech; in the MelOnlyRecognizesNods group, 10 of 14 did so; and in the MelNodsBack group, 12 of 15 did so. However, when using normalized nod rates for this behavior for each group, there is no statistical difference. This lack is certainly partially due to the small number of times subjects did this: the vast majority of participants (44/49) did a nod without speech only 1, 2, or 3 times during the entire conversation. A larger data set might show more variation, but we believe it would take a great deal of conversation to produce this effect. We conclude that nods without speech are a natural conversational behavior. Hence using vision techniques with a robot that capture this conversation behavior is valuable to any human–computer conversation, whether with a robot or with an onscreen agent.

6.6 Lessons Learned

Looking at another person, looking at objects when one mentions them, especially initially, nodding appropriately in conversation, and recognizing the nods of others are all gestures that people typically produce in interactions with others. While looking gestures are clearly indicative of engagement, nodding gestures may seem less so since their most obvious function is to supplement and sometimes replace utterances such as "okay", "yes", "uh-huh". In this respect nodding may be akin to iconic gestures, which some researchers believe supplement or provide additional semantic content during speaking (McNeill 1992).

However, nodding gestures are principally feedback to the speaker (see also Chapter 5). They work even if the hearer is not looking at the speaker and does not say anything. In our observations of human–human interactions (Sidner *et al.* 2005), we observed many occasions where a hearer nodded (sometimes without speech) to provide feedback to the speaker that they were following the speaker's utterances. And as we discovered, people naturally nod at robots during conversation. In our human–robot studies, our human participants nodded both when looking at the robot and when looking at demo objects.

This feedback serves to signal engagement in subtle ways. While speakers will proceed without a nod or a "yes" when they have indicated that they expect such feedback,[1] it is unlikely they typically will do so for a long stretch of conversation. Speakers find they must know if their conversational partners are attending to them. When a hearer does not provide that feedback on a regular basis, their behavior indicates a failure to be tracking the speaker and thereby failing to be engaged in the interaction. Conversely, there are speakers who disregard the lack of feedback from their hearers. While this behavior is not the normative way to behave, it still occurs. Importantly, speakers who speak for long stretches of conversation and ignore the lack of feedback from hearers are generally judged as holding the floor of the conversation in an unacceptable way.

The lack of feedback from a person to a robot as well as the robot's failure to recognize that feedback limits the robot's assessment of what is happening in the interaction. The studies undertaken so far provide only initial observations about how people engage with robots and how the interaction changes, as the robot knows more about engagement behaviors. However, these first steps indicate that human–machine conversations must include and account for these behaviors if robots are to function fully in conversation.

---

[1] The linguistic means for signaling that feedback is called for is not completely understood. However, a typical means of signaling feedback is to bring an utterance to a close and to use end-of-utterance prosodic effects to indicate that the speaker is expecting a response from the hearer.

Without the ability to perceive human engagement signals, robots will misinterpret what is occurring in the interaction. Without producing proper engagement signals, robots may misinform human partners about their role in the interaction.

6.7 Future Directions

The research reported in this article addresses the maintenance of engagement during an interaction. Equally interesting to us is the process by which two interactors find each other and begin to interact. Research on "catching the eye of another person" (Miyauchi *et al.* 2004) illustrates ways that robots can undertake one part of this process. Much remains to be done to understand this phenomenon and provide algorithms for the robot to perform such activities.

As was mentioned earlier, body stance serves to indicate one's focus of attention. When a conversational participant stands so that his or her body is pointed away from the conversational partner, there is an apparent conflict between engaging the partner and focusing on something in the direction of the body. How this conflict is managed and how body stance is combined with looking and gazing are not well understood. Yet for many activities one's body must be pointed in another direction than the one of the conversational participant. We plan to investigate the issue of body stance as an attentional mechanism and its relation to engagement and to looking and gazing.

Acknowledgments

This work has been accomplished in collaboration with a number of colleagues over the years who have worked with us on Mel at various stages of the research. Our thanks to Myrosia Dzikovska, Clifton Forlines, Cory Kidd, Neal Lesh, Max Makeev, Louis-Philippe Morency, Chuck Rich, and Kate Saenko. We are indebted to them for their contributions to the development of Mel, and to the studies reported here.

This work was undertaken and completed while the authors were at Mitsubishi Electric Research Laboratories, Cambridge, MA 02139 (www.merl.com).

This work is based on an earlier work: "The effect of head-nod recognition in human–robot conversation", in *Proceedings of the 1st ACM SIGCHI/SIGART Conference on Human–Robot Interaction*, pages 290–296; ⓒ ACM, 2006. see http://portal.acm.org/citation.cfm?doid=1121241.1121291

References

Argyle M. and Cook M. (1976) *Gaze and Mutual Gaze*. Cambridge University Press, New York.

Bickmore T. (2002) Towards the design of multimodal interfaces for handheld conversational characters. In *Proceedings of the CHI: Extended Abstracts on Human Factors in Computing Systems Conference*, Minneapolis, MN, ACM Press, pp. 788–789.

Breazeal C. and Aryananda L. (2002) Recognizing affective intent in robot directed speech. *Autonomous Robots* **12** (1), 83–104.

Breazeal C., Brooks A., Gray J., Hoffman G., Kidd C., Lee H., Lieberman J., Lockerd A. and Chilongo D. (2004) Tutelage and collaboration for humanoid robots. *International Journal of Humanoid Robotics* **1** (2), 315–348.

Cassell J. (2000) Nudge nudge wink wink: elements of face-to-face conversation for embodied conversational agents. In J. Cassell J., Sullivan S., Prevost and E. Churchill (eds), *Embodied Conversational Agents*. MIT Press, Cambridge, MA, pp. 1–18.

Cassell J., Vilhjlmsson H.H. and Bickmore T. (2001) BEAT: the behavior expression animation toolkit. In E. Fiume (ed.), *SIGGRAPH 2001, Computer Graphics Proceedings*. ACM Press, pp. 477–486.

Clark H.H. (1996) *Using Language*. Cambridge University Press, Cambridge.

Dietz P.H., Leigh D.L. and Yerazunis W.S. (2002) Wireless liquid level sensing for restaurant applications. *IEEE Sensors* **1**, 715–720.

Duncan S. (1973) On the structure of speaker–auditor interaction during speaking turns. *Language in Society* **3**, 161–180.

Goodwin C. (1986) Gestures as a resource for the organization of mutual attention. *Semiotica* **62** (1/2), 29–49.

Ishiguro H., Ono T., Imai M. and Kanda T. (2003) Development of an interactive humanoid robot "Robovie": an interdisciplinary approach. In R.A. Jarvis and A. Zelinsky (eds), *Robotics Research*. Springer, pp. 179–191.

Kendon A. (1967) Some functions of gaze direction in two person interactions, *Acta Psychologica* **26**, 22–63.

Kozima H., Nakagawa C. and Yano H. (2003) Attention coupling as a prerequisite for social interaction. In *Proceedings of the 2003 IEEE International Workshop on Robot and Human Interactive Communication*. IEEE Press, New York, pp. 109–114.

McClave E.Z. (2000) Linguistic functions of head movements in the context of speech. *Journal of Pragmatics* **32**, 855–878.

McNeill D. (1992) *Hand and Mind: What gestures reveal about thought*. University of Chicago Press.

Minato T. MacDorman K., Simada M., Itakura S., Lee K. and Ishiguro H. (2002) Evaluating humanlikeness by comparing responses elicited by an android and a person. In *Proceedings of Second International Workshop on Man–Machine Symbiotic Systems*, pp. 373–383.

Miyauchi D., Sakurai A., Makamura A. and Kuno Y. (2004) Active eye contact for human–robot communication. In *Proceedings of CHI 2004: Late Breaking Results*, CD Disc 2. ACM Press, pp. 1099–1104.

Morency L.-P., Lee C., Sidner C. and Darrell T. (2005) Contextual recognition of head gestures. In *Proceedings of Seventh International Conference on Multimodal Interfaces* (ICMI'05), pp. 18–24.

Morency L.-P., Rahimi A., Checka N. and Darrell T. (2002) Fast stereo-based head tracking for interactive environment. In *Proceedings of the International Conference on Automatic Face and Gesture Recognition*, pp. 375–380.

Nakano Y., Reinstein G., Stocky T. and Cassell J. (2003) Towards a model of face-to-face grounding. In *Proceedings of the 41st Meeting of the Association for Computational Linguistics*, Sapporo, Japan, pp. 553–561.

Rich C., Sidner C.L. and Lesh N.B. (2001) Collagen: applying collaborative discourse theory to human–computer interaction. *Artificial Intelligence Magazine* **22** (4), 15–25.

Sakamoto D., Kanda T., Ono T., Kamashima M., Imai M. and Ishiguro H. (2005) Cooperative embodied communication emerged by interactive humanoid robots. *International Journal of Human–Computer Studies* **62** (2), 247–265.

Sidner C., Lee C., Kidd C., Lesh N. and Rich C. (2005) Explorations in engagement for humans and robots. *Artificial Intelligence* **166** (1–2), 140–164.

Sidner C., Lee C., Morency L. and Forlines, C. (2006) The effect of head-nod recognition in human–robot conversation. In *Proceedings of the ACM Conference on Human–Robot Interaction*, pp. 290–296.

Viola P. and Jones M. (2001) Rapid object detection using a boosted cascade of simple features. In *Proceedings of the IEEE Conference on Computer Vision and Pattern Recognition*, Hawaii, pp. 905–910.

7

Dialogue Context for Visual Feedback Recognition

Louis-Philippe Morency, Candace L. Sidner, and Trevor Darrell

7.1 Introduction

During face-to-face conversation, people use visual feedback to communicate relevant information and to synchronize rhythm between participants. A good example of nonverbal feedback is head nodding and its use for visual grounding, turn-taking and answering yes/no questions. When recognizing visual feedback, people use more than their visual perception. Knowledge about the current topic and expectations from previous utterances help guide our visual perception in recognizing nonverbal cues. Our goal is to equip an embodied conversational agent (ECA) with the ability to use contextual information for performing visual feedback recognition much in the same way people do.

In the last decade, many ECAs have been developed for face-to-face interaction. A key component of these systems is the dialogue manager, which usually provides a history of the past events, the current state, and an agenda of future actions. The dialogue manager uses these contextual information sources to decide which verbal or nonverbal action the agent should perform next. This is called context-based synthesis.

Contextual information has proven useful for aiding speech recognition. In Lemon *et al.* (2002), the grammar of the speech recognizer dynamically changes depending on the agent's previous action or utterance. In a similar fashion, we want to develop a context-based visual recognition module that builds upon the contextual information available in the dialogue manager to improve performance.

The use of dialogue context for visual gesture recognition has, to our knowledge, not been explored before for conversational interaction. In this chapter we present a prediction frame work for incorporating dialogue context with vision-based head gesture recognition. The contextual features are derived from the utterances of the ECA, which is readily available from the dialogue manager. We highlight four types of contextual features: lexical, prosodic, timing, and gesture, and select a subset for our experiment that were topic independent. We use a discriminative approach to predict head nods and head shakes from a small set of recorded interactions. We then combine the contextual predictions with a vision-based recognition algorithm based on the frequency pattern of the user's head motion. Our context-based recognition framework allows us to predict, for example, that in certain contexts a glance is not likely

Conversational Informatics: An Engineering Approach Edited by Toyoaki Nishida
© 2007 John Wiley & Sons, Ltd

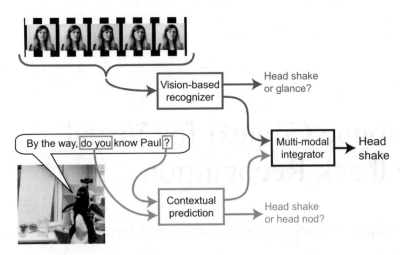

Figure 7.1 Contextual recognition of head gestures during face-to-face interaction with a conversational robot. In this scenario, contextual information from the robot's spoken utterance helps disambiguating the listener's visual gesture

whereas a head shake or nod is (as in Figure 7.1), or that a head nod is not likely and a head nod misperceived by the vision system can be ignored.

The following section describes related work on gestures with ECAs. Section 7.3 presents a general discussion on how context can be used for different types of visual feedback. Section 7.4 describes the contextual information available in most embodied agent architectures. Section 7.5 presents our general framework for incorporating contextual information with visual observations. Section 7.6 shows how we automatically extract a subset of this context to compute lexical, prosodic, timing, and gesture features. Finally, in section 7.7, we describe context-based head gesture recognition experiments, performed on 16 video recordings of human participants interacting with a robot.

7.2 Background and Related Research

There has been considerable work on gestures with ECAs. Bickmore and Cassell (2004) developed an ECA that exhibited many gestural capabilities to accompany its spoken conversation and could interpret spoken utterances from human users. Sidner *et al.* (2005) have investigated how people interact with a humanoid robot. They found that more than half their participants naturally nodded at the robot's conversational contributions even though the robot could not interpret head nods. Nakano *et al.* (2003) analyzed eye gaze and head nods in computer–human conversation and found that their subjects were aware of the lack of conversational feedback from the ECA. They incorporated their results in an ECA that updated its dialogue state. Numerous other ECAs (e.g., de Carolis *et al.* 2001; Traum and Rickel 2002) are exploring aspects of gestural behavior in human–ECA interactions. Physically embodied ECAs – for example, ARMAR II (Dillman *et al.* 2004, 2002) and Leo (Breazeal *et al.* 2004) – have also begun to incorporate the ability to perform articulated body tracking and recognize human gestures.

Head pose and gesture offer several key conversational grounding cues and are used extensively in face-to-face interaction among people. Stiefelhagen (2002) developed several successful systems for tracking face pose in meeting rooms and has shown that face pose is very useful for predicting turn-taking.

Takemae *et al.* (2004) also examined face pose in conversation and showed that if tracked accurately, face pose is useful in creating a video summary of a meeting. Siracusa *et al.* (2003) developed a kiosk front end that uses head pose tracking to interpret who was talking to who in conversational setting. The position and orientation of the head can be used to estimate head gaze which is a good estimate of a person's attention. When compared with eye gaze, head gaze can be more accurate when dealing with low-resolution images and can be estimated over a larger range than eye gaze (Morency *et al.* 2002).

Kapoor and Picard (2001) presented a technique to recognize head nods and head shakes based on two Hidden Markov Models (HMMs) trained and tested using 2D coordinate results from an eye gaze tracker . Fujie *et al.* (2004) also used HMMs to perform head nod recognition. In their paper, they combined head gesture detection with prosodic recognition of Japanese spoken utterances to determine strongly positive, weak positive, and negative responses to yes/no type utterances.

Context has been previously used in computer vision to disambiguate recognition of individual objects given the current overall scene category (Torralba *et al.* 2003). While some systems (Breazeal *et al.* 2004; Nakano *et al.* 2003) have incorporated tracking of fine motion actions or visual gesture, none has included top-down dialogue context as part of the visual recognition process.

7.3 Context for Visual Feedback

During face-to-face interactions, people use knowledge about the current dialogue to anticipate visual feedback from their interlocutor. Following the definitions of Cassell and Thorisson (1999) for nonverbal feedback synthesis, we outline three categories for visual feedback analysis: (1) content-related feedback, (2) envelope feedback, and (3) emotional feedback. Contextual information can be used to improve recognition in each category.

- *Content-related feedback.* This is concerned with the content of the conversation. For example, a person uses head nods or pointing gestures to supplement or replace a spoken sentence. For this type of feedback, contextual information inferred from speech can greatly improve the performance of the visual recognition system. For instance, to know that the embodied agent just asked a yes/no question should indicate to the visual analysis module a high probability of a head nod or a head shake.
- *Envelope feedback.* Grounding visual cues that occur during conversation fall into the category of envelope feedback. Such visual cues include eye gaze contact, head nods for visual grounding, and manual beat gestures. Envelope feedback cues accompany the dialogue of a conversation much in the same way audio cues like pitch, volume and tone envelope spoken words. Contextual information can improve the recognition of envelope visual feedback cues. For example, knowledge about when the embodied agent pauses can help to recognize visual feedback related to face-to-face grounding.
- *Emotional feedback.* Emotional feedback visual cues indicate the emotional state of a person. Facial expression is an emotional feedback cue used to show one of the six basic emotions (Ekman 1992) such as happiness or anger. For this kind of feedback, contextual information can be used to anticipate a person's facial expression. For example, a person smiles after receiving a compliment.

In general, our goal is to efficiently integrate dialogue context information from an embodied agent with a visual analysis module. We define a visual analysis module as a software component that can analyze images (or video sequences) and recognize visual feedback of a human participant during interaction with an embodied agent. The next step is to determine which information already exists in most ECA architectures.

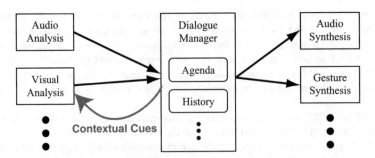

Figure 7.2 Simplified architecture for embodied conversational agent. Our method integrates contextual information from the dialogue manager inside the visual analysis module

7.4 Context from Dialogue Manager

Figure 7.2 is a general view of the architecture for an embodied conversational agent.[1] In this architecture, the dialogue manager contains two main subcomponents, an agenda and a history. The agenda keeps a list of all the possible actions the agent and the user (human participant) can do next. This list is updated by the dialogue manager based on its discourse model (prior knowledge) and on the history. Some useful contextual cues can be estimated from the agenda:

- What will be the next spoken sentence of our embodied agent?
- Are we expecting some specific answers from the user?
- Is the user expected to look at some common space?

The history keeps a log of all the previous events that happened during the conversation. This information can be used to learn some interesting contextual cues:

- How did the user answer previous questions (speech or gesture)?
- Does the user seem to understand the last explanation?

Based on the history, we can build a prior model about the type of visual feedback shown by the user. Based on the agenda, we can predict the type of visual feedback that will be shown by the user.

The simplified architecture depicted in Figure 7.2 highlights the fact that the dialogue manager already processes contextual information in order to produce output for the speech and gesture synthesizer. The main idea is to use this existing information to predict when visual feedback gestures from the user are likely. Since the dialogue manager is already merging information from the input devices with the history and the discourse model, the output of the dialogue manager will contain useful contextual information.

We highlight four types of contextual features easily available in the dialogue manager:

- *Lexical features*. These features are computed from the words said by the embodied agent. By analyzing the word content of the current or next utterance, one should be able to anticipate certain visual feedback. For example, if the current spoken utterance started with "Do you", the interlocutor will most likely answer using affirmation or negation. In this case, it is also likely to see visual feedback like a head nod or a head shake. On the other hand, if the current spoken utterance started with "What", then it's

---

[1] In our work we use the COLLAGEN conversation manager (Rich *et al.* 2001), but other dialogue managers provide these components as well.

unlikely to see the listener head shake or head nod – other visual feedback gestures (e.g., pointing) are more likely in this case.

- *Prosody and punctuation.* Prosody can also be an important cue to predict gesture displays. We use punctuation features output by the dialogue system as a proxy for prosody cues. Punctuation features modify how the text-to-speech engine will pronounce an utterance. Punctuation features can be seen as a substitute for more complex prosodic processing that are not yet available from most speech synthesizers. A comma in the middle of a sentence will produce a short pause, which will most likely trigger some feedback from the listener. A question mark at the end of the sentence represents a question that should be answered by the listener. When merged with lexical features, the punctuation features can help recognize situations (e.g., yes/no questions) where the listener will most likely use head gestures to answer.

- *Timing.* This is an important part of spoken language and information about when a specific word is spoken or when a sentence ends is critical. This information can aid the ECA to anticipate visual grounding feedback. People naturally give visual feedback (e.g., head nods) during pauses of the speaker as well as just before the pause occurs. In natural language processing (NLP), lexical and syntactic features are predominant, but for face-to-face interaction with an ECA, timing is also an important feature.

- *Gesture display.* Gesture synthesis is a key capability of ECAs and it can also be leveraged as a context cue for gesture interpretation. As described in Cassell and Thorisson (1999), visual feedback synthesis can improve the engagement of the user with the ECA. The gestures expressed by the ECA influence the type of visual feedback from the human participant. For example, if the agent makes a deictic gesture, the user is more likely to look at the location that the ECA is pointing to.

The following section presents our framework for integrating contextual information with the visual observations, and section 7.6 describes how we can automatically extract lexical, prosodic, timing, and gesture features from the dialogue system.

7.5 Framework for Context-based Gesture Recognition

We use a two-stage discriminative classification scheme to integrate interaction context with visual observations and detect gestures. A two-stage scheme allows us the freedom to train the context predictor and vision-based recognizer independently, potentially using corpora collected at different times. Figure 7.3 depicts our complete framework.

Our context-based recognition framework has three main components: vision-based recognizer, contextual predictor, and multimodal integrator. In the vision-based gesture recognizer, we compute likelihood measurements of head gestures. In the contextual predictor, we learn a measure of the likelihood of certain visual gestures given the current contextual feature values. In the multimodal integrator, we merge context-based predictions with observations from the vision-based recognizer to compute the final recognition estimates of the visual feedback.

The input of the contextual predictor is a feature vector $\mathbf{x}_j$ created from the concatenation of all contextual features at frame j. Each contextual value is a real value encoding a specific aspect of the current context. For example, one contextual feature can be a binary value (0 or 1) telling if the last spoken utterance contained a question mark. The details on how these contextual features are encoded are described in section 7.6.

The contextual predictor should output a likelihood measurement at the same frame rate as the vision-based recognizer so the multimodal integrator can merge both measurements. For this reason, feature vectors $\mathbf{x}_j$ should also be computed at every frame j (even though the contextual features do not directly depend on the input images). One of the advantages of our late-fusion approach is that, if the contextual information and the feature vectors are temporarily unavailable, then the multimodal integrator can

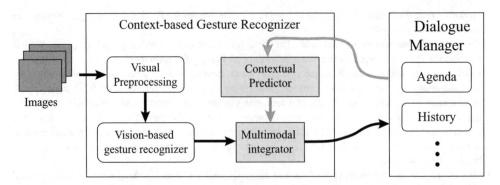

Figure 7.3 Framework for context-based gesture recognition. The contextual predictor translates contextual features into a likelihood measure, similar to the visual recognizer output. The multimodal integrator fuses these visual and contextual likelihood measures. The system manager is a generalization of the dialogue manager (conversational interactions) and the window manager (window system interactions)

recognize gestures using only measurements made by the vision-based recognizer. It is worth noting that the likelihood measurements can be a probabilities or a "confidence" measurement (as output by SVMs).

As shown in Figure 7.3, the vision-based gesture recognizer takes inputs from a visual preprocessing module. The main task of this module is to track head gaze using an adaptive view-based appearance model (see Morency *et al.* (2003) for details). This approach has the advantage of being able to track subtle movements of the head for a long periods of time. While the tracker recovers the full 3D position and velocity of the head, we found features based on angular velocities were sufficient for gesture recognition.

The multimodal integrator merges context-based predictions with observations from the vision-based recognizer. We adopt a late fusion approach because data acquisition for the contextual predictor is greatly simplified with this approach, and initial experiments suggested performance was equivalent to an early, single-stage integration scheme. Most recorded interactions between human participants and conversational robots do not include estimated head position; a late fusion framework gives us the opportunity to train the contextual predictor on a larger data set of linguistic features.

Our integration component takes as input the margins from the contextual predictor described earlier in this section and the visual observations from the vision-based head gesture recognizer, and recognizes whether a head gesture has been expressed by the human participant. The output from the integrator is further sent to the dialogue manager or the window manager so it can be used to decide the next action of the ECA.

In our experiments, we used Support Vector Machines (SVMs) to train the contextual predictor and the multimodal integrator. We estimate the likelihood measurement of a specific visual gesture using the margin of the feature vector $\mathbf{x}_j$. During training, the SVM finds a subset of feature vectors, called support vectors, that optimally define the boundary between labels. The margin $m(\mathbf{x}_j)$ of a feature vector $\mathbf{x}_j$ can be seen as the distance between the $\mathbf{x}_j$ and the boundary, inside a kernel space K. The margin $m(\mathbf{x}_j)$ can easily be computed given the learned set of support vectors $\mathbf{x}_k$, the associated set of labels y_k and weights w_k, and the bias b:

$$m(x) = \sum_{k=1}^{l} y_k w_k K(\mathbf{x}_k, \mathbf{x}_j) + b \qquad (7.1)$$

where l is the number of support vectors and $K(\mathbf{x}_k, \mathbf{x}_j)$ is the kernel function. In our experiments we used a radial basis function (RBF) kernel:

$$K(\mathbf{x}_k, \mathbf{x}_j) = e^{-\gamma \|\mathbf{x}_k - \mathbf{x}_j\|^2} \tag{7.2}$$

where γ is the kernel smoothing parameter learned automatically using cross-validation on our training set.

7.6 Contextual Features

In this section we describe how contextual information is processed to compute feature vectors $\mathbf{x}_j$. In our framework, contextual information is inferred from the input and output events of the *dialogue manager* (see Figure 7.3 and section 7.5).

We tested two approaches to send event information to the contextual predictor: (1) an active approach where the system manager is modified to send a copy of each relevant event to the contextual predictor, and (2) a passive approach where an external module listens to all the input and output events processed by the system manager and a copy of the relevant events is sent to the contextual predictor. In the contextual predictor, a preprocessing module receives the contextual events and outputs contextual features. Note that each event is accompanied by a time-stamp and optionally a duration estimate.

In our framework, complex events are split into smaller sub-events to increase the expressiveness of our contextual features and to have a consistent event formatting. For example, the next spoken utterance event sent from the conversational manager will be split into sub-events including words, word pairs, and punctuation elements. These sub-events will include the original event information (time-stamp and duration) as well as the relative timing of the sub-event.

The computation of contextual features should be fast so that context-based recognition can happen online in real time. We use two types of functions to encode contextual features from events: (1) binary functions and (2) ramp functions.

A contextual feature encoded using a binary function will return 1 when the event starts and 0 when it ends. This type of encoding supposes that we know the duration of the event or that we have a constant representing the average duration. It is well suited for contextual features that are less time-sensitive. For example, the presence of the word pair "do you" in an utterance is a good indication of a yes/no question but the exact position of this word pair is not as relevant.

A ramp function is a simple way to encode the time since an event happened. We experimented with both negative slope (from 1 to 0) and positive slope (from 0 to 1) but did not see any significant difference between the two types of slopes. A ramp function is well suited for contextual features that are more time-sensitive. For example, a grounding gesture such as a head nod is most likely to happen closer to the end of a sentence than the beginning.

The following sub-section gives specific examples of our general framework for contextual feature encoding applied to conversational interfaces.

Conversational Interfaces

The contextual predictor receives the avatar's spoken utterance and automatically processes them to compute contextual features. Four types of contextual features are computed: lexical features, prosody and punctuation features, timing information, and gesture displays. In our implementation, the lexical feature relies on an extracted word pair (two words that occur next to each other, and in a particular order) since they can efficiently be computed given the transcript of the utterance.

While a range of word pairs may be relevant to context-based recognition, we currently focus on the single phrase "do you". We found this feature is an effective predictor of a yes/no question in many of our training dialogues. Other word pair features will probably be useful as well (e.g., "have you", "will you", "did you"), and could be learned from a set of candidate word pair features using a feature selection algorithm.

We extract word pairs from the utterance and set the following binary feature:

$$f_{\text{"do you"}} = \begin{cases} 1 & \text{if word pair "do you" is present} \\ 0 & \text{if word pair "do you" is not present} \end{cases}$$

The punctuation feature and gesture feature are coded similarly:

$$f_? = \begin{cases} 1 & \text{if the sentence ends with "?"} \\ 0 & \text{otherwise} \end{cases}$$

$$f_{\text{look\_left}} = \begin{cases} 1 & \text{if a "look left" gesture happened during the utterance} \\ 0 & \text{otherwise} \end{cases}$$

The timing contextual feature f_t represents proximity to the end of the utterance. The intuition is that verbal and nonverbal feedback most likely occurs at pauses or just before. This feature can easily be computed given only two values: t_0, the utterance start-time, and δ_t, the estimated duration of the utterance. Given these two values for the current utterance, we can estimate f_t at time t using:

$$f_t(t) = \begin{cases} 1 - \left| \frac{t-t_0}{\delta_t} \right| & \text{if } t \leq t_0 + \delta_t \\ 0 & \text{if } t > t_0 + \delta_t \end{cases}$$

We selected our features so that they are topic independent. This means that we should be able to learn how to predict visual gestures from a small set of interactions and then use this knowledge on a new set of interactions with a different topic discussed by the human participant and the ECA. However, different classes of dialogues might have different key features, and ultimately these should be learned using a feature selection algorithm (this is a topic of future work).

The contextual features are evaluated for every frame acquired by the vision-based recognizer module. The lexical, punctuation and gesture features are evaluated based on the current spoken utterance. A specific utterance is active until the next spoken utterance starts, which means that in-between pauses are considered to be part of the previous utterance. The top three graphs of Figure 7.4 show how two sample utterances from our user study (described in section 7.7) will be coded for the word pair "do you", the question mark, and the timing feature.

A total of 236 utterances were processed to train the multiclass SVM used by our contextual predictor. Positive and negative samples were selected from the same data set based on manual transcription of head nods and head shakes. Test data was withheld during evaluation in all experiments in this chapter.

Figure 7.4 also displays the output of our trained contextual predictor for anticipating head nods and head shakes during the dialogue between the robot and a human participant. Positive margins represent a high likelihood for the gesture. It is noteworthy that the contextual predictor automatically learned that head nods are more likely to occur around the end of an utterance or during a pause, while head shakes are more likely to occur after the completion of an utterance. It also learned that head shakes are directly correlated with the type of utterance (a head shake will most likely follow a question), and that head nods can happen at the end of a question (i.e., to represent an affirmative answer) and can also happen at the end of a normal statement (i.e., to ground the spoken utterance).

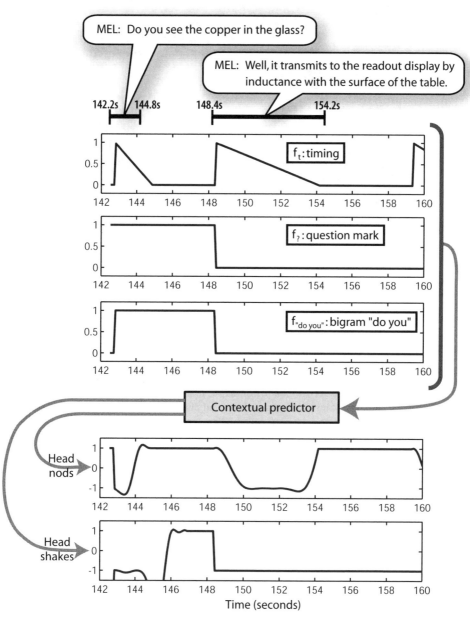

Figure 7.4   Prediction of head nods and head shakes based on three contextual features: (1) distance to end-of-utterance when ECA is speaking, (2) type of utterance, and (3) lexical bigram feature. We can see that the contextual predictor learned that head nods should happen near or at the end of an utterance or during a pause while head shakes are most likely at the end of a question

Figure 7.5   Mel, the interactive robot, can present the iGlassware demo (table and copper cup on its right) or talk about its own dialogue and sensorimotor abilities

7.7 Context-based Head Gesture Recognition

The following experiment demonstrates how contextual features inferred from an agent's spoken dialogue can improve head nod and head shake recognition. The experiment compares the performance of the vision-only recognizer with the context-only prediction and with multimodal integration.

For this experiment, a first data set was used to train the contextual predictor and the multimodal integrator (the same data set as described in section 7.5), while a second data set with a different topic was used to evaluate the head gesture recognition performance. In the training data set, the robot interacted with the participant by demonstrating its own abilities and characteristics. This data set, called Self, contains seven interactions. The test data set, called iGlass, consists of nine interactions of the robot describing the iGlassware invention (~340 utterances).

For both data sets, human participants were video recorded while interacting with the robot (Figure 7.5). The vision-based head tracking and head gesture recognition was run online (~18 Hz). The robot's conversational model, based on COLLAGEN (Rich *et al.* 2001), determines the next activity on the agenda using a predefined set of engagement rules, originally based on human–human interaction (Sidner *et al.* 2005). Each interaction lasted between 2 and 5 minutes.

During each interaction, we also recorded the results of the vision-based head gesture recognizer as well as the contextual cues (spoken utterances with start-time and duration) from the dialogue manager. These contextual cues were later automatically processed to create the contextual features (see section 7.6) necessary for the contextual predictor (see section 7.5).

For ground truth, we hand-labeled each video sequence to determine exactly when the participant nodded or shook his/her head. A total of 274 head nods and 14 head shakes were naturally performed by the participants while interacting with the robot.

Results

Our hypothesis was that the inclusion of contextual information within the head gesture recognizer would increase the number of recognized head nods while reducing the number of false detections. We tested three different configurations: (1) using the vision-only approach, (2) using only the contextual

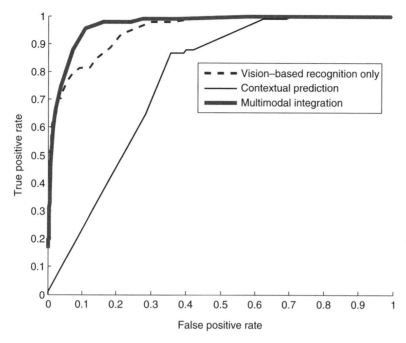

Figure 7.6   Head nod recognition curves when varying the detection threshold

information as input (contextual predictor), and (3) combining the contextual information with the results of the visual approach (multimodal integration).

Figure 7.6 shows head nod detection results for all nine subjects used during testing. The ROC curves present the detection performance of each recognition algorithm when varying the detection threshold. The areas under the curve for each technique are 0.9482 for the vision-only, 0.7691 for the predictor, and 0.9678 for the integrator.

Figure 7.7 shows head shake detection results for each recognition algorithm when varying the detection threshold. The areas under the curve for each technique are 0.9780 for the vision-only, 0.4961 for the predictor, and 0.9872 for the integrator.

Table 7.7 summarizes the results from Figures 7.6 and 7.7 by computing the true positive rates for the fixed negative rate of 0.1. Using a standard analysis of variance (ANOVA) on all the subjects, results on the head nod detection task showed a significant difference among the means of the three methods of detection: $F(1, 8) = 62.40$, $p < 0.001$, $d = 0.97$. Pairwise comparisons show a significant difference between all pairs, with $p < 0.001$, $p = 0.0015$, and $p < 0.001$ for vision-predictor, vision-integrator, and predictor-integrator respectively. A larger number of samples would be necessary to see the same significance in head shakes.

We computed the true positive rate using the following ratio:

$$\text{True positive rate} = \frac{\text{Number of detected gestures}}{\text{Total number of ground truth gestures}}$$

A head gesture is tagged as detected if the detector triggered at least once during a time window around the gesture. The time window starts when the gesture starts and ends k seconds after the gesture. The parameter k was empirically set to the maximum delay of the vision-based head gesture recognizer

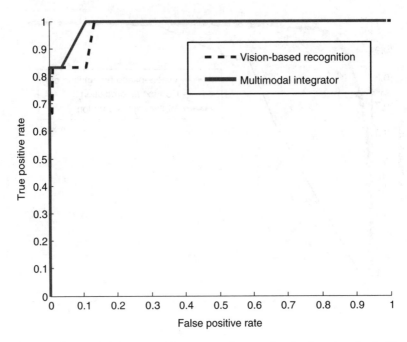

Figure 7.7 Head shake recognition curves when varying the detection threshold

(1.0 second). For the iGlass dataset, the total numbers of ground truth gestures were 91 head nods and 6 head shakes.

The false positive rate is computed at a frame level:

$$\text{False positive rate} = \frac{\text{Number of falsely detected frames}}{\text{Total number of non-gesture frames}}$$

A frame is tagged as falsely detected if the head gesture recognizer triggers and if this frame is outside any time window of a ground truth head gesture. The denominator is the total of frames outside any time window. For the iGlass data set, the total number of non-gestures frames was 18 246 frames and the total number of frames for all nine interactions was 20 672 frames.

Figure 7.8 shows the head nod recognition results for a sample dialogue. When only vision is used for recognition, the algorithm makes a mistake at around 101 seconds by false detecting a head nod. Visual grounding is less likely during the middle of an utterance. By incorporating the contextual information, our context-based gesture recognition algorithm is able to reduce the number of false positives. In Figure 7.8 the likelihood of a false head nod happening is reduced.

Table 7.1   True detection rates for a fix false positive rate of 0.1

	Vision	Predictor	Integrator
Head nods	81%	23%	**93%**
Head shakes	83%	10%	**98%**

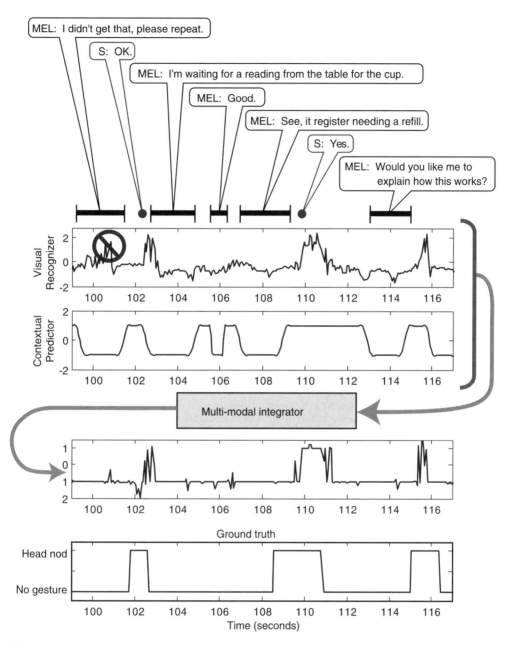

Figure 7.8 Head nod recognition results for a sample dialogue. The last graph displays the ground truth. We can observe at around 101 seconds (circled and crossed in the top graph) that the contextual information attenuates the effect of the false positive detection from the visual recognizer

7.8 Conclusions

Our results show that contextual information can improve user gesture recognition for interactions with embodied conversational agents. We presented a prediction framework that extracts knowledge from the spoken dialogue of an embodied agent to predict which head gesture is most likely. By using simple lexical, prosodic, timing, and gesture context features, we were able to improve the recognition rate of the vision-only head gesture recognizer from 81% to 93% for head nods and from 83% to 98% for head shakes. As future work, we plan to experiment with a richer set of contextual cues including those based on gesture display, and to incorporate general feature selection to our prediction framework so that a wide range of potential context features can be considered and the optimal set determined from a training corpus.

References

Bickmore T. and Cassell J. (2004) Social dialogue with embodied conversational agents. In J. van Kuppevelt, L. Dybkjaer and N. Bernsen (eds), *Natural, Intelligent and Effective Interaction with Multimodal Dialogue Systems*. Kluwer Academic.

Breazeal C., Hoffman G. and Lockerd A. (2004) Teaching and working with robots as a collaboration. *Proceedings Third International Conference on Autonomous Agents and Multi-Agent Systems* AAMAS, ACM Press, pp. 1028–1035.

Cassell J. and Thorisson K.R. (1999) The poser of a nod and a glance: envelope vs. emotional feedback in animated conversational agents. *Applied Artificial Intelligence*.

de Carolis B., Pelachaud C., Poggi I. and de Rosis F. (2001) Behavior planning for a reflexive agent. *Proceedings of IJCAI*, Seattle.

Dillman R., Ehrenmann M., Steinhaus P., Rogalla O. and Zoellner R. (2002) Human friendly programming of humanoid robots: the German Collaborative Research Center. *Third IARP Intenational Workshop on Humanoid and Human-Friendly Robotics*, Tsukuba Research Centre, Japan.

Dillman R., Becher R. and Steinhaus P. (2004) ARMAR II: a learning and cooperative multimodal humanoid robot system. *International Journal of Humanoid Robotics* 1 (1), 143–155.

Ekman P. (1992) An argument for basic emotions. *Cognition and Emotion* 6 (34), 169–200.

Fujie S., Ejiri Y., Nakajima K., Matsusaka Y. and Kobayashi T. (2004) A conversation robot using head gesture recognition as para-linguistic information. In *Proceedings of 13th IEEE International Workshop on Robot and Human Communication* (RO-MAN), pp. 159–164.

Kapoor A. and Picard R. (2001) A real-time head nod and shake detector. In *Proceedings of the Workshop on Perspective User Interfaces*.

Lemon O., Gruenstein A. and Peters S. (2002) Collaborative activities and multi-tasking in dialogue systems. *Traitement Automatique des Langues (TAL), special issue on dialogue* 43 (2), 131–154.

Morency L.P., Rahimi A., Checka N. and Darrell T. (2002) Fast stereo-based head tracking for interactive environment. In *Proceedings of International Conference on Automatic Face and Gesture Recognition*, pp. 375–380.

Morency L.P., Rahimi A. and Darrell T. (2003) Adaptive view-based appearance model. In *Proceedings of IEEE Conference on Computer Vision and Pattern Recognition*, vol. 1, pp. 803–810.

Nakano Y., Reinstein G., Stocky T. and Cassell J. (2003) Towards a model of face-to-face grounding. In *Proceedings of Annual Meeting of the Association for Computational Linguistics*, Sapporo, Japan.

Rich C., Sidner C. and Lesh N. (2001) Collagen: applying collaborative discourse theory to human–computer interaction. *AI Magazine, special issue on intelligent user interfaces* 22 (4), 15–25.

Sidner C., Lee C., Kidd C.D., Lesh N. and Rich C. (2005) Explorations in engagement for humans and robots. *Artificial Intelligence* 166 (1–2), 140–164.

Siracusa M., Morency L.P., Wilson K., Fisher J. and Darrell T. (2003) Haptics and biometrics: a multimodal approach for determining speaker location and focus. In *Proceedings of 5th International Conference on Multimodal Interfaces*.

Stiefelhagen R. (2002) Tracking focus of attention in meetings. In *Proceedings of International Conference on Multimodal Interfaces*.

Takemae Y., Otsuka K. and Mukaua N. (2004) Impact of video editing based on participants' gaze in multiparty conversation. *Extended Abstract of CHI'04*.

Torralba A., Murphy K.P., Freeman W.T. and Rubin M.A. (2003) Context-based vision system for place and object recognition. *IEEE Intl. Conference on Computer Vision* (ICCV), Nice, France.

Traum D. and Rickel J. 2002 Embodied agents for multi-party dialogue in immersive virtual world. In *Proceedings of the International Joint Conference on Autonomous Agents and Multi-agent Systems* (AAMAS 2002), pp. 766–773.

8

Trading Spaces: How Humans and Humanoids Use Speech and Gesture to Give Directions

Stefan Kopp, Paul A. Tepper, Kimberley Ferriman, Kristina Striegnitz, and Justine Cassell

8.1 Introduction

When describing a scene, or otherwise conveying information about objects and actions in space, humans make frequent use of gestures that not only supplement, but also complement the information conveyed in language. Just as people may draw maps or diagrams to illustrate a complex spatial layout, they may also use their hands quite spontaneously to represent spatial information. For example, when asked how to find a particular address, it is common to see a direction-giver depicting significant landmarks with the hands – the fork where one road joins another, the shape of remarkable buildings, or their spatial relationship to one another.

Figure 8.1 shows just such an example of spontaneous gesture in direction-giving. Here, the speaker has just said "If you were to go south", then, while making this two-handed gesture, he says "there's a church". The gesture imparts visual information to the description, the shape of the church (left hand) and its location relative to the curve of a road (represented by the right arm). This meaning is instrumental to the understanding of the scene, and the listener took the gesture as intrinsic to the description.

But how do these gesture–speech combinations communicate? Unlike language, coverbal gesture does not rely upon a lexicon of stable form–meaning pairings. And in virtual humans that are programmed to give directions to humans, how do we generate these same kinds of direction-giving units of speech and gesture? To date, research in this area has only scratched the surface of this problem, since previous ECAs have drawn upon a "gestionary" or lexicon of predefined gestures. While this approach can allow for the automatic generation of utterances, complemented by coordinated gestures, it provides neither an explanatory model of the behavior, nor the aspect of gesture that makes it so valuable – the communicative power and flexibility of our ever-present, visual, spatial modality.

Conversational Informatics: An Engineering Approach   Edited by Toyoaki Nishida
© 2007 John Wiley & Sons, Ltd

Figure 8.1   Coverbal gesture on "There's a church". Reproduced by permission of the ACM, Kopp *et al.,* 2004

In this chapter, we start with a discussion of the role of the two generative systems of speech and gesture for giving directions. We will show that there are well-known, fundamental differences between the kinds of information expressed by the two modalities (i.e., between the semantics of natural language and the meaning of speech-accompanying gestures), and that these differences must be taken into account in order to understand and model the behavior. As posited by other researchers working on spatial language, we agree that this distinction necessitates the addition of an additional level of representation for spatial and visual information, beyond the two-level models of form and meaning seen in language. We report on an empirical study on spontaneous gesture in direction-giving, whose results provide evidence for the existence of patterns in the way humans compose their representational gestures out of morphological features, encoding meaningful geometric and spatial properties themselves. It also suggests a qualitative, feature-based framework for describing these properties at the newly introduced level, which we call the *image description feature* level (IDF). We then show how this approach allows us to model multimodal generation in conversational agents by extending existing techniques for natural language generation. It lends itself to a smooth integration of gesture generation into a system for *microplanning* of language and gesture, wherein linguistic meaning and structure can be coordinated with gesture meaning and structure at various levels. This approach allows us to capture the differences between the two systems, while simultaneously providing a means to model gesture and language as two intertwined facets of a single communicative system.

8.2   Words and Gestures for Giving Directions

Spatial descriptions including speech and gesture figure in descriptions of motion (Cassell and Prevost 1996), in descriptions of houses (Cassell *et al.* 2000), in descriptions of object shape (Sowa and Wachsmuth 2003), in descriptions of routes (Tversky and Lee 1998, 1999), and in descriptions of assembly procedures (Daniel *et al.* 2003; Heiser *et al.* 2004; Heiser and Tversky 2003; Heiser *et al.* 2003; Lozano and Tversky 2004). The kinds of gestures involved in these descriptions are mainly *iconic gestures*, where the form of the gesture bears a visual resemblance to what is referred to in speech and gesture, and *deictic gestures* that point out or indicate a path. Both kinds of gestures belong to the class of coverbal gestures, hand movements that don't have a meaning independent of the speech with which they occur in temporal coordination (McNeill 1992). In this respect they are fundamentally different from emblematic gestures, which are conventionalized and differ from culture to culture, or sign language movements, whose form–meaning relation is standardized, determined by convention, and independent of the context. In this chapter we focus on iconic gestures, which are often an intrinsic part of a spatial description.

Across all kinds of spatial description, researchers have found evidence that communicative behaviors portray a single underlying conceptual representation. For example, in an investigation of manner-of-motion verbs and gesture in describing motion events, Cassell and Prevost (1996) found ample evidence that the same concept may, in different situations, result in different realizations at the level of lexical items and paired gestures (e.g., "walked" vs. "went"+gesture). This suggests that communicative content may be conceived of in terms of semantic components that can be distributed across the modalities. In this study, roughly 50% of the semantic components of the described events were observed to be encoded redundantly in gestures *and* speech, while the other 50% were expressed non-redundantly either by speech or by gesture.

Similarly, an experiment on house descriptions (Cassell *et al.* 2000) demonstrated that properties like shape, location, relative position, and path could be discerned in both speech and gesture. The particular distribution of properties, however, appeared to depend not only on the nature of the object described but also on the discourse context. While the location of an object was redundantly conveyed by speech and gesture, properties such as contrast between two objects, their shape, location, relative position, and path through space most frequently occurred only in gesture while the existence of the objects was conveyed in co-occurring speech (e.g., "there was a porch" with a gesture describing the curved shape of the porch).

Route directions are a particular kind of spatial description designed to assist a traveler in finding a way from point A to point B in an unknown environment. A route description is typically organized into a set of route segments that connect important points, and a set of actions – reorientation (turning), progression, or positioning – one of which is taken at the end of each segment. In order to ensure that the listener will be able to recognize the segments and accomplish the correct actions, the speaker refers to significant *landmarks* (Denis 1997). Landmarks are chosen for mention based on perceptual and conceptual salience (Conklin and McDonald 1982), informative value for the actions to be executed, as well as visibility, pertinence, distinctiveness, and permanence (Couclelis 1996).

When describing routes through environments too large to be taken in at a single glance, speakers adopt either a *survey* or a *route* perspective, or a mixture of both (Taylor and Tversky 1992, 1996). These two kinds of perspectives in direction-giving result in two different patterns of gestures. When giving direction in the *route* perspective, speakers take their listeners on an imaginary tour of the environment, describing the locations of landmarks in terms of relative directions with respect to the traveler's changing position; left, right, front, and back (e.g., "you walk straight ahead"). In the *survey* perspective, speakers adopt a bird's eye viewpoint, and locate landmarks with respect to one another, in terms of an extrinsic or absolute reference frame, typically cardinal diections; north, south, east, west (e.g., "the house is south of the bridge"). Gestures during a *route* description tend to be in an upright plane in front of the body whereas gestures during a *survey* description are on a table-top plane (Emmorey *et al.* 2001).

In all of these cases, however, iconic gestures do not communicate independent of speech, for their meaning depends on the linguistic context in which they are produced. And listeners are unable to remember the form of gestures that they have seen in conversation (Krauss *et al.* 1991), although they do attend to the information conveyed in gesture, and integrate it into their understanding of what was said (Cassell *et al.* 1999). How then, do we understand gesture?

8.2.1 Arbitrary and Iconic Signs

Despite their variety and complexity, when the findings described above have been applied to embodied conversational agents, researchers have followed the gestionary approach; that is, ECAs have been limited to a finite set of gestures, each of which conveys a predefined meaning. This is equivalent to a lexicon, wherein each word conveys a predefined meaning. The first step to move beyond this approach, both in understanding human behavior, and in computational models of virtual humans, is to realize that there are fundamental differences between the way iconic gestures and words convey meaning – differences that already form the basis of the study of signs by semioticians like Saussure (1985) and Peirce (1955).

Words are *arbitrarily* linked to the concepts they represent. They can be used to convey meaning because they are conventionalized symbols ("signifiers"), just like the manual signs of a sign language, agreed upon by members of a linguistic community. Conversely, iconic gestures communicate through *iconicity*; that is, in virtue of their resemblance to the information they depict. These two "semiotic vehicles" also bear a markedly different relation to the context in which they occur. No matter what the context or the particular lexical semantics, a word always has a limited number of specific meanings. In contrast, an iconic gesture is *underspecified* (or indeterminate) from the point of view of the observer. That is, an iconic gesture has a potentially countless number of interpretations, or images that it could depict (Poesio 1996) and is almost impossible to interpret outside of the context of the language it co-occurs with. For example, even limiting it to depictions of the concrete, the gesture shown in Figure 8.1 can be used to illustrate the vertical movement of an object, the shape of a tower, the relative location of two objects, or a reenactment of a character performing some action. Clearly, it does not make sense to say that a gesture – observed as a stand-alone element separate from the language it occurs with – has semantics in the same way as language does when interpreted within linguistic context.

Even if an iconic gesture by itself does not uniquely identify an entity or action in the world, it depicts (or specifies) features of an image through some visual or spatial resemblance. As a result, we refer to the gesture as underspecified, and it is precisely this underspecification that is missing from most models of gestural meaning.[1] To account for how iconic gestures are able to express meaning, one must have accounts of both *how images are mentally linked to entities* (referents) and *how gestures can depict images*. Therefore, to provide a way to link gestures to their referents, a third, intermediate level of abstraction and representation that accounts for a context-independent level of visuo-spatial meaning is required. While its application to gesture is novel, the idea of such a spatial representation is not new. For example, Landau and Jackendoff (1993) discuss it as "a level of mental representation devoted to encoding the geometric properties of objects in the world and the spatial relationships among them" (p. 217). Further, it has been suggested that this cognitive representation is modality-independent (amodal); that is, it underlies the processing of spatial information in different modalities, being drawn upon by language, gesture and the motor system more generally.

8.2.2 Systematicity from Iconicity

If iconic gestures are indeed communicative, people must be able to recover and interpret their meaning and there must be a process by which we encode and decode information in gesture. A reliable system for this requires some systematicity in the way gesture is used to depict, and the evidence from previous literature in several domains indeed suggests that patterns exist in the form and function of iconic gestures with respect to expressing spatial information and communicating meaning more generally. Sowa and Wachsmuth (2003) report that one can find consistencies in the ways the fingers are used to trace a shape and that, for example, both palms may be held facing each other to depict an object's extent. Unlike language, in gesture multiple form features may be combined to express multiple spatial aspects (e.g., extent and shape) simultaneously. Emmorey *et al.* (2001) observed that depictions of complex spatial structures are broken down into features that are then built up again by successive gestures. The fact that a single spatial structure is referred to across gestures (e.g., a winding road) is signaled by spatial coherence; that is, the gestures employ the same viewpoint, size scale, and frame of reference, as indicated by a constancy of hand shape, trajectory, and position in space. Sometimes, the frame of reference (e.g., relative to the winding road) is explicitly anchored in gesture space by one hand, and then held throughout while the other hand describes additional landmarks at appropriate relative locations. McNeill and Levy (1982) found positive and negative correlations for the association of distinct "kinesic"

---

[1] See Poesio (2005) for an approach to natural language processing based on underspecification.

features in gesture, like fingers curled, palm down, or motion upwards, with semantic features of the motion verbs the gestures co-occurred with. For example, verbs with a horizontal meaning feature tended to co-occur with gestures with a sideways movement, but almost never with downward motion.

These results suggest that there are prevalent patterns in the ways the hands and arms are used to create iconic, gestural images of the salient, visual aspects of objects/events, and that such patterns may account for the ways human speakers derive novel gestures for objects they are describing for the first time. Further, the generativity that human gesture displays suggests that such patterning or commonality pertains not to the level of gestures as a whole, but to subparts – *features* of shape, spatial properties, or spatial relationships that are associated with more primitive *form features* of gesture morphology, like hand shapes, orientations, locations, movements in space, or combinations thereof. Consequently, it can be hypothesized that the intermediate level of meaning, which explicates the imagistic content of an iconic gesture, consists of separable, qualitative features describing the meaningful geometric and spatial properties of entities. We call these descriptors *image description features* (henceforth, IDFs).

Landau and Jackendoff (1993), as well as several others in this area, posit a range of geometric entities and spatial relations that seem to exist in such a mental spatial representation, based on analytical studies of linguistic data (Herskovits 1986; Talmy 2000). They show that the semantics of linguistic structures, such as prepositions, dimensional terms or named objects, can be described in terms of entities, axes, and spatial relations. These abstract spatial and visual features form the basis for the kinds of IDFs we have proposed. The basic assumption is that, although human mental representations are not necessarily feature-based, a qualitative feature-based approach will be adequate for describing these spatial and imagistic aspects. Such techniques have also been successfully employed in research on spatial reasoning systems in the artificial intelligence literature (Forbus 1983). This approach has two advantages over a quantitative, analog representation, for example, using 3D graphics. First, using quantitative features requires that analog models be annotated or linked to logical or semantic representations of what they depict. This is similar to the arbitrary way in which words are linked to their meanings in language. Second, our approach uses the same symbols at every level of the representation, so that processing of natural language semantics, gesture meaning, and knowledge representation can all be carried out in a single, underlying representation. These advantages will be exploited in the computational model described below.

IDFs can be considered links between gestures and the images they depict. They are features, which can be used to describe the visuo-spatial aspects of both a gesture's morphology and the entities to which a gesture can refer. So just as certain objects can be abstracted away from into classes based on visual and spatial features, certain hand shapes can be grouped into iconic categories, describing the image that they can carry. For example, just as *lakes, roads* and *walls* fall into the abstract class of *surfaces* or *planes*, likewise, certain hand shapes can be thought of as flat (2D), while others seem to have volume (3D). This allows for treating iconic gestures as composed of sets of one or more morphological features that convey sets of one or more IDFs.

To illustrate this view, let us return to the utterance example in Figure 8.1. The subject's right hand is held in place from the previous utterance and represents the curve in a road, anchoring the frame of reference. In the left-hand gesture, there are three morphological features: the flat hand shape with slightly bent fingers, the vertical linear trajectory, and the hand location relative to the right-hand gesture. Supposing that each form feature corresponds to one or more IDFs in virtue of the resemblance of the former to the latter, the gesture can be analyzed as follows: the relatively flat hand shape resembles, in descriptive spatial terms, a two-dimensional, planar shape in a certain orientation. The vertical, linear trajectory shape corresponds to a feature that marks a vertical extent. Finally, the gesture location corresponds to a relative spatial location. All three IDFs in combination define an upright plane with a significant vertical extent, in a particular orientation and location. However, this inherent content of the iconic gesture does not suffice for a successful interpretation. Only when the gesture is placed in linguistic context does the set of possible interpretations of the IDFs become so constrained as to make it unique. In this example, the indefinite noun phrase "a church" implies that the IDFs represent spatial information about the referent of the expression, namely a church. Linking the underspecified, imagistic features to this specific referent makes a successful interpretation possible, and we arrive at what McNeill (1992) deems the *global-synthetic* property of gesture, namely, that "the meanings of the parts of the gesture are determined by the whole (*global*), and different meaning segments are synthesized into a single gesture (*synthetic*)" (p. 41). The synthetic aspect is captured in some detail by the IDF approach; the global meaning of the gesture implies that the depicted upright plane becomes the wall of the church, viewed relative to the location of the road, and the vertical trajectory emphasizes the salient, vertical dimension, now corresponding to the height of the wall. Overall, we conclude that the communicative intention of the speaker was to introduce a church, which has a tall, upright wall, and which is located near the curve of the road.

Importantly, the IDF approach does not assume a gestionary or lexicon of gestures that consistently refer to objects or actions in the world. On the contrary, it is suggesting that *features* of gestures – like handshapes, particular kinds of trajectories through space, palm orientations – refer to *features* of referents in the world – like flatness in the horizontal plane, small extent, roundness. It is this level of granularity that allows for describing how gestures can communicate, without requiring standards of form or consistent form–meaning pairings. Mappings from IDFs onto form features, however, should be shared among gestures that depict different, but visually similar things. Evidence for this hypothesis, or the systematic use of certain visually similar classes of gestures to depict visually similar classes of referents, would be adduced if one could find correlations between the morphological features that were generally used to depict similar visual or spatial properties. In the next section we describe an exploratory study aimed at providing this evidence.

8.3 Relationship between Form and Meaning of Iconic Gestures in Direction-giving

To date, no literature describes in detail the interaction between iconic gestures and spatial language in route directions. Thus, to test the IDF hypothesis, we collected video and audio recordings of 28 dyads (more than five hours of dialogue) engaging in direction-giving. In each dyad, one person described, without any external aids such as maps, a route from point A to point B on the Northwestern University campus to another person, who was unfamiliar with the campus. As mentioned above, in such route descriptions, speakers refer to a constrained set of entities, including actions like progression (continuing along a path) and reorientation (turning), and objects like landmarks, their parts, shapes, and spatial configurations (Denis 1997). The Northwestern campus is ideal for this task as it provides numerous examples of objects (buildings, gates, bridges, etc.) that can serve as landmarks, while at the same time necessitating extensive and detailed instructions due to its size and complexity. This direction-giving task thus demanded the speaker to communicate complex spatial and visual information only by means of the natural modalities. Consequently, all subjects used coverbal iconic gestures to create representations of the spatial and visual information about landmarks and actions they needed to describe.

In order to examine whether the IDF level of analysis is valid, we *independently* (a) coded gestures into features of their morphology, (b) coded referents in the world into their visuo-spatial features, and then (c) examined if correlations existed between the two sets of features. Since all the directions were given around Northwestern campus, we were able to trace each route through the campus. Maps and photographs of the campus provided an independent source of information about the context of the utterances and gestures. This information allowed us to determine what seemed to be the real-world referent for each phrase (e.g., a specific landmark, or aspects or subparts of it) and to determine its visual, spatial, or geometric properties. To narrow down the scope of our exploratory study, we focused on gestures that seem to depict aspects of the shape of concrete objects, i.e. landmarks, parts of landmarks, streets and paths,[2] as opposed to abstract entities like actions.

---

[2] By path here we mean real paved or dirt paths around campus, not abstract paths of motion or trajectory.

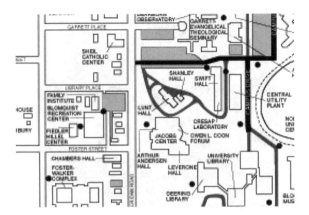

Figure 8.2   Segment of the campus map given to direction-giver for preparation

8.3.1 Method

Twenty-eight undergraduates (11 male and 17 female, all native speakers of English), who were familiar with Northwestern University campus, participated individually as direction-givers. Three undergraduates unfamiliar with the campus participated as direction-followers; in the other 25 dyads, a confederate acted as naïve participant receiving the directions. At the beginning of each experiment, the direction-giver was given a list of ten routes, each of which consisted of five locations on campus to be visited in the order of listing. The starting point was always the building in which the experiment took place, and the first segment, from the starting point to the first waypoint, was identical in each route. The direction-giver was asked to check off every route on the list s/he felt comfortable to give directions for. The subject was provided with the campus map in case s/he needed to look up names of waypoints or locations (Figure 8.2).

One of the checked routes was randomly chosen and assigned to the direction-giver to describe. In order to guarantee comparable conditions, the participant was instructed to familiarize herself with that particular route by then walking it herself. After the direction-giver returned, she was seated face-to-face with the second subject (the direction-follower) in a quiet room. Both participants were instructed to make sure that the direction-follower understood the directions, and they were informed that the follower would have to find the route on her own, right after they concluded the session. Audio and video recordings of four different, synchronized camera views were taken of the dyad, to ensure that detailed coding of language and gesture is possible (Figure 8.3). No time limits were imposed on the dyads.

8.3.2 Coding of Gesture Morphology

The audio and video data was annotated in separate independent passes by a team of coders, using the *PRAAT*[3] and *TASX Annotator* software (Milde and Gut 2002). In the first pass, the words of the direction-giver were transcribed. The next pass was for segmentation, in which the expressive, meaning-bearing phase of each gesture was spotted, and the gesture was classified according to the categories *iconic*, *deictic*, or *iconic+deictic* (all other types of gestures were ignored). In the final pass, the morphology of each included gesture was coded, using a scheme based on the McNeill Coding Manual (McNeill 1992), refined for the purpose of our study. As shown in Figure 8.4, the TASX annotator software was

---

[3] See http://www.fon.hum.uva.nl/praat/

Figure 8.3 Sample of the video data gathered (the arrow is inserted here to indicate the direction of movement)

adapted to allow separate descriptions of the shape, orientation, and location of each hand involved in the gesture:

- *Hand shape* was denoted in terms of ASL (American Sign Language) shape symbols, optionally modified with terms like "loose", "bent", "open", or "spread".
- *Hand orientation* was coded in terms of the direction of an axis orthogonal to the palm, and the direction the fingers would point in if they were extended (Figure 8.5). Both were coded in terms of six speaker-centric, base- or half-axes (Herskovits 1986), namely forward, backward, left, right, up and down. Assuming the left hand in Figure 8.5 is being held straight out in front of the body, it would be described by extended finger direction forward (away from the body) and palm facing left.

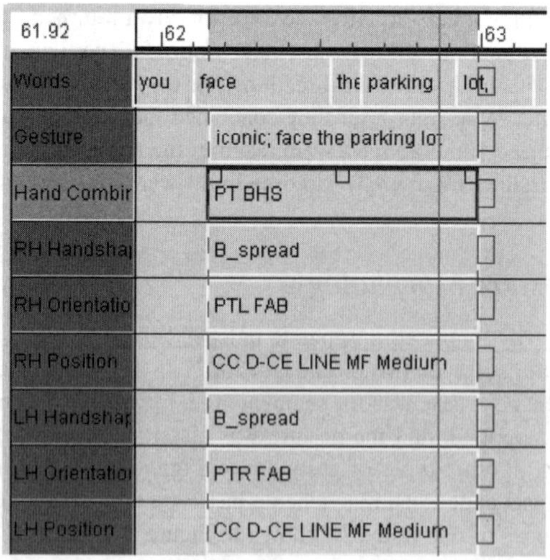

61.92		62			63
Words	you	face	the parking	lot	
Gesture		iconic; face the parking lot			
Hand Combir		PT BHS			
RH Handsha|		B_spread			
RH Orientatio		PTL FAB			
RH Position		CC D-CE LINE MF Medium			
LH Handsha|		B_spread			
LH Orientatio|		PTR FAB			
LH Position		CC D-CE LINE MF Medium			

Figure 8.4 The morphological features of the gesture shown in Figure 8.3 are coded in an annotation window

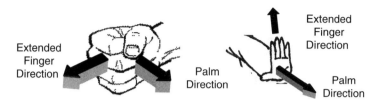

Figure 8.5 Hand orientation defined in terms of extended finger direction and palm direction

Combinations of these features were used to code diagonal or mixed directions (e.g. forward and to the left).

• *Hand location* was described relative to a zoning of the space in front of the gesturer as suggested by McNeill (1992), which determines a position in the frontal body plane. An additional symbol was used to denote the distance between the hand and the body (in contact, between contact and elbow, between elbow and knee, or outstretched).

Change or movement in any of the three features was described using symbols to denote shape (line, arc, circle, chop, or wiggle), direction, and extent. Again, directions were restricted to the six relative directions in space, or pairwise combinations of them. In addition to the three features for each hand, two-handed configurations (e.g., palms together, or finger tips of one hand touching the palm of the other) as well as movements of one hand relative to the other were explicitly denoted (e.g., hands move as mirror images, or one hand is held to anchor a frame of reference while the other hand is active). Figure 8.4 shows the results of morphology coding for the gesture in Figure 8.3: the palms are touching each other ("PT"); the hands are mirror images, yet moving in the same direction ("BHS"); the hands are shaped flat (ASL shape "B_spread"); the fingers are pointed away from the body ("FAB"); the palms are facing toward right/left ("PTR", "PTL"); the hands are positioned at the center of gesture space ("CC"), at a distance between contact and elbow ("D-CE"); a linear movement forward of medium extent is performed ("LINE MF Medium"). Gesture morphology was coded for the first four minutes or more of ten dyads, giving a total of 1171 gestures.

Inter-rater Reliability

It is important in any investigation into the relationship between speech and gesture to be certain that the analysis is not circular, and that coding is rigorous. As far as the first question is concerned, as described above, speech and gesture were coded independently so that the content of the speech did not influence the coding of the morphology of the gesture. In addition, the coding of the morphology was carried out independently from the coding of the referents in the real world, such that neither influenced the other.

As far as the second question is concerned, inter-rater reliability is extremely hard to assess for coding in which there are as many as 12 sub-parts to each coding decision (extent of gesture, kind of gesture, shape of right hand, trajectory of right hand, shape of left hand, location in space of left hand, etc.). We therefore depended on two methods to ensure rigor and assess accuracy. First, all coding, both of gesture and of referents in the world, was carried out by a minimum of two coders, with any disagreements resolved by discussion. Second, accuracy was assessed by an additional experiment, where four subjects were asked to *reproduce* 75 randomly chosen gestures from the data set, solely on the basis of the morphology codes (one subject 15 gestures, three subjects 20 gestures each). All of the participants were generally familiar with gesture, but none had seen any of the original movies or dealt with this data set. After a short practice session to understand our coding manual (five gestures, solely verbal instructions), each participant was videotaped while reproducing the test gestures from their morphology annotations. The video recordings were compared with the original data to assess similarity between the original and

Table 8.1   Criteria used for reproducibility of morphological coding

Hand shape	1: same shape; 2: same shape, but wrongly modified (e.g., "open"); 3: different, but similar shape (e.g., ASL "5" and "B"); 4: otherwise
Hand orientation	1: same orientation; 2: extended finger directions or palm directions differ less than 45°; 3: directions differ more than 45°, but less than 90°; 4: otherwise
Hand position	1: same position; 2: different, but still in the same gesture space region and distance; 3: in adjacent regions; 4: farther away
Movement	1: same movement; 2: same direction and plane of movement, but slightly different extent or shape (e.g., arc, but stronger curved); 3: movements differ considerably in either shape, extent, direction, or plane of movement; 4: movements differ considerably in at least two of the four criteria

the reproduced gestures. Similarity was rated from 1 (identical) to 4 (completely different), separately for hand shape, hand orientation, and hand location (plus movement), based on feature-specific criteria as listed in Table 8.1.

Three gestures, in which subjects chose the wrong hand compared to the original gesture, were excluded from analysis leaving a total of 72 gestures, of which 15 were static postures and 57 included movement (19 linear, 23 arcs, 10 chops, 3 circles). Similarity was judged for each feature of each gesture independently and the arithmetic mean was calculated. The resulting average value across all gesture ratings was 1.54 (SD = 0.44), with static gestures being more accurate than gestures with movement. Table 8.2 shows the results for single features of different kinds of gesture. Overall, the results of the assessment test indicate that the morphology codes specify almost all of the information needed to reproduce a gesture that is largely identical to the original. Even more importantly, the number of errors in our form coding proves to be well within acceptable limits.

8.3.3 Referent Coding

Using independent information about the campus from maps, photographs, and trips across campus, three of the experimenters named each place on campus that was referred to in the first four minutes of each of the complete set of direction-giving episodes. Those episodes yielded 195 unique, concrete objects or subparts of objects, including parking lots, signs, buildings, lakes and ponds, etc.

Then, several visuo-spatial features based upon Landau and Jackendoff (1993) and Talmy (1983) were chosen and used to mark each referent as to whether or not we perceived the feature as salient when looking at it from various angles in the route perspective (i.e., not from above). Paths and roads were marked as "sideways planes", which refer to objects that have two parallel sides or borders, and which seem to be conceptualized as one-dimensional lines or two-dimensional planes. In general, every building was marked as having vertically oriented plane or surface features, corresponding to its walls. Every lake and parking lot was marked with horizontal planes, corresponding to the surface of the region. Many

Table 8.2   Quality of morphology coding, values from 1 (correct) to 4 (completely different)

	Average	Static	Dynamic	Line	Chop	Arc	Circle
Hand shape	1.28	1.54	1.20	1.15	1.06	1.36	1
Orientation	1.42	1.34	1.45	1.15	1.26	1.53	2
Position	1.86	1.34	2.04	1.69	1.93	2.4	2
Average (SD)	1.54 (0.44)	1.43 (0.26)	1.57 (0.47)	1.33 (0.38)	1.52 (0.46)	1.73 (0.51)	1.54 (0.44)

subparts of buildings were referred to (e.g., windows), which were marked with vertical planes, likewise for various kinds of signs. A few buildings were also marked with horizontal plane features; for example, Northwestern University's Central Utility Plant, which is a low, long building – when one walks by the building, the surface of the roof is a salient characteristic.

8.3.4 Pre-study

In order to determine the feasibility of the approach, in a preliminary analysis (Kopp *et al.* 2004) we selected several hundred gestures illustrating several particular morphological features, and combinations thereof, and evaluated what they referred to. The results suggested a correspondence between combinations of morphological features and the visual or geometrical similarity of referents. For example, we found that 67% of the gestures with a linear trajectory ($N = 48$) referred to objects, and that the gestures tended to depict a significant axis with the linear movement (e.g., length of a street, extent of a field, transverse overpass). Likewise, 80% of gestures with a flat hand shape and palm oriented vertically ($N = 45$) referred to objects whose shape comprised an upright plane (walls, signs, windows). In addition, 85% of the gestures with a flat hand shape and the palm oriented sideways ($N = 61$) referred to directed actions (go, keep on, take a, look, etc.), and they always finished with the fingers pointing in the direction of the action. These gestures seemed to fuse iconic with deictic aspects, in that the trajectory of the hands depicted the concrete path or direction of the action described, while hand shape and palm orientation appeared to be more "conventionalized" in their indexing of locations and directions. These results provided confidence to continue the analysis on the full set of 1000 gestures.

8.3.5 Analysis of Morphology Features and Visual Features of Gesture Referents

The analysis looked at simple morphological configurations to see if they correlate with two particular IDF sets. The morphological configurations comprise combinations of hand shape features that appear relatively "flat" (e.g., "5", "B", and their loose and open variants, according to the ASL alphabet), in several orientations in space. We looked at flat hand shapes oriented *vertically* (with fingers pointing up) and *horizontally* (with palm pointing down) to see if they correlated positively with referents that possess salient visual characteristics corresponding to vertical and horizontal planes. Using maps, photographs and trips across campus, each landmark (including landmark aspects and sub-parts) was annotated with information on whether it had these salient characteristics. For example, the walls of a tall building contain vertically oriented planes (surfaces) and a parking lot contains a horizontal plane (surface). In order to relate them to gesture morphology, these two features must be formalized using IDFs; e.g., a vertical plane as *building(ams)* $\wedge$ *has_part(wall1,ams)* $\wedge$ *isa(wall1,surface)* $\wedge$ *orientation(wall1, vertical)*. It is important to note that these features are not mutually exclusive. For example, some buildings seem to possess both vertical surfaces (walls) and horizontal planes (their wide footprint, overhanging roof, etc.). Figure 8.6 illustrates (from left to right) the hypothesized correspondences, between a gesture with a flat,

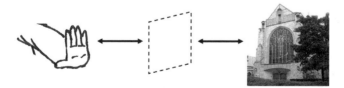

Figure 8.6   (From left to right) Gesture morphology, corresponding salient visual characteristic (IDFs), and a corresponding concrete referent

Table 8.3 *Vertical*: Flat hand + vertical orientation + extended-finger-direction up or down

| Palm dir. (P) | Extended finger direction (F) | |
	Weak	Strong
Weak	F = Up or Down	F = Up or Down F ≠ Left, Right, Forward, or Backward
Strong	F = Up or Down P ≠Up or Down	F = Up or Down F ≠ Left, Right, Forward, or Backward P ≠ Up or Down

vertically oriented morphology, a vertical surface definable in terms of IDFs, and a building (wall) that has the vertical plane feature.

Because gestures do not fall naturally into such discrete categories, various levels of strictness for their high-level categorization must be chosen. Specifically, for deciding whether a gesture should be categorized morphologically as having a flat shape plus a vertical or horizontal orientation, we defined a *weak* and a *strong* version for each of our sets of criteria. Since we have two classes of orientation features (finger and palm) we required a 2 × 2 classification system for each feature, as shown in Table 8.3.

Each cell of this table contains a set of criteria, or constraints, that specify variations on a gesture with a flat hand shape, extended finger direction up or down, and palms facing any direction but up or down. Abbreviating finger direction as F, palm direction as P, weak as w and strong as s, the box for $FwPw$ contains the constraint "F = Up or Down", meaning that the extended finger direction must have an up or down aspect. This can be combined with other features, such as up and to the left, overall a weak version of pointing up. The strong F cells include additional negative constraints to ensure that the F feature is purely up and down.

Overall, three basic morphological configurations including a flat hand shape were distinguished in terms of the following criteria: *vertical* means that the extended finger direction is up or down (Table 8.3); *sideways*, where the palm is oriented towards the left or right, and the thumb is pointing up or down (Table 8.4); and *horizontal* means that the fingers are pointing left, right, forward, or backward, with the palm oriented downwards (Table 8.5). With these specifications in place, one can analyze which of the four levels of classification stringency provides the highest degree of correspondence between the occurrence of shape features of the referent and the morphological features in the gesture. As there

Table 8.4 *Sideways*: Flat hand + vertical orientation + extended finger direction forward, backward, right or left

| Palm dir. (P) | Extended finger direction (F) | |
	Weak	Strong
Weak	F = (Left or Right and/or Forward or Backward) P = (Left or Right and/or Forward or Backward)	F ≠ Up or Down P = (Left or Right and/or Forward or Backward)
Strong	F = (Left or Right and/or Forward or Backward P ≠ Up or Down P = (Left or Right and/or Forward or Backward)	F ≠ Up or Down P ≠ Up or Down P = (Left or Right and/or Forward or Backward)

Table 8.5 *Horizontal*: Flat hand + horizontal orientation

Palm dir.	Extended finger direction	
	Weak	Strong
Weak	P = Up or Down	F ≠ Up or Down P = Up or Down
Strong	P = Up or Down P≠ Left or Right P ≠ Forward or Backward	F ≠ Up or Down P = Up or Down P≠ Left or Right P≠ Forward or Backward

are potentially infinite features of referents about which to gesture, we examined the extent to which shape features of gesture morphology predict the existence of corresponding features in the referent. Dichotomous outcome decision tables were used with gestures classified as possessing horizontal plane, vertical plane, and sideways plane features, each at our four levels of stringency (12 decision tables in total). Table 8.6 shows the decision table for the *FwPw* (Finger-weak Palm-weak) classification of the flat *vertical* gestures.

In this case, 77 of the 123 gestures that have the flat vertical morphology also refer to a landmark that has the predicted vertical plane feature, while the other 46 do not. Of the 315 gestures without flat vertical morphology, 178 also do not refer to a landmark with the vertical plane feature; the referents of the other 137 gestures have a vertical plane feature. Using Paul Barrett's DICHOT[4] program, the chi-squared value (the *p*-value of which indicates the likelihood that the results occur by chance) and the phi coefficient (the correlation between predictors and outcomes, or the validity of the gesture morphology as a predictor) of the data in this and the rest of the decision tables were calculated.

8.3.6 Results

The results of the analysis are summarized in Tables 8.7–8.9. Of the three gesture morphology types, *horizontal* (Table 8.7) is the weakest predictor overall, while the Finger-strong + Palm-weak (*FsPw*) stringency level provides the best predictions across all three morphological types. For example, in flat *horizontal* gestures (Table 8.7), *FsPw* predicts horizontal planes in the referent with a small but statistically significant validity (Φ) of 0.1194. That is, when extended finger direction contains no up or down features and the palm is not necessarily perfectly horizontal, we are most likely to find that the referent contains a salient horizontal plane. Note, however, that about one-fifth of these gestures depict buildings that do not have the horizontal plane feature.

For the prediction of the existence of vertical planes in the referent by occurrence of flat *vertical* morphology in the gestures (Table 8.8), we find that all levels of stringency are statistically significant. Again, *FsPw* stringency is the strongest predictor overall with a somewhat larger validity (Φ) of 0.2071, significant at the 0.001 level. Note also that the Finger-strong + Palm-weak and Finger-strong + Palm-strong stringency levels return the same number of gestures. Finally, for prediction of the existence of "ribbonal planes" in the referent by occurrence of flat *sideways* morphology in the gestures (Table 8.9), all levels but the weakest are statistically significant. Here, the best predicting *FsPw* level reaches a validity (Φ) of 0.2002, significant at the 0.001 level.

---

[4] Version 3.0 (http://www.pbmetrix.com/)

Table 8.6   Example of a frequency table used in this study

FwPw	Occurrence of a vertical plane in the referent	No occurrence of a vertical plane in the referent	
Flat vertical gesture morphology (Predicted occurrence of vertical plane in the referent)	77	46	123
No flat vertical gesture morphology (No predicted occurrence of vertical plane in the referent)	137	178	315
	214	224	438

8.4 Discussion of Empirical Results

The analysis yielded evidence for a systematic relationship between visual characteristics of the gesture form and the spatial features of the entities they refer to, for every morphological class investigated. Of the three morphological classes, the flat *vertical* type was the best predictor of referent form. However, for all three morphological classes, a significant correlation exists between a "more-or-less-flat" hand shape with perfectly-oriented extended finger direction (*FsPw*) and the spatial features of the referent.

There are, however, several caveats. First of all, it should be noted that there are quite a number of false negatives in each frequency table. That is, although it is frequently the case that a flat vertical hand shape refers to a flat vertical image description feature linked to, for example, a tall building, it is also quite often the case that a tall flat building is not described using a flat handshape. From the perspective of gesture analysis, this is quite understandable. Route directions include reference to landmarks as a way of ensuring that the listener is on the correct path. It therefore stands to reason that landmarks must be described in such a way as to disambiguate them from the other buildings and objects that are nearby. If this is the case, then gesture may participate in the act of disambiguating a referent from a kind of distractor set – a quite different task than simply accomplishing similarity with a building in and of itself. In order to determine whether gesture is serving this function, the referents would need to be coded for

Table 8.7   Percentage of flat *gorizontal* gestures depicting referents with horizontal planes and other features (numbers in parentheses)

Stringency of morphology categorization	Referent feature					Test validity (Φ)
	Horizontal plane	No horizontal plane		Total	χ^2	
		Building	Misc. referent			
FwPw	50.6% (49)	21.6% (21)	27.8% (27)	100% (97)	3.1153	0.0843
FsPw	57.4% (35)	18.0% (11)	24.6% (15)	100% (61)	6.2449*	0.1194*
FwPs	55.6% (25)	20.0% (9)	24.4% (11)	100% (45)	3.3909	0.088
FsPs	55.6% (25)	20.0% (9)	24.4% (11)	100% (45)	3.3909	0.088

   *$p < 0.05$

Table 8.8 Percentage of flat *vertical* gestures depicting referents with vertical planes and other features (numbers in parentheses)

Stringency of morphology categorization	Referent feature		Total	χ^2	Test validity (Φ)
	Horizontal plane	No vertical plane			
FwPw	63% (77)	37% (46)	100% (123)	12.9279*	0.1718*
FsPw	79% (37)	21% (10)	100% (47)	18.7934*	0.2071*
FwPs	68% (59)	32% (28)	100% (87)	15.615*	0.1888*
FsPs	79% (37)	21% (10)	100% (47)	18.7934*	0.2071*

$^*p < 0.001$

the visuo-spatial features that serve to differentiate them from their neighbors, rather than the features that are generally visually salient.

Second, the results did not – as we had expected – adduce evidence for stronger morphological features being more strongly linked to features of referents. The effect of strictness criteria for palm and extended finger direction is significant, but a clear pattern of predictive power does not emerge. In fact, for every morphological class, the most predictive stringency level was strong extended finger direction and weak palm direction. This result may derive from the different ways in which gestures are produced depending on (a) the response of the listener, and (b) the discourse function of the entity being referred to, the placement of the reference in the set of directions. That is, gestures might be made more forcefully the first time an entity is introduced, or when an entity is more important, or when the listener shows signs of not following. In a sense, weak features may act like proforms – like the schwa in "the" – once a referent has been established with a stronger hand shape. This hypothesis certainly bears further investigation.

Finally, all referents were coded as individual entities. Future work should attempt to identify systematicity in the features regarded as salient for each referent or referent class. Thus, buildings may always be portrayed in gesture with horizontal flat features. This could be carried out by examining the correlations between morphological configurations within a group of gestures that all refer to the same thing, such as a particular landmark.

However, when generating multimodal directions in a virtual human system, the information about referent classes is needed for use only in content planning of natural language and gesture, which selects the IDFs and semantics for inclusion in the goals it sends to a "microplanner". The microplanner takes these goals and turns them into coordinated language and gesture. For modeling this problem, one needs evidence of the existence of correlations (systematicity), which we have found. And so, with this evidence in hand, we now turn to the task of modeling direction-giving with embodied conversational agents.

Table 8.9 Percentage of flat *sideways* gestures depicting referents with sideways planes and other features (numbers in parentheses)

Stringency of morphology categorization	Referent feature		Total	χ^2	Test validity (Φ)
	Sideways plane	No sideways plane			
FwPw	46% (100)	54% (117)	100% (217)	2.0188	0.0679
FsPw	56% (86)	44% (67)	100% (153)	17.5546***	0.2002***
FwPs	58% (28)	42% (20)	100% (48)	5.3891*	0.1109*
FsPs	61% (28)	39% (18)	100% (46)	6.9399**	0.1259**

$^*p < 0.05;\ ^{**}p < 0.01;\ ^{***}p < 0.001$

8.5 Generating Directions with Humanoids

These empirical findings can lead to embodied conversational agents that can perform appropriate speech and novel gestures in direction-giving conversation with real humans. In the following sections we describe our implementation of an integrated microplanner that derives the form of both natural language and gesture directly from communicative goals. The computational model is based upon the theoretical assumptions of the IDF approach, as well as the empirically suggested patterns in the mapping of features of visuo-spatial meaning onto features of gesture morphology.

Although much existing work has addressed the automatic generation of coordinated language and visualization for complex spatial information (e.g., Towns *et al.* 1998; Kerpedjiev *et al.* 1998; Green *et al.* 1998), little of this research has addressed coordinated generation of speech and gesture for spatial tasks. Traum and Rickel (2002) present a model of dialogue acts for spoken conversation that incorporates nonverbal behavior into its representation as well as accounting for a representation of the discourse state of these dialogue acts. This work is related in that it deals with the discourse state of nonverbal behavior (Rickel *et al.* 2002), but it does not consider questions of generating these behaviors. Nijholt *et al.* (2005) discuss architectural issues for multimodal microplanning and the factors influencing modality choice, but adhere in their proposed model to selecting iconic and deictic gestures from a lexicon; the issues of iconicity of gesture and their underlying semantics are not considered. The REA system (Cassell *et al.* 2000) represents one of the most elaborated works on the automatic generation of natural language and gesture in embodied conversational agents (ECAs). Using the SPUD system (Stone *et al.* 2003) for planning natural language utterances, REA was able to successfully generate context-appropriate language and gesture, relying upon empirical evidence (Cassell and Prevost 1996; Yan 2000) that communicative content can be defined in terms of semantic components, and that different combinations of verbal and gestural elements represent different distributions of these components across the modalities. This approach was able to account for the fact that iconic gestures are not independent of speech but vary with the linguistic expression they accompany and the context in which they are produced, being sometimes redundant and sometimes complementary to the information conveyed in words. However, whole gestures were treated exactly like words, associated to syntactic trees by a specific grammatical construction, the SYNC structure, and gesture planning only extended as far as the selection of a complete gesture from a library and its context-dependent coordination with speech. This does not allow for the expression of new content in gestures, as is possible in language with a generative grammar. Gao (2002) extended the REA system to derive iconic gestures directly from a 3D graphics scene. He augmented the VRML scene description with information about 3D locations of objects and their basic shapes (boxes, cylinders, spheres, user-defined polygons, or composites of these), which were mapped onto a set of hand shapes and spatial hand configurations. This method allows for deriving a range of new gesture forms, but it does not provide a unified way of representing and processing the knowledge underlying coordinated language and gesture use.

The fact that previous systems usually draw upon a lexicon of self-contained gestures is also a consequence of the use of canned animations. Although previous systems, such as BEAT (Cassell *et al.* 2001), were able to create nonverbal as well as paraverbal behaviors – eyebrow raises, eye gaze, head nods, gestures, and intonation contours – and to schedule those behaviors with respect to synthesized speech output, the level of animation was always restricted to predefined animations. Sometimes, motor primitives were used that allowed for some open parameters (e.g., in the STEVE system (Rickel *et al.* 2002) or REA (Cassell *et al.* 2000)), were adjustable by means of procedural animation (EMOTE (Chi *et al.* 2000)), or could be combined to form more complex movements (e.g., Perlin and Goldberg 1996). Kopp and Wachsmuth (2004) presented a generation model that assembles gestural motor behaviors entirely based on specifications of their desired overt form. This method allows for greater flexibility with respect to the producible forms of gesture, which is a prerequisite for the level of gesture generation targeted here, but it does not determine the morphology of the gesture from the communicative intent.

What is missing in previous systems is a model to plan a detailed morphology of a gesture from a given set of communicative goals. Here we explore how this problem can be addressed by following our work on IDFs and form–meaning pairings. To this end, we extend the standard Natural Language Generation (NLG) model to allow for the integrated generation of both natural language and iconic gesture (henceforth, NLGG). NLG systems commonly have a modular, pipeline architecture, broken down into three subtasks – content planning, microplanning, and surface realization (in that order; Reiter and Dale 2000). In ordinary language, the work done by these three subsystems boils down to figuring out what to say, figuring out how to say it, and finally, saying it, respectively. We concentrate on microplanning, the second stage of the NLGG pipeline, where domain knowledge must be recoded into linguistic and gesture form, although we will first outline some prerequisites to be met by the other stages for this.

8.5.1 Modeling Content and Spatial Information

Content planning involves the selection and structuring of domain knowledge into coherent directions. We do not discuss the details of this process here, as it is beyond the scope of this chapter; but see Striegnitz *et al.* (to appear), Guhe *et al.* (2003), and Young and Moore (1994). However, we must note that our NLGG model requires a rich representation of domain knowledge that pays attention to the affordances of both language and gesture as output media. In the direction-giving scenario, most of this content is spatial information about actions, locations, orientations, and shapes of landmarks, and a representation is needed powerful enough to accommodate all information expressible in spatial language and gesture. Two levels of abstraction with corresponding layers of formal representation are required to model spatial language. For example, for an NLG system to refer to an object as "tall", first, the concept or property of tallness must be formalized. This can be done as a simple logical formula like $tall(X)$, where *tall* is a predicate symbol representing the concept, and X is an open variable which can be bound to another ground symbol, representing a particular discourse referent, such as $tall(church)$ or $tall(john)$. Second, this formula must be associated with the string "tall" representing the word itself. For an iconic gesture, a more fine-grained specification in terms of the intrinsic spatial nature of this property is required. For example, tallness can be described as holding of an object when the extent of its vertical axis is longer than its other axes, or more likely it is long relative to the vertical axes of some other relevant objects (e.g., a man might be tall relative to some other men standing nearby), or relative to some stereotype. The intermediate IDF level represents such spatial properties that can be displayed by gesture. If the concept of tallness is represented as $tall(X)$, and its spatial description is represented as a set of IDFs, these IDFs can then be mapped onto form features, and this iconic gesture can be used to refer to the concept.

This example motivates IDFs and conceptual/semantic knowledge as different kinds of knowledge with different levels of abstraction, needed to exploit the representational capabilities of the two modalities, and meriting separation into two ontologically distinct levels. However, at the same time, we follow the ideas of one amodal, common representation of – even spatial – content (e.g., Landau and Jackendoff 1993) that both language and gesture utilize, working together to express information as parts of one communicative system (McNeill 1992). Our model thus maintains a single, common representation system encompassing all the kinds of domain knowledge needed, formalized in terms of qualitative, logical formulae. It is based on a formal, extensible ontology that encompasses objects (buildings, signs, etc.), regions (parking lots, lake, etc.), and actions (go, turn, etc.). In addition, it defines IDF-related symbols for basic shapes, locations, directions, or qualitative extents (long, short, large, tall, narrow, etc.). When ontologically sound, entities are connected using taxonomic (is-a), partonomic (part-of), and spatial relations (in, on, left-of, etc.). The ontology thus provides for IDFs and lays down their assignment to concrete objects or actions. In other words, it provides for the link between an entity and a mental image thereof, the latter formalized in terms of IDFs. Such an ontology has been built for parts of Northwestern University campus, including all landmarks that were referred to in the analyzed route descriptions. Content plans are then specified in terms of these entities and relations. Figure 8.7 shows

```
instruction(e2). see(e2,user,cook,future,place(on,right)).
tense(e2,future). name(cook,cook_hall). type(cook,building).
place(on,right). rel_loc(cook,user,right).
shape(dim,vert,cook). shape(primary_dim(longit,cook)).
shape(dim,longit,cook).
```

Figure 8.7   Content plan in logics notation (propositions are delimited by points)

an example content plan that comprises all kinds of qualitative knowledge (including IDFs) required for employing language and gesture in instructing someone that she will see a particularly shaped building ("Cook Hall") on her right.

8.6 Multimodal Microplanning

The multimodal microplanner must link domain-specific representations of meaning like the ones shown in Figure 8.7 to linguistic form and gesture form. As language and gesture require different kinds of information, provide different representational capacities, and convey information in different ways, NLGG calls for specific models of how each modality encodes content. Thus, a new component, the gesture planner (GP), was added to the microplanning stage of NLGG, as illustrated in Figure 8.8. It is responsible for planning the morphology of a gesture appropriate to encode a set of one or more input IDFs. That is, the GP is itself a microplanner, addressing the problem of recoding content into form, but this time on a feature level, from IDFs to morphological features.

8.6.1 Grammer-based Sentence Planner

To connect content to linguistic forms, we employ a grammar-based sentence planner, SPUD (Stone *et al.* 2003). SPUD takes a uniform approach to microplanning, framing it as a search task wherein utterances are iteratively constructed from an input specification of a set of resources and a knowledge base (KB) that

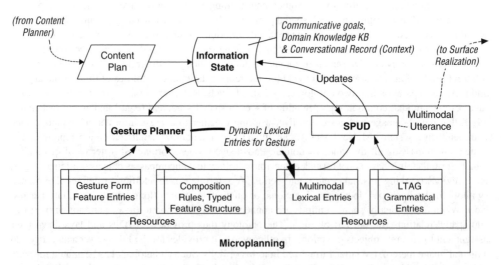

Figure 8.8   Overview of the multimodal microplanning process. Reproduced by permission of the ACM, Kopp *et al.*, 2004

contains, among others, the facts to be communicated (communicative effects). All facts are explicitly labeled with information about their conversational status (e.g., whether the fact is private or shared), constraining decisions about what information the system must assert as new to the user, and what it can presuppose as information in the common ground (Clark 1996). The grammar includes a set of lexical entries and a set of syntactic constructions, formalized using Lexicalized Tree Adjoining Grammar (Joshi 1987), and optionally containing pragmatic constraints on use of this construction relative to discourse context. Each lexical entry consists of a lexical item (word), a set of logical formulae defining its semantics and, again, pragmatic conditions for use in conversation. SPUD works towards a complete, grammatical structure by iteratively selecting words and syntactic structures to add, such that their semantics allows for the assertion of the maximum number of communicative effects to be achieved per state (simulating an economical, Gricean approach to generation). Additionally, for each state, the system maintains a representation of the utterance's intended interpretation, or communicative intent, a record of inferential links made in connecting the semantics and pragmatics, associated with linguistic terms, to facts about referents in the world, as recorded in the knowledge base.

In previous work (Cassell *et al.* 2000), SPUD's linguistic resources have already been extended to include a set of predefined gestures, from which it drew upon to express its communicative goals. We follow this same strategy here, using SPUD to compose full, multimodal utterances via a single, uniform algorithm. But, instead of drawing upon a static set of predefined gestures, the GP is added into the pipeline: before calling SPUD, the GP plans iconic gestures that express some or all of the given IDFs. The planned gestures are then dynamically incorporated into SPUD's (now multimodal) resources and utilized in the same way as described in Cassell *et al.* (2000).

8.6.2 Gesture Planning and Integration

Similar to the sentence planner SPUD, the Gesture Planner system draws upon a bipartite input specification of domain knowledge, plus a set of entries to encode the connection between semantic content and form. Using such data structures, the same kind of close coupling between gesture form and meaning can be achieved, allowing for efficient, incremental construction of gestures and maintenance of inferential links from abstract meaning (logical) formulae to specific discourse referents. For the GP, the form–meaning coupling is formalized in a set of "form feature entries", data structures that connect IDFs to morphological features. Form feature entries implement the patterns that were found in the empirical data (see section 8.3), and may contain "clusters" of features on either side – conjunctions of IDFs as well as combinations of morphological features. Also, these entries encode the ways in which the function of a gesture (e.g., deictic) influences its morphology (e.g., hand shape) through conventionalized patterns, again, as evidenced by our empirical data.

When receiving a set of IDFs as input, the desired communicative effects, the GP searches for all combinations of form feature entries that can realize them. Because iconic gesture is not governed by a hierarchical system of well-formedness, feature structure unification is employed to combine morphological features, whereby any two features may combine provided that the derived feature structure contains only one of any feature type at a time. Through iterative application of this operation, the GP builds up gestures incrementally until all the desired communicative effects are encoded. Figure 8.9 shows a state in the generation of a gesture, being composed to depict the IDFs from the content plan in Figure 8.7. Location and hand shape have already been inserted according to patterns observed in the data, such as the use of a flat hand shape (ASL sign 5) for depicting the wall of the Cook building. A related pattern now informs the palm orientation, together with the location of the object (Cook) to be depicted. Note that the GP may output an underspecified gesture if a morphological form feature does not meaningfully correspond to any of the disposed IDFs; that is, it remains undefined by the selected patterns.

Similar to SPUD's pragmatic constraints on the way language is used in context, the GP process can be guided by composition constraints on all possible ways to combine a set of form features into a feature

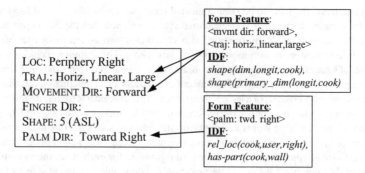

Figure 8.9 Example of form features entries filling a gesture feature structure (Kopp et al. 2004; reproduced by permission of ACM)

structure that defines a realizable gesture. Such composition constraints could formalize restrictions over the ways in which different form features combine, and could, for example, be utilized to favor the reuse of feature structures that have been successfully employed before to express a common set of semantic formulae. This would require comparison to the KB's record of context, and allows for simulation of what McNeill (1992) has called catchments, the maintenance of a certain gesture morphology to indicate cohesion with what went before.

The GP is implemented (in Prolog) to return all possible combinations of morphology features; that is, it delivers all gestures that could take on communicative work by encoding some or all of the desired communicative effects. Each dynamically planned gesture is added to SPUD's resources, which also contain a set of dedicated SNYC constructions that state the possible ways in which language and gesture can combine. Each SYNC construction pairs a certain syntactic constituent and a gesture feature structure under the condition that their predicate arguments are connected to the same discourse referents, achieving coordination of meaning in context. In addition, it imposes a constraint of temporal surface synchrony between both elements. The SPUD algorithm chooses the gesture feature structure and the construction that, when combined with appropriate words, allow for the most complete intended interpretation in context; see Cassell *et al.* (2000) for details. Figure 8.10 shows how a gesture feature structure, derived from the IDFs in the content plan shown in Figure 8.7, is combined with a linguistic tree to form a multimodal utterance. Finally, the tree of the resulting multimodal utterance is converted into an XML description, containing the textually defined words along with the feature structure for the gesture. This tree is passed on to the next and final stage of the NLGG pipeline, surface realization.

8.7 Surface Realization

Starting out with the XML specification output from the microplanner, surface realization turns this tree into multimodal output behaviors to be realized by our embodied conversational agent NUMACK (the Northwestern University Multimodal Autonomous Conversational Kiosk). This is done in two steps, behavior augmentation and behavior realization.

8.7.1 Behavior Augmentation

The XML specification coming in from microplanning amounts to communicatively *intended* behaviors – words and gestures that directly derive from communicative goals. To achieve natural multimodal output, the surface realization stage imparts to the utterance additional nonverbal and paraverbal behaviors

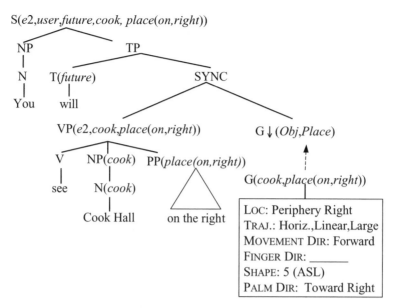

Figure 8.10   Insertion of the gesture into the utterance tree (Kopp et al. 2004; reproduced by permission of ACM)

like intonation, eyebrow raise, head nod, or posture shifts. These behaviors do not encode explicit communicative goals, but are essential to meaning as they, for example, mark the central parts of the utterance.

   This task is accomplished by the BEAT system (Cassell *et al.* 2001), which first suggests all plausible behaviors, and then uses filters to trim overgenerated behaviors down to a set appropriate for a particular character. The behavior suggestion draws upon information about grammatical units (clause structure), information structure (theme and rheme), word newness, and contrast, each of which is represented by dedicated tags in the XML tree. This tree gets augmented with appropriate behaviors by application of an extensible set of rule-based generators to each node (Cassell *et al.* 2001). When a node meets the criteria specified by a generator, a behavior node is suggested and inserted independent of any other. Its position in the tree defines the time interval the behavior is supposed to be active; namely, synced with all words contained in its sub-tree. Note that the information at disposal for behavior suggestion is limited to clause structure and information structure, the latter being always set to "rheme", as NUMACK currently lacks a discourse planner for composing multi-clause utterances with thematic parts. After behavior suggestion, filters are applied to the tree, which delete all inserted behaviors that cannot physically co-occur, or whose assigned priority falls below a pre-specified threshold. Figure 8.11 shows a complete behavior augmentation example, in which gaze and intonation behaviors are added to the utterance "Make a left". In addition, the clause is turned into a "chunk" node that demarcates a unit for speech-gesture realization to follow.

8.7.2 Behavior Realization

Upon completion of behavior augmentation, the XML tree with its morphologically and phonologically specified behaviors is turned into synthesized speech and intonation, expressive gestures, and other animations for a graphical avatar body, all being scheduled into synchronized multimodal output. The ACE system (Kopp and Wachsmuth 2004) is employed for this, which provides modules for synthesizing

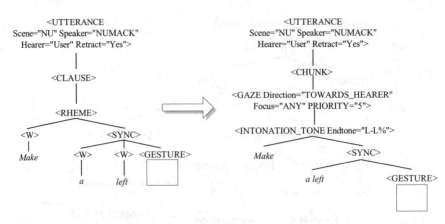

Figure 8.11   Insertion of non-/paraverbal bejaviors in a multimodal utterance tree during behavior augmentation

gestural, verbal, and facial behaviors, and embeds them in an incremental process model to schedule and link their output into synchronized, fluent utterances.

Before behavior realization starts, the XML gesture description by the microplanner, stating a typed feature structure that contains a possibly incomplete set of form features, is converted into a format that can be processed by ACE's gesture generation module. As illustrated in Figure 8.12, the feature structure is translated into MURML (Multimodal Utterance Representation Markup Language; Kopp and Wachsmuth 2004), an XML-conforming, feature-based representation that denotes a hand–arm configuration in terms of the same features as the original feature structure, but using a slightly different set of descriptive symbols based on *HamNoSys*, a notation system for German sign language (Prillwitz 1989). A gesture is described as a combination of separate, yet coordinated postures or sub-movements within the features, the latter being defined as a sequence of elements each of which is piecewise guiding it. To capture the gesture's inner structure, features are arranged in a constraint tree by combining them with dedicated nodes for expressing simultaneity, posteriority, symmetry, and repetition of sub-movements (for example, the PARALLEL node in Figure 8.12 denotes simultaneity of its child features). Form features that have been left open by the microplanner are either set to default values, or remain undefined when no action needs to be taken about them. For example, a gesture whose location has not been laid down during microplanning will per default be performed in the center of gesture space.

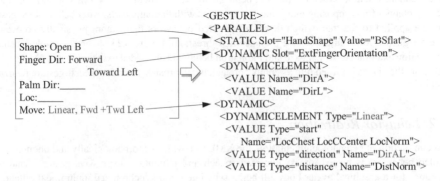

Figure 8.12   Conversion of a feature structure into a MURML specification

The complete multimodal utterance tree is finally processed by the realization model, which schedules, synthesizes, and executes all verbal and nonverbal behaviors. Its approach is based on the assumption that continuous speech and gesture are co-produced in successive chunks, which pair one intonation phrase with one synchronized gesture (as, for example, the tree in Figure 8.11), and that the modalities are coordinated at two different levels of an utterance. Within a chunk, the gesture is timed such that its meaning-bearing stroke phase starts before or with the affiliated linguistic elements and spans them. Consequently, the intonation phrase along with its pitch accents is synthesized in advance using the Festival system for text-to-speech conversion (Black and Taylor 1997). Timing constraints for coverbal gestural or facial behaviors are drawn from the phoneme information obtained. Second, in-between two successive chunks, the synchrony between speech and gesture in the forthcoming chunk is anticipated by the movements between the gestures (co-articulation), ranging on demand from the adoption of an intermediate rest position to a direct transition movement. Likewise, the duration of the silent pause between two intonation phrases varies according to the required duration of gesture preparation.

Both effects are determined and simulated at a time when the next chunk is ready for being uttered ("lurking"), while the former is "subsiding"; that is, done with executing all mandatory parts (intonation phrase and gesture stroke). At this time, intra-chunk synchrony is defined and reconciled with the onsets of phonation and movement, and animations are created that satisfy the movement and timing constraints now determined. This method rests upon the ability of ACE to generate all animations required to drive the skeleton as specified, in real time and from scratch (Kopp and Wachsmuth 2004), which is also indispensable for the level of generativity targeted here. Following a biologically motivated decomposition of motor control, a motor planner breaks down the control problem – to steer the control variables such that the resulting movement meets all constraints – into sub-problems that get solved by specialized modules for the hands, the wrists, and the arms. Their solutions are local motor programs (LMPs) that utilize suitable computer animation techniques to control sub-movements that affect a limited number of joints for a limited period of time. To create the whole gesture, the LMPs run concurrently and synchronized in more abstract motor control programs, in which they autonomously (de-)activate themselves as well as other LMPs. As a result, different, yet coordinated motion generators create different parts of a gesture and automatically create context-dependent gesture transitions.

8.7.3 Generation Examples

Based on the described theoretical and technical approaches, NUMACK is able to produce a considerable range of directions, using semantically coordinated language and gesture. Figure 8.13 demonstrates two

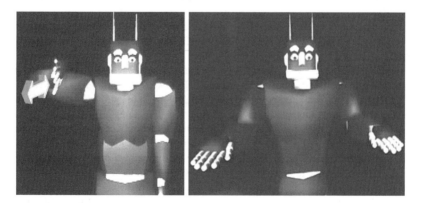

Figure 8.13 NUMACK generation examples: *"You will see Cook Hall on your right"* (left; Reproduced by permission of the ACM, Kopp *et al.*, 2004), and *"You will see the Lake ahead"* (right)

example utterances. The one in the left picture was generated from the content plan in Figure 8.7. Surface realization is in real time, in that the time to produce an utterance is less than the natural pause between two utterances in dialogue. Together with the lightweight implementation of the microplanner in Prolog, NUMACK is the first system that creates a multimodal utterance from a given set of communicative goals (including all factual and spatial knowledge about the referents) in less than one second on a standard PC.

8.8 Discussion of Generation Results

As described in the previous sections, the NUMACK embodied conversational agent was implemented on the basis of theoretical assumptions and corresponding empirical results. Consequently, NUMACK is able to realize direction-giving in quite different ways from that of other ECAs, and other non-embodied dialogue systems. However, "different" does not equal "better" and it is therefore important to assess the fit of the NUMACK system both as a cognitive model of the empirical results described above, and in its role as an autonomous direction-giving system. In the first instance, the question is how similar NUMACK's performance comes to human direction-giving. In the second instance, we ask how effective NUMACK is in guiding people to their destination, and how NUMACK compares to other direction-giving devices, such as maps. In particular, how effective are the advances that we made in this system – the generation of direction-giving gestures, and the use of landmarks in route descriptions – to actual human use of the system?

With respect to the first question, concerning NUMACK as a model of human direction-giving, we rely on our own evaluation of NUMACK's performance, and our comparison of NUMACK with the human direction-givers whom we have examined. NUMACK displays some natural hand shapes in describing landmarks, and is certainly capable of a wider variety of direction-giving gestures than previous ECAs that have relied on gesture libraries. However, the ways in which NUMACK is *not* human-like are perhaps more salient than NUMACK's successes. Here we notice that NUMACK tends to give directions all in one go – from point A to point B, without asking the direction-follower if s/he can remember this much information at once. This behavior is striking, and allows us to realize that people must use some heuristic to *chunk* direction-giving into segments. In human direction-giving segments might be separated by explicit requests for feedback ("are you still following me") or perhaps even followed by a suggestion to ask another passerby ("at that point, you might want to ask somebody else where to go"). The study of route direction chunking, and whether it is based on the speaker's *a priori* beliefs about how long directions should be, or on cues that are emitted by the listener, is therefore one of our topics for future research.

We likewise notice that NUMACK uses the left and right hand interchangeably in pointing out the route or describing landmarks, based on whichever hand is closest to the side he is referring or pointing to. Something about this seems unnatural, leading us to think that direction-givers must use one hand or the other time after time as a way of marking cohesion among direction-giving segments. In addition there is something unnatural about the way in which NUMACK uses himself as the recipient of his direction-giving, as if he is walking through the scene. Looking at his performance, one is led to think that humans might use their hands to follow an imaginary walker along a route. This too is a topic for future research. In each of these instances, it should be noted that only because we have NUMACK as an instantiation of our theory of gesture and speech in direction-giving are we even able to evaluate the completeness of the theory, and the places in which we have omitted pertinent analysis of the data.

A second and quite separate topic concerns NUMACK's potential role as a direction-giving kiosk. Is NUMACK more effective than the display of a map? Are NUMACK's directions more effective with gestures, or do the gestures not add much at all? And, are the descriptions of landmarks useful, or are they unnecessary? These three questions lead to six different conditions of an experiment to test how people assess the naturalness and effectiveness of direction-giving. The experimental design, with six

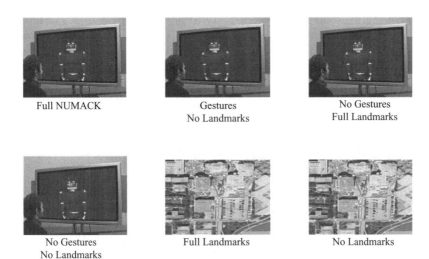

Full NUMACK   Gestures   No Gestures
                               No Landmarks   Full Landmarks

No Gestures   Full Landmarks   No Landmarks
No Landmarks

Figure 8.14   Experimental design with six conditions for the evaluation of NUMACK

conditions – an outgrowth of $2 \times 2 \times$ (gestures vs. no gestures; landmarks vs. no landmarks; NUMACK vs map only) – is represented in Figure 8.14. This experiment is currently under way.

8.9 Conclusions

It has been proposed that the different media that participate in direction-giving – diagrams, maps, speech, and gesture – all convey one single mental model. And yet, such a claim does not account for the actual form of each medium of expression – why do we use a flat hand with the palm facing inward and the thumb up to represent a path, and a flat hand with the palm facing outward and the palm down to represent walking past a building? And, even more perplexingly, given the lack of standards of form in the gestural medium – even the lack of codified symbols such as those used in maps – how do listeners interpret gesture? In this article we suggest that gesture interpretation and gesture production are facilitated by a layer of meaning that mediates between the morphology of the gesture, and the visuo-spatial attributes of the thing in the world that gesture represents. This level of meaning we call the *image description feature*, and it allows gestures themselves to remain underspecified (lacking consistent form–meaning pairings) and yet meaningful in the context of the speech with which they co-occur.

Analysis of 28 direction-giving sessions allowed for adducing evidence for the image description framework. An analysis of flat horizontal, flat vertical, and flat sideways morphological configurations revealed that there was a significant correlation between these features of gesture morphology and similar features in the objects in the world to which these gestures referred in context. In fact, such a result may seem trivial at best – *of course* iconic gesture resembles that which they are iconic of. And yet most researchers have looked for resemblances at the level of the whole gesture, and have not found it (apart from a very small number of culturally specific gestures called *emblems*). And, were the connection so obvious, we would not have found so many instances of false negatives – cases where the visuo-spatial feature was not represented by a similar feature in gesture. Part of the issue may come from the simultaneity of gesture; unlike phonemes, for example, units of meaning do not line up neatly one after another in gesture, but occur simultaneously in packages – flat handshape *and* movement away from the body *and* bouncing manner. In addition, IDFs are chosen for their salience *in a particular context*. That is, the same object in the world may very well be conveyed quite differently, depending on what aspect of it is salient in different contexts. Or particular morphological features may be carried over from previous gestures, as part of catchments. Or particular gestures may last longer, and be made more clearly, as a function of the retrievability of the referent. It is clear that we still have a long way to go in order to understand the relationship between gesture and language in particular discourse and pragmatic contexts.

The empirical results to date nevertheless are strong enough to support the concept of a mediating level of meaning that links gestural features and meaning. This framework has allowed us to model direction-giving in an embodied conversational agent, and escape the gestionary approach to gesture generation. To this end, we have implemented an integrated, on-the-fly microplanner that derives coordinated surface forms for both modalities from a common representation of context and domain knowledge. In extending the SPUD microplanning approach to gesture planning, lexical entries were replaced with form feature entries; LTAG trees were replaced with feature structures, more closely resembling the global and synthetic nature of gesture; and pragmatic constraints were carried over to guide gesture use in context.

In following projects, we continue to analyze empirical data to refine the model, and to find further patterns in the way iconic gesture expresses visual domain knowledge in order to extend the system's generation capabilities. We believe that the approach to microplanning described here is one step closer to a psychologically realistic model of a central step in utterance formation. However, a range of open questions still need to be investigated, and evaluation of our system will help us shed light on some of these. Such questions include whether a higher degree of interaction might be necessary between the two separate, but interacting, planning processes for language and gesture; or whether one unified, qualitative, logic-based representation is sufficient to represent the knowledge required for planning the surface structure of both modalities. In the meantime, we have established a relationship between words and images, language and gesture, where the latter remains underspecified until joined to the former in a particular pragmatic context, and where taken together, words and gestures provide a window onto the representation of space by both humans and humanoids.

Acknowledgments

The authors gratefully thank Matthew Stone for valuable input, Marc Flury and Joseph Jorgenson for technical assistance, and Jens Heemeyer for help with data annotation. This work was supported by a post-doctoral fellowship from the Postdoc-Program of the German Academic Exchange Service (DAAD).

References

Black A. and Taylor P. (2002) *The Festival Speech Synthesis System: System documentation*. Retrieved. August 2007 from http://festvox.org/docs/manual-1.4.3.

Cassell J. and Prevost S. (1996) Distribution of semantic features across speech and gesture by humans and computers. Paper presented at the Workshop on the Integration of Gesture in Language and Speech, Newark, DE.

Cassell J., McNeill D. and McCullough K.-E. (1999) Speech–gesture mismatches: evidence for one underlying representation of linguistic and nonlinguistic information. *Pragmatics and Cognition* 7 (1), 1–33.

Cassell J., Stone M. and Yan H. (2000) Coordination and context-dependence in the generation of embodied conversation. Paper presented at the INLG 2000, Mitzpe Ramon, Israel.

Cassell J., Vilhjálmsson H. and Bickmore T. (2001) BEAT: the behavior expression animation toolkit. Paper presented at the SIGGRAPH 01, Los Angeles, CA.

Chi D., Costa M., Zhao L. and Badler N. (2000) The EMOTE model for effort and shape. Paper presented at the SIGGRAPH '00, New Orleans, LA.

Clark H.H. (1996) *Using Language*. Cambridge University Press, Cambridge.

Conklin E.J. and McDonald D.D. (1982) Salience: The key to the selection problem in natural language generation. Paper presented at the 20th Annual Meeting of the Association for Computational Linguistics, Toronto.

Couclelis H. (1996) Verbal directions for way-finding: space, cognition, and language. In J. Portugali (ed.), *The Construction of Cognitive Maps*. Kluwer Academic Publishers, Boston, pp. 133–153.

Daniel M.P., Heiser J. and Tversky B. (2003) Language and diagrams in assembly instructions. Paper presented at the European Workshop in Imagery and Cognition, Pavia, Italy.

Denis M. (1997) The description of routes: a cognitive approach to the production of spatial discourse. *Current Psychology of Cognition* **16**, 409–458.

Emmorey K., Tversky B. and Taylor H.A. (2001) Using space to describe space: perspective in speech, sign, and gesture. *Spatial Cognition and Computation* **2** (3), 1–24.

Forbus K. (1983) Qualitative reasoning about space and motion. In D. Gentner and A.L. Stevens (eds), *Mental Models*. Erlbaum, Hillsdale, NJ, pp. 53–73.

Gao Y. (2002) *Automatic Extraction of Spatial Location for Gesture Generation*. MIT Press, Cambridge, MA.

Green N., Carenini G., Kerpedjiev S. and Roth S.F. (1998) A media-independent content language for integrated text and graphics generation. Paper presented at the Workshop on Content Visualization and Intermedia Representations at COLING and ACL '98, University of Montreal, Quebec.

Guhe M., Habel C. and Tschander L. (2003) Incremental production of preverbal messages with in C. Paper presented at the 5th International Conference on Cognitive Modeling, Bamberg, Germany.

Heiser J. and Tversky B. (2003) Characterizing diagrams produced by individuals and dyads. Paper presented at the Workshop on Interactive Graphics, London.

Heiser J.L., Tversky B., Agrawala M. and Hanrahan P. (2003) Cognitive design principles for visualizations: Revealing and instantiating. Paper presented at the Cognitive Science Society Meetings.

Heiser J., Phan D., Agrawala M., Tversky B. and Hanrahan P. (2004) Identification and validation of cognitive design principles for automated generation of assembly instructions. Paper presented at the Advanced Visual Interfaces '04.

Herskovits A. (1986) *Language and Spatial Cognition: An interdisciplinary study of prepositions in English*. Cambridge University Press, Cambridge.

Jackendoff R. and Landau B. (1991) Spatial language and spatial cognition. In D.J. Napoli, J.A. Kegl *et al.* (eds), *Bridges between Psychology and Linguistics: A Swarthmore Festschrift for Lila Gleitman*. Lawrence Erlbaum Associates, Hillsdale, NJ, pp. 145–169.

Joshi A. (1987) The relevance of tree adjoining grammar to generation. In G. Kempen (ed.), *Natual Language Generation: new results in artificial and intelligence, psychology and linguistics*. Kluwer Academic Publishers, Boston, pp. 233–252.

Kerpedjiev S., Carenini G., Green N., Moore J. and Roth S. (1998) Saying It In Graphics: from intentions to visualizations. *Proceedings of the IEEE Symposium on Information Visualization*, pp. 97–101.

Kopp S. and Wachsmuth I. (2004) Synthesizing multimodal utterances for conversational agents. *Journal of Computer Animation and Virtual Worlds* **15** (1), 39–52.

Kopp S., Tepper P. and Cassell J. (2004) Towards integrated microplanning of language and iconic gesture for multimodal output. *Proceedings of International Conference* on Multimodal Interfaces (ICMI), ACM Press, pp. 97–104.

Krauss R.M., Morrel-Samuels P. and Colasante C. (1991) Do conversational hand gestures communicate? *Journal of Personality and Social Psychology* **61** (5), 743–754.

Landau B. and Jackendoff R. (1993) "What" and "where" in spatial language and spatial cognition. *Behavioral and Brain Sciences* **16** (2), 217–265.

Lozano S. and Tversky B. (2004) Communicative gestures benefit communicators. Paper presented at CogSci 2004, Chicago, IL.

McNeill D. (1992) *Hand and Mind: What gestures reveal about thought*. University of Chicago Press, Chicago.

McNeill D. and Levy E. (1982) Conceptual representations in language activity and gesture. *Speech, Place, and Action*, 271–295.

Milde J.-T. and Gut U. (2002) The TASX-environment: an XML-based toolset for time aligned speech corpora. Paper presented at the 3rd International Conference on Language Resources and Evaluation (LREC), Las Palmas.

Nijholt A., Theune M. and Heylen D. (2005) Embodied language generation. In O. Stock and M. Zancanaro (eds), *Intelligent Information Presentation*, Springer, vol. 27, pp. 47–70.

Peirce C. (1955) *Philosophical Writings of Peirce*. Dover Publications, New York.

Perlin K. and Goldberg A. (1996) Improv: a system for scripting interactive actors in virtual worlds. Paper presented at the SIGGRAPH '96, New Orleans, LA.

Poesio M. (1996) Semantic ambiguity and perceived ambiguity. In K. van Deemter and S. Peters (eds), *Ambiguity and Underspecification*. CSLI, Palo Alto.

Poesio M. (2005) *Incrementality and Underspecification in Semantic Processing*. CSLI, Palo Alto.

Prillwitz S. (1989) *HamNoSys: Hamburg Notation System for Sign Languages: an introductory guide* (ver. 2.0 ed., Vol. 5). Signum Press, Hamburg.

Reiter E. and Dale R. (2000) *Building Natural Language Generation Systems*. Cambridge University Press, Cambridge.

Rickel J., Marsella S., Gratch J., Hill R., Traum D. and Swartout W. (2002) Toward a new generation of virtual humans for interactive experiences. *IEEE Intelligent Systems* **17** (4), 32–38.

Saussure F.d. (1985) *Cours de linguistique générale*. Payot, Paris.

Sowa T. and Wachsmuth I. (2003) Coverbal iconic gestures for objects descriptions in virtual environments: an empirical study. Paper presented at the Gestures, Meaning and Use Conference, Edicoes Fernando Pessoa.

Stone M., Doran C., Webber B., Bleam T. and Palmer M. (2003) Microplanning with communicative intentions: the SPUD system. *Computational Intelligence* **19** (4), 311–381.

Striegnitz K., Tepper P., Lovett A. and Cassell J. (to appear). Knowledge representation for generating locating gestures in route directions. In K. Coventry, T. Tenbrink and J. Bateman (eds), *Spatial Language in Dialogue*. Oxford University Press.

Talmy L., University of California Berkeley, Institute of Cognitive Studies (1983). *How Language Structures Space*. Cognitive Science Program, Institute of Cognitive Studies, University of California at Berkeley.

Talmy L. (2000) *Toward a Cognitive Semantics*. MIT Press, Cambridge, MA.

Taylor H.A. and Tversky B. (1992) Spatial mental models derived from survey and route descriptions. *Journal of Memory and Language* **31**, 261–282.

Taylor H.A. and Tversky B. (1996) Perspective in spatial descriptions. *Journal of Memory and Language* **35**, 371–391.

Towns S., Callaway C. and Lester J. (1998) Generating coordinated natural language and 3D animations for complex spatial explanations. Paper presented at the AAAI-98.

Traum D. and Rickel J. (2002) Embodied agents for multi-party dialogue in immersive virtual worlds. Paper presented at the Autonomous Agents and Multi-Agent Systems, Melbourne.

Tverskey B. and Lee P. (1998) How space structures language. In C. Freksa, C. Habel and K.F. Wender (Eds), *Spatial Cognition: An Interdisciplinary Approach to Representation and Processing of Spatial Knowledge*. Springer, Berlin, 57–75.

Tverskey B. and Lee P. (1999) Pictorial and verbal tools for conveying routes. In C. Freksa and D. Mark (Eds), *Spatial Information Theory: Cognitive and Computational Foundations of Geographic Information Science*. Springer, Berlin, 51–64.

Yan H. (2000) *Paired Speech and Gesture Generation in Embodied Conversational Agents*. Unpublished Masters of Science thesis, MIT, Cambridge, MA.

Young R.M. and Moore J.D. (1994) DPOCL: a principled approach to discourse planning. Paper presented at the 7th International Workshop on Natural Language Generation, Kennebunkport, ME.

9

Facial Gestures: Taxonomy and Application of Nonverbal, Nonemotional Facial Displays for Embodied Conversational Agents

Goranka Zoric, Karlo Smid, and Igor S. Pandžić

9.1 Introduction

Communication is the information exchange process. Humans communicate in order to share knowledge and experiences, to tell who they are and what they think, or to cooperate with each other. Communication implies bidirectional interchange of messages. One-way transfer of information is inefficient. In one-way communication there is no proof that what is heard is what is intended. It is necessary to have interaction between people trying to communicate. Therefore, a feedback from the listener is required in order to avoid misunderstanding.

To transfer their meaning to another person, humans use different methods or channels. Communication can be verbal or nonverbal, and often mixed. Albert Mehrabian (1971), a psychologist, came to the conclusion that the nonverbal accounts for 93% of the message while words account for 7%. The disproportionate influence of the nonverbal becomes effective only when the communicator is talking about their feelings or attitudes, since experiments were conducted dealing with communications of feelings and attitudes (i.e., like–dislike). Anthropologist Ray Birdwhistell, who studied nonverbal communication extensively in the 1960s, claimed that in conversation about 35% of the message was carried in the verbal modality and the other 65% in the nonverbal (Knapp 1978).

Although psychologists still argue about the percentage of information nonverbally exchanged during a face-to-face conversation, it is clear that the nonverbal channel plays an important role in understanding human behavior. If people rely only on words to express themselves, that can lead to difficulties in communication. Words are not always associated with similar experiences, similar feelings, or even meaning by listeners and speakers. That is why nonverbal communication is important. A situation when there is no consistency (i.e., one signal is being said and another is shown) can be very misleading to another person. When listeners are in doubt they tend to trust the nonverbal message since disturbances in nonverbal communication are "more severe and often longer lasting" than disturbances in verbal language (Ruesch 1966).

Nonverbal communication refers to all aspects of message exchange without the use of words. It includes all expressive signs, signals, and cues (audio, visual, etc.) apart from manual sign language and speech. Nonverbal communication has multiple functions. It can repeat the verbal message, accent the verbal message, complement or contradict the verbal message, regulate interactions, or substitute for the verbal message (especially if it is blocked by noise, interruption, etc.).

There are a number of categories into which nonverbal communication can be divided:

- *Kinesics* (body language). Nonverbal behavior related to movement of the body. Includes facial expressions, eye movements, gestures, posture, and the like.
- *Oculesics* (eye contact). Influence of visual contact on the perceived message that is being communicated.
- *Haptics* (touch). Touching behavior.
- *Proxemics* (proximity). Concerned with personal space usage.
- *Paralanguage* (paralinguistics). Non-word utterances and other nonverbal clues relatively closely related to language use.
- *Chronemics.* Use of time, waiting, pausing.
- *Silence.* Absence of sound (muteness, stillness, secrecy).
- *Olfactics* (smell).
- *Vocalics* (vocal features of speech). Tone of voice, timbre, volume (loudness), speed (rate of speech).
- *Physical appearance and artifacts.* Physical characteristics of body, clothing, jewellery, hairstyle, etc.
- *Symbolism* (semiotics). Meaning of signs and symbols.

Nonverbal communication can be both conscious and subconscious. Culture, gender, and a social status might influence the way nonverbal communication is used. There are certain rules that apply to nonverbal communication. It has a form, a function, and a meaning, all of which may be culturally specific (Knapp 1978); for example, some cultures forbid direct gaze while others find gaze aversion an offense (Cassell 2000). Facial expressions of primary emotions (e.g., disgust, surprise) are universal across cultures as the psychologist Paul Ekman and his colleagues have shown.[1]

Nonverbal cues may be learned, innate, or mixed. Some of them are clearly learned (e.g., eye wink, thumbs-up) and some are clearly innate (e.g., eye blink, facial flushing). Most nonverbal cues are mixed (e.g., laugh, shoulder shrug), because they originate as innate actions, but cultural rules and environment shaped their timing, energy, and use.

What arises from the above is that, in a face-to-face conversation, much can be said even without words. This fact should be kept in mind when generating computer correspondents of humans – ECA. Just as with a human–human interaction, a human–computer interaction should consist of two-way communication channel for a verbal and a nonverbal message exchange. By building the human–computer communication on the rules of human–human communication, the ECA behavior will be more like human behavior. Therefore, humans are more likely to consider computers human-like and use them with the same efficiency and smoothness that characterizes their human dialogues (Cassell 1989).

In this chapter we attempt to give a complete survey of facial gestures that can be useful as guidelines for their implementation in an ECA. Facial gestures include all facial displays except explicit verbal and emotional displays. A facial gesture is a form of nonverbal communication made with the face or head, used continuously instead of or in combination with verbal communication. There are a number of different facial gestures that humans use in everyday life. While verbal and emotional displays have been

---

[1] See http://www.paulekman.com/

investigated substantially, existing ECA implementations typically concentrate on some aspects of facial gestures but do not cover the complete set. We will first describe existing systems that introduce one or more aspects of facial gesturing, and then we will attempt to specify a complete set of facial gestures with its usage, causes, and typical dynamics.

Related Work

- The Autonomous Speaker Agent (Smid *et al.* 2004) performs dynamically correct gestures that correspond to the underlying text incorporating head movements (different kinds of nods, swing), eye movements (movement in various directions and blinking), and eyebrow movements (up and down and v.v.).
- Albrecht *et al.* (2002) introduce a method for automatic generation of the following nonverbal facial expressions from speech: head and eyebrow raising and lowering dependent on the pitch; gaze direction, movement of eyelids and eyebrows, and frowning during thinking and word search pauses; eye blinks and lip moistening as punctuators and manipulators; random eye movement during normal speech. The intensity of facial expressions is additionally controlled by the power spectrum of the speech signal, which corresponds to the loudness and intensity of the utterance.
- Poggi and Pelachaud (2002) focus on the gaze behavior, analyzing each single gaze in terms of a small set of physical parameters like eye direction, humidity, eyebrow movements, blinking, pupil dilatation, etc. In addition, they tried to find the meanings that gaze can convey.
- The Eyes Alive system (Lee *et al.* 2002) reproduces eye movements that are dynamically correct at the level of each movement, and that are also globally statistically correct in terms of the frequency of movements, intervals between them, and their amplitudes. Although the speaking and the listening modes are distinguished, movements are unrelated to the underlying speech contents, punctuation, accents, etc.
- Cassell *et al.* (1994) automatically generate and animate conversations between multiple human-like agents including intonation, facial expressions, lip motions, eye gaze, head motion, and hand gestures.
- The BEAT system (Cassell *et al.* 2001) controls movements of hands, arms and the face and the intonation of the voice, relying on rules derived from extensive research into human conversational behavior.
- Graf *et al.* (2002) analyze head and facial movements that accompany speech and investigate how they relate to the text's prosodic structure. They concluded that despite large variations from person to person, patterns correlated with the prosodic structure of the text.
- Pelechaud *et al.* (1996) report result from a program that produces animation of facial expressions and head movements conveying information correlated with the intonation of the voice. Facial expressions are divided by its function (determinant) and the algorithm for each determinant is described, with special attention to lip synchronization and coarticulation problems.

9.2 Facial Gestures for Embodied Conversational Agents

Facial gestures are driven by (see also Cassell *et al.* 2000):

- *Conversational function of speech.* We unconsciously use facial gestures to regulate the flow of speech, accent words, or segments and punctuate speech pauses.
- *Emotions.* They are usually expressed with facial gestures.
- *Personality.* It can often be read through facial gestures.
- *Performatives.* For example, advice and order are two different performatives and they are accompanied with different facial gestures.

The same facial gesture can have different interpretations in different conditions. For example, blinking can be the signal of a pause in the utterance, or can serve to wet the eyes. In the context of conversational speech, all facial gestures can be divided into four categories according to the function (usually called *determinant*) they have (Pelachaud *et al.* 1996):

- *Conversational signals.* They correspond to the facial gestures that clarify and support what is being said. These facial gestures are synchronized with accents or emphatic segments. Facial gestures in this category are eyebrow movements, rapid head movements, gaze directions, and eye blinks.
- *Punctuators.* They correspond to the facial gestures that support pauses. These facial gestures group or separate the sequences of words into discrete unit phrases, thus reducing the ambiguity of speech. Examples are specific head motions, blinks, or eyebrow actions.
- *Manipulators.* They correspond to the biological needs of a face, such as blinking to wet the eyes or random head nods, and have nothing to do with the linguistic utterance.
- *Regulators.* They control the flow of conversation. A speaker breaks or looks for an eye contact with a listener. He turns his head towards or away from a listener during a conversation. We have three regulator types: *speaker-state-signal* (displayed at the beginning of a speaking turn), *speaker-within-turn* (a speaker wants to keep the floor), and *speaker-continuation-signal* (frequently follows speaker-within-turn). The beginning of themes (previously introduced utterance information) is frequently synchronized by a gaze-away from a listener, and the beginning of rhemes (new utterance information) is frequently synchronized by a gaze-toward a listener.

In addition to characterizing facial gestures by their function, we can also characterize them by the amount of time that they last (Cassell *et al.* 2000). Some facial gestures are linked to personality and remain constant across a lifetime (e.g., frequent eye blinking). Some are linked to emotional state, and may last as long as the emotion is felt (e.g., looking downward and frowning in case of sadness). And some are synchronized with the spoken utterance and last only a very short time (e.g., eyebrow raising on the accented word).

All these functions are supported by a fairly broad repertoire of facial gestures. What follows is a detailed description of all facial gestures. Anatomic parts of the face we take into account are: head, mouth (including lips, tongue, and teeth), eyebrows (inner, medial, outer), eyelids (upper, lower), eyes, forehead, nose, and hair. Within each part, different gestures are recognized. Each facial gesture class is described with its attributes, parameters (aspects, actions, and presence/absence), and the function it can serve. According to the function a single facial gesture can have, it is additionally described by its causes and usage, level of synchronization, and typical dynamics and amplitudes.

An overview of facial gestures is shown in the Table 9.1, which is organized as follows.

- The first column on each page contains the anatomic part of the face that takes part in facial gesture creation. The regions that are used during speech, but not in any other facial gesture, are not included in the table (e.g., cheek or chin). Similarly, the regions that are not connected with any facial gesture (at least according to the available literature) are not included (e.g., iris or ears).
- Facial gestures recognized within each part of the face are given in the second column, *Gesture*. For some facial regions there are several known gestures (such as different kinds of nods) while some regions are characterized with only one gesture (e.g., hair with the hairline motion).
- The column *Description* gives a brief description of each facial gesture.
- The column *Attributes* contains characteristics important for a single gesture or a group of gestures. If the single gesture is not characterized with any attribute, the cell is left empty and shaded (this is the case when the gesture is described unambiguously without any additional parameters – e.g., nose wrinkling).
- The column *Parameters* provides values that attributes can achieve.

- A function that a facial gesture can have is given in the column *Function*. According to the previous division, C stands for conversational signal, P for punctuator, M for manipulator, and R for regulator. Some facial gestures can have more than one function (e.g., frowning serves as a conversational signal as well as being a punctuator). For several facial gestures, the function is not assigned and that field is left empty and shaded, since those gestures are connected to personality (e.g., teeth gnashing while listening or talking).
- The column *Usage/causes* provides information about each gesture, according to the function it has and concerning typical usage scenarios. It gives information about situations in which a certain facial gesture might appear and the meaning it has. Also, any available knowledge important for facial gesture understanding and use in ECA systems is given here.
- Verbal and nonverbal signals are synchronized. Synchrony occurs at all levels of speech: phonemic segment, word, syllable or long utterance, as well as at pauses. The column *Level of synchronization* specifies on which level the gesture is synchronized with the underlying speech.
- The column *Typical dynamics and amplitude* gives available knowledge on typical dynamics and amplitudes of the gesture, including known rules for its application in an ECA. For example, head nods are characterized by small amplitude and the direction up–down or left–right. Another example is saccadic eye movements which are often accompanied by a head rotation.

Cells left empty but not shaded in the table indicate fields for which these is insufficient information in the studied literature. The focus of this chapter is nonverbal and nonemotional facial displays, so the influence of affects is not described in detail. However, in some cases their presence is stated in order to make clear the usage and causes of facial gestures.

9.2.1 The Head

Head movements are frequently used. Attributes and parameters that characterize head movements are: direction (left, right, up, down, forward, backward, diagonal), amplitude (wide, small), and velocity (slow, ordinary, rapid). Amplitude and velocity seem to be inversely related. Movement with big amplitude is rather slow. Different combinations of these parameters define several head movement types (Pelachaud *et al.* 1996; Smid *et al.* 2004).

- *Nod.* An abrupt swing of the head with a similarly abrupt motion back. Nod can be used as a conversational signal (e.g., nodding for agreement/disagreement or to accentuate what is being said), synchronized at the word level, or as a punctuation mark. Typically, a nod is described as a rapid movement of small amplitude with four directions: left and right, right and left, up and down, and down and up.
- *Postural shifts.* Linear movements of wide amplitude often used as a regulator. Postural shifts occur at the beginning of speech, between speaking-turns, and at grammatical pauses maintaining the flow of conversation. Synchronization with verbal cues is generally achieved at pauses of speech.
- *Overshoot nod.* Nod with an overshoot at the return. The pattern looks like an "S" lying on its side. It is composed of two nods: the first one is with bigger amplitude starting upwards, while the second one is downwards with smaller amplitude.
- *Swing.* An abrupt swing of the head without the back motion. Sometimes the rotation moves slowly, barely visible, back to the original pose; sometimes it is followed by an abrupt motion back after some delay. Possible directions are up, down, left, right, and diagonal. It occurs at increased speech dynamics (when the pitch is also higher) and on shorter words.
- *Reset.* Sometimes follows swing movement; returns head to central position. Reset is a slow head movement. It can be noticed at the end of a sentence: the sentence finishes with a slow head motion coming to rest.

Table 9.1 Facial gestures

Facial region	Gesture	Description	Attributes	Parameters
HEAD	nod	An abrupt swing of the head with a similarly abrupt motion back.	direction	left, right, up, down, forward, backward, diagonal
	postural shift	Linear movements of big amplitude (i.e. they change the axis of motion).	amplitude	big, small
	swing	An abrupt swing of the head without the back motion.	velocity	slow, ordinary, rapid
	reset	Sometimes follows swing movement. Returns head in central position.		
	overshoot nod	Nod with an overshoot at the return (i.e. the pattern looks like an 'S' lying on its side).		
FOREHEAD	frown	Wrinkling (contracting) the brow.	intensity	strong, normal, weak
EYEBROWS (left, right) (inner, medial, outer)	raise	Eyebrows go up and down.	direction	up, central, down
	frown	Eyebrows go down and up.	amplitude	wide, small, slow
			velocity	ordinary, rapid
EYELIDS (left, right) (upper, lower)	blinking	Periodic or voluntary eye blink (closing and opening one or both of the eyes rapidly).	velocity frequency	rapid frequent, normal, rarely
	winking	The eyelid of one eye is closed and opened deliberately.	side	left, right
EYES GAZE	eye avoidance	Aversion of gaze – the speaker looking away from the listener.	duration	
	eye contact	The speaker is steadily looking toward to the listener for a period of time.		
	lowered gaze	The level of gaze falls.		
	rising gaze	The level of gaze rises.		
	saccade	A rapid intermittent eye movements from one gaze position to another.	velocity direction magnitude duration inter-saccadic interval	
PUPIL	dilation	The pupil condition of being expanded or stretched.	diameter centre	default, dilated, narrow
HAIR	hairline motion	Moving hairline (the outline of the growth of hair on the head).	direction	up, down
NOSE	wrinkling	Nose wrinkling in order to show an emotional state.		
LIPS	wetting	Periodic moistening of the lips done by passing one lip over another (upper and lower) wihtout using a tongue.	frequency velocity	frequent, normal, rarely slow, ordinary, rapid
TONGUE	lips licking	Passing the tongue over or along the lips.	*the same as lips wetting*	
TEETH/JAW	lips biting	Biting one's lips.		
	gnashing	Grinding or striking the teeth together.		

Table 9.1   (*Continued*)

Facial region	Function*	Usage/causes	Level of synchronization	Typical dynamics and amplitude
HEAD	C	agreement/disagreement, emphatic discourse, accent	word	Small amplitude, left-right or up-down
	P	– as punctuation mark	pause	Small amplitude, left-right or up-down
	R	at the beginning of the speech, between speaking-turns, at the grammatical pauses	pause	High velocity, big amplitude
	C	– at increased speech dynamics (higher pitch) – on shorter words	word	Sometimes the rotation moves slowly, barely visible, back to the original pose sometimes it is followed by an abrupt motion back after some delay; up, down, left, right, diagonal
	C	– the sentence finishes with slow head motion coming to rest	word	Slow head movement
	C	– as a swing nod, but it happens less frequent	word	Two nods – the first one is with bigger amplitude starting upwards; the second one is downwards with smaller amplitude
FOREHEAD	P	– mark a period – when thinking or in search pauses	pause	
	C	– showing feelings such as dislike, displeasure	word	
EYEBROWS (left, right) (inner, medial, outer)	P	– as punctuation mark (e.g. when asking a question), when thinking	pause	
	C	to accentuate a word, showing affirmation (yes) or not sure (perhaps)	word	
	P	as punctuation mark, in word search pauses	pause	
	C	– when speaker experiences difficulties, distress, doubt or sadness	word	
EYELIDS (left, right) (upper, lower)	M	– wet the eye (period of occurrence is affect dependent)	phoneme or pause	They appear every 4.8s and last 1/4s with 1/8s closure time 1/24s of closed eyes and 1/12s opening time.
	P	– mark a pause	pause	
	C	to emphasize the speech, to accentuate a word	word or syllable	
	C	– to convey a message, signal, or suggestion	word	
EYES GAZE	R	– at the beginning of an utterance, signalling that a person is thinking – looking down when answering questions (e.g. someone might look away when asked a question as they compose their response) – at hesitation pause when thinking what to say (looking up)	pause	
	C	– while speaking as opposed to listening	word	
	R	at the end of an utterance (when passing speaking turn) during pauses in speech, at the beginning of a phrase boundary pause (the pause between two grammatical phrases of speech), when asking questions	pause or word	
	R	– at the hesitation pause (delays that occur when the speaker is unsure of what to say next), which requires more thinking	pause	
	C	– during discussion of cognitively difficult topics	word	
	R	at the end of an utterance in order to collect feedback from the listner	word	
	C	– clarifies what is being said – often accompanied by a head rotation	word	Natural saccade magnitude – less than 15 degrees, direction-up-down, left–right, duration – 40 deg/sec.;
PUPIL		Pupil changes occur during affectual experiences. Pupil dilation expresses a level of interest.		Pupil dilation is followed by constriction during "happiness" and "anger" and remains dilated during "fear" and "sadness". When we find a particular subject fascinating. Our pupils are unconsciously dilated, as to opening up to the speaker. Pupil dilation correlates quite highly with heart-rate.
HAIR	C	– to accentuate a word	word	
NOSE	C	– showing feelings such as disgust, dislike disdain	word	
LIPS	M	– during long speech periods, due to biological need	pause	During pauses where thinking or word search expression is exhibited, the tongue/lip motion is slower, because the speaker is concentrating entirely on what to say next.
	P	– during thinking or word search pauses	pause	
TONGUE				
TEETH/JAW	P	at the hesitation pause (delays that occur when the speaker is unsure of what to say next). which requires more thinking	pause	
	C	– showing nervousness	utterance	
		Related to personality.		

* C – conversational signal, P – punctuator, M – manipulator, R – regulator

Another issue is the base head position or orientation which can be towards or away from a listener, up or down, etc. The head direction may depend on affect (e.g., the speaker–listener relationship) or can be used to point at something. For example, if the utterance is a statement, the head is positioned to look down as the speaker reaches the end of the sentence.

9.2.2 The Mouth

The parts of the mouth, including lips, tongue, jaw, and teeth, take part in speech production. Their shape or position depends on the articulated phoneme and forms a *viseme*, which is the visual representation of a phoneme (Benoit *et al.* 1992). According to the MPEG-4 standard we can distinguish only 15 different visemes, including a neutral face (Pandžić and Forchheimer 2002).

Openness of the lips is dependent not only on the articulated speech: emotional state can also influence how wide the lips will be open. Also, intensity of the lip shape action decreases during fast speech.

In speech production, the lips, tongue and teeth are used in various ways:

- *Lip wetting*. Periodic moistening of the lips done by passing one lip over another (upper and lower) without using the tongue. It occurs during long speech periods due to biological need (serving as manipulator), but also during thinking or word search pauses (serving as punctuator). Lip wetting is characterized by the frequency and velocity attributes. During pauses where thinking or word search expression is exhibited, the tongue/lip motion is slower, because the speaker is concentrating entirely on what to say next. Frequency of occurrence also depends on the personality and the outside conditions.
- *Lip licking*. Passing the tongue over or along the lips. The function, usage and causes are the same as for lip wetting (the same result is obtained in two different ways).
- *Lip biting*. Biting one's lip is often a sign of nervousness or insecurity. It might occur in hesitation pauses, when the speaker is thinking what to say next (serving as punctuator).
- *Teeth gnashing*. Grinding or striking the teeth together is a gesture related to personality and not to the context of speech.

9.2.3 The Eyebrows

Eyebrow movements appear frequently as conversational signals or punctuators. When serving as punctuator, they are used to mark a period for thinking and word search pauses.

Eyebrow raise (eyebrows go up and down) is often used to accentuate a word or a sequence of words as well as to show affirmation (yes) or insecurity (perhaps) (Poggi and Pelachaud 2002) . Eyebrow frown (eyebrows go down and up) might appear when the speaker experiences difficulties, distress, or doubt.

Eyebrow movement also has amplitude and velocity, and is closely related to pitch contour: eyebrows are raised for high pitch and lowered again with the pitch (Albrecht *et al.* 2002).

9.2.4 The Eyelids

Eyelids determine the openness of eyes. Closure of the eyes happens quite frequently due to eye blinks, described as rapid closing and opening of one or both eyes which might happen in frequent, normal, or rare periods.

- *Periodic blinks* serve the physical need to keep the eyes wet. Periodic eye blinks are manipulators. On average they appear every 4.8 seconds but their occurrence is dependent on affect. The eye blink consists of three components: closure time (about 1/8th of a second), closed eyes ($\sim$ 1/24th second), and opening time, ($\sim$ 1/12th second). The total duration is therefore about one-quarter of a second.

- *Voluntary blinks* appear in two roles, as punctuators (to mark a pause), synchronized with a pause, or as conversational signals (to emphasize speech or to accentuate a word), synchronized with a word or syllable.

Eye openness varies also depending on the affect. For "surprise" and "fear" the eyes are wide open and they are partially closed during "sadness", "disgust", and "happiness" (Pelachaud *et al.* 1996).

Another facial gesture performed by eyelids is eye winking (the eyelid of one eye is closed and opened deliberately). It is quite common in everyday communication and is used to convey a message, signal, or suggestion. This conversational signal, is synchronized with the word spoken.

9.2.5 *The Eyes*

Eyes play an essential role as a major channel of nonverbal communicative behavior. Different expressions can be reflected in eyes. Eyes can be in tears, red or dry, open or closed, showing clearly the state of our mind. Eyes can slightly narrow when adding more precise information, or widen when asking for a speaking turn (Poggi and Pelachaud 2002).

Pupil changes occur during emotional experiences. They may be dilated or narrow, centered or not. Pupil dilation is followed by constriction during "happiness" and "anger" but remains during "fear" and "sadness". Pupil dilation also expresses a level of interest: when we find a particular subject fascinating, our pupils are unconsciously dilated, as if opening up to the speaker (Pelachaud *et al.* 1996).

Eyes interact in face-to-face communication through gaze direction or intensity and *saccade*. Saccade is a rapid intermittent eye movement from one gaze position to another executed voluntarily. It has several attributes: *direction*; *velocity*; *magnitude or amplitude* (the angle through which the eyeball rotates as it changes fixation from one position to another); *duration* (the time that the movement takes to execute, typically determined using a velocity threshold); and *Intersaccadic interval* (the time which elapses between the termination of one saccade and the beginning of the next one). Natural saccade movement (usually up–down, left–right) rarely have a magnitude greater than 15 degrees, while the duration and velocity are functions of its magnitude (Lee *et al.* 2002) .

Eye gaze is used to signal the search for feedback during an interaction, look for information, express emotion, influence another person's behavior, or help regulate the flow of conversation (Cassell *et al.* 1994). However, some cultural differences are found in the amount of gaze allowed. Gaze can be classified into four primary categories depending on its role in the conversation (Cassell *et al.* 1994):

- *Planning* corresponds to the first phase of a turn when the speaker organizes thoughts.
- *Comment* accompanies speech, by occurring in parallel with accent and emphasis.
- *Control* controls the communication channel and functions as a synchronization signal.
- *Feedback* is used to collect and seek feedback.

According to Poggi and Pelachaud (2002), two broad types of meanings of gaze can be distinguished: information on the world and information on the sender's mind. The first class includes places, objects and times to which we refer; the second class contains sender's beliefs, goals and emotions (e.g., words giving information <*of course, no*> are spoken with eyebrows central and down).

Aversion of gaze happens at the beginning of an utterance, signaling that a person is thinking what to say. Eye contact occurs at the end of an utterance (when passing the speaking turn), during pauses in speech, or at the beginning of a phrase boundary pause (the pause between two grammatical phrases of speech). Eye avoidance and eye contact follow the same rules as head movements for speaking turns. The level of gaze falls at the hesitation pause or during discussion of cognitively difficult topics, while it rises at the end of an utterance in order to collect feedback from the listener.

9.2.6 The Forehead

Wrinkles often appear on the forehead. Vertical or horizontal, curved or oblique, they are positioned central, lateral or all along the forehead, giving a special note to one's personality. During speech, depending on context and affect, wrinkles often deepen or change in direction and shape.

Another gesture connected with the forehead is a frown. It is used during the search and thinking pauses and to mark a period (serving as a punctuator) or to show feelings such as dislike or displeasure. Intensity of the frown varies from strong to weak, depending on the context.

9.2.7 Other Facial Parts

- Some people move their hairline (the outline of the growth of hair on the head) up and down to accentuate what is being said. This conversational signal is synchronized on the word level.
- Nose wrinkling is used to show an emotional state and feelings such as disgust, dislike, or disdain.
- Color of the cheeks is connected with the emotional state, outside conditions, or personality. The human face can loose color, blush, or stay as it is.
- There are some other facial gestures not directly affecting visual output, but as other nonverbal facial displays they complete the verbal output. Examples are: heavy gulp, rapid breathing, and dry cough.

9.2.8 Combinations of Facial Gestures

Table 9.2 gives a few representative examples of how groups of facial gestures might be used together to convey different meanings. This is not meant to give a complete overview, but only to illustrate the idea that facial gestures often come in groups.

Table 9.2 Frequently used combinations of facial gestures

Combination of facial gestures	Meaning / function / context
Head aside Eyebrows up	Performative eyes (I suggest)
Looking up Eyebrows raise Raise head Frowning	Thinking
Avoidance of gaze Head of the speaker turns away from listener	Hesitation pause (speaker is concentrating on what to say)
Look down at the end of the sentence Head down	Statement
Eyebrows raise Head raising at the end Gaze toward to listener High pitch at the end of the utterance	Question
Head nodding Eyebrows raise	Affirmation (yes)

Figure 9.1 Model of the Autonomous Speaker Agent

9.3 Example of a Practical System Implementation

In this section we present an example of a practical ECA system. Our Autonomous Speaker Agent (ASA) implements a subset of a full set of ECA functionality, so we first give the scope of the system. Then the technologies and architecture we have used are explained in detail. Finally, we present empirical results along with real users' impressions of our system implementation.

9.3.1 System Scope

ASA is a presentation system in the ECA domain. Since its function is only presentation of content, the first simplification is that we implemented only a subset of facial gestures. That means that our ASA is presented with a model of a human head (Figure 9.1).

We focus our ASA implementation on a subset of the interactional functions of speech. It supports conversational signals, punctuators, and manipulators. The final simplification is the set of facial gestures (Table 9.3).

Table 9.3 Implemented facial gestures

Facial display	Type	Direction
Head	Nod	Up Down Left Right
	Overshoot nod	Up then down
	Swing	Up Down Left Right Diagonal
	Reset	To central
Eyes	Simple gaze Blink	
Eyebrows	Raise	^^

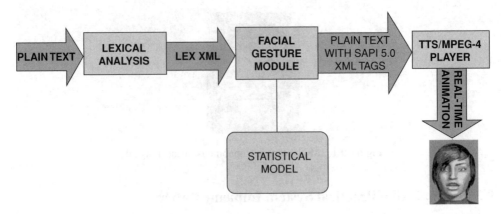

Figure 9.2   ASA system modules

9.3.2 System Overview

We will first give brief ASA system overview with a basic explanation of data flow between system modules. After that, each system module is described in detail along with technology and standards used in its implementation.

The ASA system consists of several modules. System input is plain English text and output is real-time animation with appropriate facial gestures and audio. Figure 9.2 shows the modules along with data flow through them. Plain English text is first processed using the Lexical Analysis module that generates English text in XML format with appropriate lexical tags (describing new/old words and punctuation marks). The Facial Gesture module then inserts appropriate gestures in the form of Microsoft Speech API[2] (SAPI 5.0) XML tags. This text with appropriate gesture tags represents input for the TTS/MPEG-4 Player module that is based on the SAPI 5.0 Text To Speech (TTS) engine and Visage SDK API.[3] While the SAPI 5.0 TTS engine generates an audio stream, the SAPI 5.0 notification mechanism is used to catch the timing of phonemes and XML tags containing gesture information. Gesture XML tags hold three gesture parameters: gesture type, amplitude, and duration in word units. In order to produce real-time animation, we used the assumption that average word duration is one second.

The Lexical Analysis module performs linguistic and contextual analysis of a text written in English with the goal of enabling the nonverbal (gestures) and verbal (prosody) behavior assignment and scheduling. The main goal of this module is to determine whether a word (or group of words) is new in the utterance context (*new*), or extends some previously used word or group of words (*old*), and to determine the punctuation marks. Punctuation mark determination is straightforward, but for the first two tasks there is a need to know morphological, syntactic and part-of-speech word information. For that purpose we used WordNet,[4] which is a lexical reference system whose design is inspired by current psycholinguistic theories of human lexical memory. English nouns, verbs, adjectives, and adverbs are organized into synonym sets, each representing one underlying lexical concept. Different relations link the synonym sets. First we need to determine word type (noun, verb, adverb, or adjective) by querying the WordNet system, using English grammatical rules and parsing input text multiple times. Each noun, verb, adverb, or adjective is tagged as new if itself or any of its synonyms (queried from the WordNet system) has not been mentioned

---

[2] Microsoft speech technologies (http://www.microsoft.com/speech/)
[3] Visage Technologies AB (http://www.visagetechnologies.com/)
[4] See http://www.cogsci.princeton.edu/~wn/

Table 9.4 Input and output data of the Lexical Analysis module

Module input	Module output

"However, figures presented by Business Unit Systems prompted more positive reactions."	`<?xml version="1.0" encoding="UTF-8"?>`
	`<UTTERANCE>`
	`  <CLAUSE>`
	`    <WORD Text="However" New="Yes"/>`
	`    <WORD Text="," />`
	`    <WORD Text="figures" New="Yes"/>`
	`    <WORD Text="presented" New="Yes"/>`
	`    <WORD Text="by"/>`
	`    <WORD Text="Business"/>`
	`    <WORD Text="Unit" New="Yes"/>`
	`    <WORD Text="Systems" New="Yes"/>`
	`    <WORD Text="prompted" New="Yes"/>`
	`    <WORD Text="more" New="Yes"/>`
	`    <WORD Text="positive" New="Yes"/>`
	`    <WORD Text="reactions" New="Yes"/>`
	`    <WORD Text="."/>`
	`  </CLAUSE>`
	`</UTTERANCE>`

in previous text. Other word classes are not considered as new. WordNet does not handle pronouns, so we developed a special algorithm for processing them. Every pronoun that is not preceded by a noun in the sentence, and is part of the set {"any", "anything", "anyone", "anybody", "some", "somebody", "someone", "something", "no", "nobody", "no-one", "nothing", "every", "everybody", "everyone", "everything", "each", "either", "neither", "both", "all", "this", "more", "what", "who", "which", "whom", "whose"}, or any pronoun that is part of the set {"I", "you", "he", "she", "it", "we", "they"}, gets the **new** tag assigned. All other pronouns that do not fulfill the above-stated requirements are tagged with **old**. Also, each pronoun, substituting a noun that appears after it, or a noun that does not appear in the text at all, needs to be tagged as **new**. Table 9.4 represents one example of input and output data of the Lexical Analysis module.

Next, the Statistical Model of facial gestures is the basic building block of the Facial Gesture module. The decision-tree algorithm uses data from the Statistical Model (explained in detail later). First, we will present the Statistical Model of facial gestures along with methods, tools, and data sets used to build it.

Figure 9.3 represents the training process of the ASA system. As an input for this process we used as a training data set a number of Ericsson's "5 minutes" video clips. These clips are published by LM Ericsson for internal usage and offer occasional in-depth interviews and reports on major events, news, or hot topics from the telecom industry and they are presented by professional newscasters. As we have already stated, we identified three facial gesture parameters: gesture type (see Table 9.3), amplitude, and duration. So, the Statistical Model consists of components for those three facial gesture parameters for every lexical context (new, old, and no lexical information). We will next explain the process of generating text annotated with gesture and lexical data.

Table 9.5 presents an extract of text annotated with gesture and lexical data. The *word* row contains an analyzed news extract separated word-by-word. The *eyes*, *head*, and *eyebrows* rows hold data about facial motion that occurred on the corresponding word: type of motion, direction, amplitude and determinant, according to the notation summarized in Table 9.6.

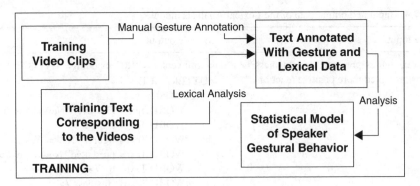

Figure 9.3   ASA training process

These basic facial motions were mapped to the facial gestures described in Table 9.3. We replayed newscaster footage to determine the facial gestures type, direction, and duration parameters. For example, the last row in Table 9.6 contains the head movement facial gesture parameters.

The *pitch* in Table 9.5 row indicates which words were emphasized by voice intonation. The *lexical* row holds information about a word's newness in the context of a text (output of Lexical Analysis module).

For data analysis it was very important to use good and valid analysis techniques. Our analysis was partly based on the work described in Lee *et al.* (2002). In that work, the authors proposed an eyes animation model. The animation was driven by conversation between two people. The data was gathered using a sophisticated eye-movement capturing device during a 9-minute conversation between two individuals, and the authors analyzed the gathered data using well-known statistical methods and functions. The statistical eye-movement animation model was then incorporated into the existing facial animation model.

The first significant difference here is that we are not analyzing a conversation between two individuals. Because our facial animation model was for the ASA, we analyzed recorded video of real speakers who were not involved in conversation. Also, we didn't use any sophisticated equipment: our tools were Movie Maker and observing the recorded video. In Movie Maker we could analyze the facial and head movements by watching the recorded video frame by frame. It was a tedious task, but the quality of analysis was not put in question.

Table 9.5   Text annotated with gesture and lexical data

Word	52		Three	arraignments	and
Eyes	3		blink::cs		blink::m
Head		|up;A=2	|d to n |d A=0.25	Id A =0.5	|up to n A=0.5
Eyebrows	2		raise::cs A=1/4		
Pitch	13		+		
Lexical	44		new	new	
			cs – conversational signal		
			p – punctuator		
			m – manipulator		
		~nod::A1=2:A2=0.5::cs			

Table 9.6 Basic facial motions triggered by words

Facial gesture	Type of motion	Description
Head movement	|up	vertical up
	|up to n	vertical up to neutral (center) position
	|down	vertical down
	|down to n	vertical down to neutral (center) position
	– to left	horizontal left
	– to right	horizontal right
	– to n	horizontal to neutral (center) position
	/ up	diagonal up from left to right[a]
	/ down	diagonal down from right to left[a]
	\ up	diagonal up from right to left[a]
	\ down	diagonal down from left to right[a]
Eyebrows movement	raised s	eyebrows going up to maximal amplitude
	raised e	eyebrows going down to neutral position

[a] From the listener's point of view.

It is important to state that in our model, the basic unit, which triggers head or facial movement, is a word. Words can be divided into phonemes or syllables, but for simplicity of data analysis we chose words as basic units for the facial animation model.

In our data analysis (Table 9.5) we first populated the word row with words along with punctuation marks from the news transcripts. Then, the first row we populated with observed data was the *pitch* row. Using headphones and playing the recorded video clips over and over again, we marked the words that the speaker had emphasized with her voice. Our primary concern was not the type of accent (Silverman *et al.* 1992), only if a particular word had been accented or not. In our table, every word that had been accented was marked with "+" in the pitch row. Every pitch accent had a duration of one word.

After analyzing the pitch information, we continued with *eye blinks*. Eye blinks started and ended on the same word or punctuation mark. Extracting data about eye blinks was the easiest part of the data analysis. We played the video clips in Movie Maker and marked the columns in the eyes row with "blink" if the speaker had blinked on the corresponding word or punctuation mark. It is important to state that eye blinks were triggered, aside from words, also by punctuation marks.

After that, the *head movement* analysis followed and it was the most tedious task of the whole data analysis process because a head has the biggest freedom of movement. We can move a head vertically up and down, horizontally left and right, diagonally, and even rotate it. An additional problem was that a head did not always start (or finish) its basic movement element in the neutral position.

According to Graf *et al.* (2002) the following are the head movement patterns:

- nod
- nod with overshoot
- swing (rapid).

Those three meta head movements have basic elements which are triggered by the uttered word or punctuation mark. Since words are the basic triggering units of our facial animation model, we had to devise symbols for the basic elements of the head movement patterns. They are listed in Table 9.6.

Our task was to map basic elements into basic head movement patterns. First we observed the video clips (average duration was about 15 seconds) using Movie Maker playing functions, and, using the frame-by-frame analysis, we marked words with corresponding basic elements of head movements. When we finished with basic elements, we mapped them into the corresponding basic patterns according to Table 9.3.

The basic elements of head movements are triggered by the uttered words. Sometimes one word triggers two basic elements. In such cases we had a nod triggered by a word. That nod was very fast with a small amplitude. Nods usually lasted through two words. Nods with overshoot had the longest duration. They were not so frequent and they lasted through four or five words.

Finally, we analyzed *eyebrow movement*. The basic eyebrow patterns are raises and frowns (see Table 9.3). Eyebrow movements also had basic elements which are described in Table 9.6. Eyebrow movements also lasted through one or more words.

The value for a particular facial motion is determined as follows. If the facial motion occurred on a punctuation mark, then the determinant for that motion was punctuator (p); if the facial motion accompanied a word that was new in the context of the uttered text, then the determinant was a conversational signal (cs); otherwise, the determinant of the facial motion was a manipulator (m).

During the data analysis we also extracted amplitudes of the head and eyebrow movements. Our goal was to propose a statistical model for the amplitudes as important components of the facial animation model. Also, amplitudes of the head and eyebrow movements were in close relation with the pitch amplitudes. The recorded video data that we had analyzed and the method used (that will be also described here) gave us a good starting point for the statistical model of amplitudes.

The raw data tables were populated during manual analysis and measurement. All amplitude values were normalized to Mouth–Nose Separation unit (MNS0) for the particular speaker. MNS0 is Facial Animation Parameter Unit (FAPU) in the MPEG-4 Face and Body Animation (FBA) standard.[5] Using MNS0 FAPU, our model could be applied to every 3D model of a speaker. Algorithms used for extracting amplitude values from observed footages are described in detail in Silverman *et al.* (1992).

In order to explain how the Facial Gesture module works we will follow the decision tree (Figure 9.4). The first branch point classifies the current input text context as either a word or a punctuation mark. Our data analysis showed that only eye blink facial gesture occurred on punctuation marks. Therefore, only the blink component of the statistical model is implemented in this context. A uniformly distributed random number between 0 and 100 is generated and two non-uniform intervals are assigned. That is, a random number between 0 and 33.68 is assigned to the eye blink, and a number between 33.68 to 100.00 to the no-blink motion. Thus, there is a 33.68% chance that an eye blink will occur, and a 66.32% chance that no eye blink motion will be generated. Words could be new or old in the context of the uttered text – this is the second branch point. All facial gestures occurred in both cases but with different probabilities. Because of that we have different components for facial gesture parameters in both cases. In the case of a new word, we first compute an eye blink motion. A uniformly distributed random number between 0 and 100 is generated and two non-uniform intervals are assigned. That is, a random number between 0 and 15.86 is assigned to the eye blink, and a number between 15.86 and 100.00 to the no-blink motion. Thus, there is 15.86% chance that an eye blink will occur, and a 84.14% chance that no eye blink motion will be generated. After eye blink, we compute the eyebrows motion parameters. The two parameters are motion amplitude and duration. Again, a uniformly distributed random number between 0 and 100 is generated and two non-uniform intervals are assigned. That is, a random number between 0 and 18.03 is assigned to the eyebrows raise, and a number between 18.03 to 100.00 to the no eyebrows raise. Thus, there is a

---

[5] ISO/IEC IS 14496-2 Visual (1999).

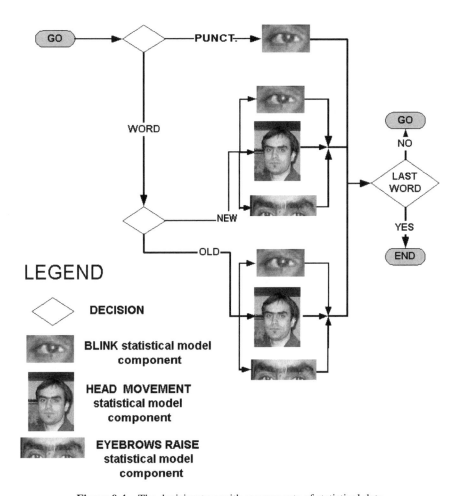

Figure 9.4 The decision tree with components of statistical data

18.03% chance that an eyebrows raise will occur, and a 81.97% chance that no eyebrows motion will be generated. After that we compute the eyebrows raise duration. A uniformly distributed random number between 0 and 100 is generated and ten non-uniform intervals are assigned. That is, a random number between 0 and 77.31 is assigned to the duration of one word, a number between 77.31 to 86.57 to the duration of two words, and so on. The eyebrows raise amplitude is dependent on the duration parameter. The duration of one word is qualified as short duration, while all other durations are qualified as long durations. A random number between 0 and 100 is generated. The number corresponds to the x-axis (percentage of frequency) in Figure 9.5.

Our decision-tree algorithm is triggered for every word. However, an eyebrows raise motion could last through more than one word, so an eyebrows raise is not calculated for words that are already covered by previous calculation for an eyebrows raise. Finally, we calculate the head motion. First, the head motion type is determined. A uniformly distributed random number between 0 and 100 is generated and five non-uniform intervals are assigned (for every head motion subtype). That is, a random number between 0 and 1.34 is assigned to the overshoot nod, a number between 1.34 and 23.71 to the nod, a number between 23.71 and 46.25 to the rapid (swing) movement, a number between 46.25 and 51.43 to the reset

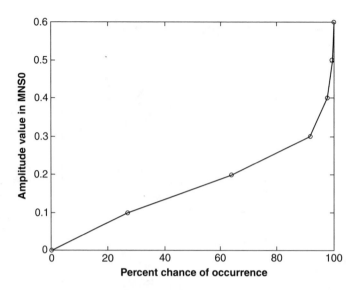

Figure 9.5 Linear approximation of the cumulative histogram for the amplitude of a short eyebrows raise

movement, and a number between 51.43 and 100.00 to the no head movement. Thus, there is a 1.34% chance that an overshoot nod will occur, a 22.37% chance that a nod will occur, a 22.54% chance that a rapid (swing) movement will occur, a 5.18% chance that a reset movement will occur, and a 48.57% chance that no head motion will be generated.

In the case of an overshoot nod, the following calculations are performed. An overshoot nod has two amplitude parameters (for first nod-up and second nod-down) and one duration parameter. We begin by computing an overshoot nod duration. A uniformly distributed random number between 0 and 100 is generated and seven non-uniform intervals are assigned. That is, a random number between 0 and 6.25 is assigned to the duration of one word, a number between 6.25 and 37.5 to the duration of two words, and so on. A random number between 0 and 100 is generated for the first and second amplitude. The random number corresponds to the x-axis (percentage of frequency) similar to Figure 9.5. Our decision-tree algorithm is triggered for every word. However, an overshoot nod could last through more than one word, so a head motion is not calculated for words that are already covered by the previous calculation for a head motion.

In the case of a nod, the following calculations are performed. A nod has a type parameter, an amplitude parameter, and one duration parameter. We begin by computing the nod type. A uniformly distributed random number between 0 and 100 is generated and four non-uniform intervals are assigned. That is, a random number between 0 and 60.15 is assigned to the nod-up, a number between 60.15 and 89.3 to the nod-down, a number between 89.3 and 97.42 to the nod-right, and a number between 97.42 and 100.00 to the nod-left. Thus, there is 60.15% chance that a nod-up will occur, a 29.15% chance that a nod-down will occur, a 8.12% chance that a nod-right will occur, and a 2.58% chance that a nod-left will be generated.

After the type parameter, we compute nod duration. A uniformly distributed random number between 0 and 100 is generated and five non-uniform intervals are assigned. That is, a random number between 0 and 62.31 is assigned to the duration of one word, a number between 62.31 and 85.45 to the duration of two words, and so on. Nod amplitude is dependent on the duration parameter. The duration of one word is qualified as short duration, and all other durations are qualified as long duration. A random number between 0 and 100 is generated for short and long amplitudes. The random number corresponds to the *x*-axis (percentage of frequency) similar to Figure 9.5. Our decision-tree algorithm is triggered for every word. However, a nod motion could last through more than one word, so a nod is not calculated for words that are already covered by previous calculation for a nod motion.

In the case of a rapid (swing) movement, the following calculations are performed. A rapid movement has a type parameter and an amplitude parameter. First we compute the rapid movement type. A uniformly distributed random number between 0 and 100 is generated and five non-uniform intervals are assigned. That is, a random number between 0 and 61.11 is assigned to the rapid down movement, a number between 61.11 and 76.3 to the rapid up motion, a number between 76.3 and 84.82 to the rapid left motion, a number between 84.82 and 95.56 to the rapid right motion, and a number between 95.56 and 100.00 to the rapid diagonal movement. Thus, there is a 61.11% chance that a rapid down motion will occur, a 15.19% chance that a rapid up motion will occur, a 8.52% chance that a rapid left motion will occur, a 10.74% chance that a rapid right motion will occur, and a 4.44% chance that a rapid diagonal motion will be generated.

A random number between 0 and 100 is generated for the amplitude calculation. The random number corresponds to the *x*-axis (percentage of frequency) similar to Figure 9.5. In the case of an old word context, the generation process for every facial display is the same as in the case of a new word context. The only differences are statistical data values.

From Figure 9.4, it is obvious that a word could be accompanied by all three facial gesture kinds. The output from the Facial Gesture module is plain English text accompanied by SAPI5 bookmarks for facial gestures. Every facial gesture has a corresponding bookmark value: Table 9.7 shows the boundary values for each bookmark. The head and eyebrows movement bookmark values not only define the type of facial gesture, but also contain the amplitude data and duration of the facial movement. For example, bookmark value 805120 (Bmk_value) defines the rapid head movement to the left (symbol L) of amplitude (A) 1.2 MNS0 and duration (D) of five words. The function for amplitudes of facial gestures L is:

$$D = (Bmk\_value - Bmk\_code)/1000.$$
$$A = ((Bmk\_value - Bmk\_code) - (D \times 1000))/100.$$

The interval for bookmark values for L is [800000, 900000> because the statistical data showed that the maximal amplitude value for facial gesture L was 2.2 MNS0, and duration was one word.

Table 9.7 Our SAPI5 bookmark codes of facial gestures

Bookmark code	Facial gesture
MARK = 1	conversational signal blink
MARK = 2	punctuator blink
MARK = 100000	eyebrows raise
MARK = 200000	nod ^
MARK = 300000	nod V
MARK = 400000	nod <
MARK = 500000	nod >
MARK = 9	rapid reset
MARK = 600000	rapid d
MARK = 700000	rapid u
MARK = 800000	rapid L
MARK = 900000	rapid R
MARK = 1000000	rapid diagonal

Head nods and eyebrows raises could last through two or more words. The statistics have shown that the maximum duration of a nod is five words, that an eyebrows raise can last through eleven words, and the maximal duration for a nod with overshoot is eight words. We code a nod with overshoot as two nods: a nod-up immediately followed by a nod-down. Every nod has its own amplitude distribution.

TTS/MPEG-4 Playing Module plays in real time, using the bookmark information, appropriate viseme and gestures model animation. The synchronization between the animation subsystem (MPEG-4 Playing) and the speech subsystem (Microsoft's TTS engine) can be realized in two-ways: with time-based scheduling and event-based scheduling. Which synchronization method will be used depends on the underling TTS engine implementation. In time-based scheduling, speech is generated before nonverbal behaviors. Event-based scheduling means that speech and nonverbal behaviors are generated at the same time. In our system we use the event-based scheduling method. We have implemented simple animation models for eye blink, simple gaze following, and head and eyebrows movement. Our system implementation is open, so every user is able to easily implement its own animation models. The animation model for head movement and eyebrows movement is based on the sine function. That means that our ASA nods his head and raises eyebrows following the sine function trajectory. In gaze following, the eye of our ASA move in the opposite direction to a head movement if the head movement amplitude is smaller than a defined threshold. This gives the impression of eye contact with ASA.

9.4 Results

We conducted a subjective test in order to compare our proposed statistical model to simpler techniques. We synthesized facial animation on our face model using three different methods. In method 1, head and eye movements were produced playing an animation sequence that was recorded by tracking movements of a real professional speaker. In method 2, we produced a facial animation using the system described in this chapter. Method 3 animated only the character's lips. We conducted a subjective test to compare those three methods of facial animation. The three characters were presented in random order to 29 subjects. All three characters presented the same text. The presentation was conducted in the Ericsson Nikola Tesla and all subjects were computer specialists. However, most of the subjects were not familiar with virtual characters. The subjects were asked the following questions:

Q1: Did the character on the screen appear interested in you (5) or indifferent to you (1)?
Q2: Did the character appear engaged (5) or distracted (1) during the conversation?
Q3: Did the personality of the character look friendly (5) or not (1)?
Q4: Did the face of the character look lively (5) or deadpan (1)?
Q5: In general, how would you describe the character?

Note that higher scores correspond to more positive attributes in a speaker. For questions 1 to 4, the score was graded on a scale of 5 to 1.

Figure 9.6 summarizes the average score and standard deviation (marked with a black color) for the first four questions. From the figure, we can remark that the character animated using method 2 was graded with the highest average grade for all questions except for Q2. The reason is that method 3 only animates lips while the head remains still. This gave the audience the impression of engagement in the presentation. A Kruskal–Wallis ANOVA indicated that the three characters had significantly different scores ($p = 0.0000$).

According to general remarks in answer to Q5, the subjects tended to believe the following:

- Type 1 looked boring and uninteresting, it seemed to have a cold personality. Also, implemented facial gestures were not related to the spoken text.

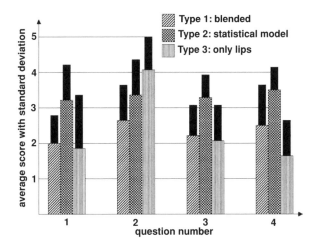

Figure 9.6   Results of subjective evaluations: average score and standard deviation

- Type 2 had a more natural facial gesturing and facial gestures were coarticulated to some extent. Head movements and eye blinks are related to the spoken text. However, eyebrow movements were with unnatural amplitudes and were not related to the spoken text.
- Type 3 looked irritating, stern, and stony. However, it appeared to be concentrated and its lips animation was the best.

9.5 Conclusions

We have tried to give a complete survey of facial gestures as a guideline for their implementation in an ECA. Specifically, we have concentrated on providing a systematically organized repertoire of common facial gestures, including information on typical usage, causes, any available knowledge on typical dynamics, and amplitude of the gestures. We have studied facial motions of head, mouth, eyebrows, eyelids, eyes, forehead, hair and nose, and within each part we have recognized different gestures. For each facial gesture class available knowledge is presented, including the function it can serve and attributes with its parameters.

Although we have attempted to cover all facial gestures with complete data that would make possible immediate implementation in an ECA system, there are still some gesture classes that are not adequately described, either because of lack of knowledge or due to complexity. Additional information needs to be added.

References

Albrecht I., Haber J. and Seidel H.P. (2002) Automatic generation of non-verbal facial expressions from speech. In *Proceedings Computer Graphics International* (CGI 2002), pp. 283–293.

Benoit C., Lallouache M.T., Abry C. (1992) A set of French visemes for visual speech synthesis. In G. Bailly and C. Benoit (eds), *Talking Machines: Theories, models and designs*. Elsevier Science Publishers, pp. 485–504.

Cassell J. (1989) Embodied conversation: integrating face and gesture into automatic spoken dialogue systems. In S. Luperfoy (ed.), *Spoken Dialogue Systems*. MIT Press, Cambridge, MA.

Cassell J. (2000) Nudge nudge wink wink: elements of face-to-face conversation for embodied conversational agents. In J. Cassell *et al.* (eds), *Embodied Conversational Agents*. MIT Press, Cambridge, MA, pp. 1–27.

Cassell J., Pelachaud C., Badler N., Steedman M., Achorn B., Becket T., Douvillle B., Prevost S. and Stone M. (1994) Animated conversation: rule-based generation of facial expressions. In *Proceedings of SIGGAPH '94*.

Cassell J., Sullivan J., Prevost S. and Churchill E. (2000) *Embodied Conversational Agents*. MIT Press, Cambridge, MA.

Cassell J., Vilhjálmsson H. and Bickmore T. (2001) BEAT: the behavior expression animation toolkit. In *Proceedings of SIGGRAPH 2001*, ACM Press, pp. 477–486.

Graf H.P., Cosatto E., Strom V. and Huang F.J. (2002) Visual prosody: facial movements accompanying speech. In *Proceedings of AFGR*, pp. 381–386.

ISO/IEC (1999) IS 14496-2 Visual.

Knapp M.L. (1978) *Nonverbal Communication in Human Interaction*, 2nd edn. Holt, Rinehart & Winston, New York.

Lee S.P., Badler J.B. and Badler N.I. (2002) Eyes alive. In *Proceedings of the 29th Annual Conference on Computer Graphics and Interactive Techniques*, San Antonio, Texas, ACM Press, pp. 637–644.

Mehrabian A. (1971) *Silent Messages*. Wadsworth, Belmont, CA.

Pandžić I.S. and Forchheimer R. (eds) (2002) *MPEG-4 Facial Animation: the standard, implementation and applications*. John Wiley.

Pelachaud C., Badler N. and Steedman M. (1996) Generating facial expressions for speech. *Cognitive Science* **20** (1), 1–46.

Poggi I. and Pelachaud C. (2002) Signals and meanings of gaze in animated faces. In P. McKevitt, S. O'Nuallàin, Conn Mulvihill (eds), *Language, Vision, and Music*. John Benjamins, Amsterdam, pp. 133–144.

Ruesch J. (1966) Nonverbal language and therapy. In A.G. Smith (ed.), *Communication and Culture: Readings in the codes of human interaction*. Holt, Rinehart & Winstone, New York, pp. 209–213. *See also* Nonverbal dictionary of gestures, signs and body language cues, http://members.aol.com/nonverbal2/diction1.htm

Silverman K., Beckman M., Pitrelli J., Osterndorf M., Wightman C., Price P., Pierrehumbert J. and Herschberg J. (1992) ToBI: a standard for labeling English prosody. In *Proceedings of Conference on Spoken Language*, Banff, Canada, pp. 867–870.

Smid K. (2004) *Simulation of a Television Speaker with Natural Facial Gestures*. Master thesis no. 03-Ac-10/2000-z, Faculty of Electrical Engineering and Computing, University of Zagreb.

Smid K., Pandžić I.S. and Radman V. (2004) Autonomous speaker agent. In *Computer Animation and Social Agents Conference* (CASA 2004), Geneva, Switzerland.

Part II

Conversational Contents

10

Conversation Quantization and Sustainable Knowledge Globe

Hidekazu Kubota, Yasuyuki Sumi, and Toyoaki Nishida

10.1 Introduction

Conversation is a primary medium for creating and sharing knowledge in many situations such as education, planning, decision-making, or even casual conversations. One of the prominent advantages of conversation as a medium is its heterogeneity, which includes both verbal and nonverbal representations. Conversation generally also includes contextual information such as circumstances, times and places, or goods and equipments concerning the conversation. This heterogeneity enables people to communicate rich information through conversation. Although the heterogeneity helps people, it makes it difficult for computers to process human conversation. Since daily conversations contain vast amounts of knowledge, computational support for them is indispensable; however, the knowledge representation used in computers is too formal for expressing them. Well-formed computational representations such as ontology do not consider the conversational nuances. Formal representations are unsuitable for describing the conversational activities of professionals. The essence of professional ideas and skills is so informal and implicit that novices can acquire professional skills through deep conversation and imitation for a long duration, as opposed to book learning. Rich information media such as video representations are popular and relatively better than formal representations from the viewpoint of expressing conversational nuances. Videos represent both the verbal and nonverbal behavior of humans and their contexts in an understandable manner. The disadvantage of video representation is the lack of interactivity. Video is a changeless stream that is not a conversational medium. The indispensable idea is to fulfill both computerization and conversational nuances.

A novel concept for supporting knowledge creating conversation is proposed in this study. The key concept is conversation quantization (Nishida 2005), a computational approach for approximating a continuous flow of conversation by using a series of conversation quanta. Conversation quantization is a feasible framework for constructing conversational knowledge and it is performed by processing a large number of conversation quanta using a computer. A conversation quantum represents a reusable knowledge material that is an individually functional, interactive, and synthesizable conversation block, similar to a Lego® toy block. In a conversation quantum, the quantum is represented in both computational and

Conversational Informatics: An Engineering Approach   Edited by Toyoaki Nishida
© 2007 John Wiley & Sons, Ltd

conversational fashions. The entire framework, problem statement, and implementation are discussed in section 10.2. A knowledge support system based on conversation quantization is described in section 10.3. This system is called "sustainable knowledge globe" (SKG) and it emphasizes the sustainability of conversational knowledge. In a conversation, sustainability implies the production of conversational materials for the next conversation. Presentation slides and meeting transcripts are good examples of conversational materials that are utilized in conversations. Three experiments are conducted using SKG and its effectiveness is discussed in section 10.4. The related works are discussed in section 10.5. Finally, the conclusions are presented in section 10.6.

10.2 Conversation Quantization

Conceptually, the framework of conversation quantization is a quantization spiral that comprises (1) quantization, (2) construction, (3) tailoring, and (4) requantization (Figure 10.1). We consider a scenario wherein people conduct conversations in many real-world situations supported by computer-supported collaborative work (CSCW) systems. Situations (A) and (B) are different in time and place. People talk about topical real-world objects or topical electronic contents. The conversation quanta are extracted from the conversation in situation (A) and are then utilized in situation (B) in the quantization spiral.

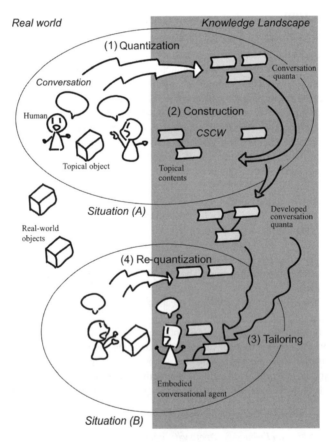

Figure 10.1   Quantization spiral

10.2.1 Quantization

The conversation quanta are quantized by identifying and encoding the segments of interactions in a conversational situation. These segments are identified by recognizing people's intentional and unintentional clues. People often explicitly record a point of conversation in a notebook. Such a process is regarded as an unsophisticated method for conversation quantization. The requirements for making it more sophisticated are to utilize a medium that is richer than text and to employ methods that are more casual than handwriting. The interaction corpus project (Sumi *et al.* 2004) is an approach to automatically capturing people's activities by using ubiquitous and wearable sensor technologies. The implemented smart room contains several video cameras, infrared signal trackers, and microphones. A person's noteworthy activities are segmented and classified into five interaction primitives, namely, "stay", "coexist", "gaze", "attention", and "facing", by categorizing the results sensed from the coexistence of LED tags. The discussion mining project (Nagao *et al.* 2004) is another approach that has not yet been fully automated; it attaches more importance to a person's intentional behavior. A point of conversation is annotated by the manipulation of colored discussion tags such as "comment", "question", and "answer". These tags are manipulated up and down by people to indicate their intentions in a formal discussion.

The problem with quantization is that the conversation models and their interpretations are not entirely clear in several daily conversational situations. We now consider knowledge-oriented conversation in a room, the goal of which is to acquire knowledge. This restriction is more informal than that of a decision-making meeting. A sample situation is a meeting room where people share and create new knowledge. An exhibition is another situation where people acquire knowledge from presenters. Approaches for expanding applicable situations into more casual situations such as chance conversations are discussed in section 10.5.

The encoding of conversation quanta into knowledge representation is another problem. Although the encoded quanta must not lack conversational nuances, they must be computationally retrievable. Using video with annotations is currently a popular approach. Several types of annotations have been proposed in previous studies, such as the time and place information, computationally recognized speech, identified speaker, interaction primitives, intention of speaker, and any other context such as manipulations of the equipment. A knowledge card using rich media with natural language annotations is another notable approach (Kubota *et al.* 2002a,b; Nakano *et al.* 2003). Rich media such as a photo image or video clip can include an extract of conversational situations. The economy of reusing knowledge cards appears to be good, although the text annotations are written by hand. The text annotations are more likely to be retrieved than rich media because of the recent evolution of text processing technologies. The text annotation also has several good conversational applications such as embodied conversational agents or communication robots.

10.2.2 Construction

The CSCW and archive systems are required to build up the store of conversational knowledge. The amount of daily conversation quanta is expected to be large. The CSCW system enables us to collaboratively arrange large contents.

We have proposed a virtual contents space called "knowledge landscape" where people can store and manage conversation quanta. The use of a spatial world is a general approach for supporting group work and managing large contents. People can look over contents if they are arranged spatially. People can also grasp content locations by using several spatial clues such as direction, distance, and arrangement.

Long-term support requires the construction of conversational knowledge because people cannot talk constructively unless they have past knowledge. The knowledge landscape provides a shape to the spatio-temporal memory model that facilitates spatial and long-term quanta management. A zoomable globe of the knowledge landscape enables the accumulation of huge quanta. Its panoramic content presentation enables people to edit and discuss the quanta according to conversational situations. The details of knowledge landscapes are described in section 10.3.

10.2.3 Tailoring

The constructed quanta are reused and tailored to people's requirements by using conversational systems such as embodied conversational agents, communication robots, or conversation support systems. The conversation quanta are designed such that they are applicable to conversational situations that differ from the original situation where they were quantized. The knowledge in the past conversation is shared and reused by people collaborating with these conversational systems. A unique issue concerning the use of conversation quanta is the coherency of quantized context. Rich media such as video includes several contexts, for example, the scenery of the conversation location, clothes of the speakers, camera angles, and so on. Various contexts imply productive conversational knowledge. The problem is that gaps between contexts in the quanta are large and it is quite difficult to modify rich media. One idea for bringing fluency to joined quanta is combining them in another coherent stream such as background music. Combining them in a coherent talk using a common agent character is another idea. This implies a TV-style program such as a news program or a talk show (Kubota et al. 2002a,b; Nakano et al. 2003).

10.2.4 Requantization

The requantization of new conversations in the applied situations is an essential process for the evolution of knowledge. People's conversations are a developed form of past conversations that are derived by collaborating with conversational systems. The conversational knowledge grows with such a quantization spiral. The point of requantization is a system design for the smooth coordination of conversational applications and the quantizing process. In fact, the spiral development of knowledge is generally observed in knowledge processes such as a conversational knowledge process (Nishida 2005) or the SECI model (Nonaka and Takevchi 1995). We realize the front-end of the knowledge spiral in the form of several conversational systems (Kubota et al. 2002a,b; Nakano et al. 2003; Kubota et al. 2005; Kumagai et al. 2006) and the back-end in the form of knowledge landscape systems (Kubota et al. 2005, 2007a).

The quantization spiral is discussed from the viewpoint of knowledge landscape in section 10.3. The spiral reuse of conversation quanta using knowledge landscape systems is discussed in section 10.4.

10.3 Knowledge Landscape

The difficulty of managing conversational quanta is their size, diversity, and situational use. In a sustainable conversation, people need to explore in a lot of past materials according to current context, and then pile new ideas upon them. It is difficult for people to manage old and new context in huge materials if they lack a good overview. Knowledge landscape is a scenic representation of the huge conversation quanta that are organized spatially and temporally. An early concept of the landscape illustrated by Nishida (2004) is an evolving memory space. The landscape enables the user to visually grasp the global nature of knowledge, explore the information space, and accommodate new information at an appropriate place.

We clear up the evolving landscape schematically in the spatio-temporal memory model (Figure 10.2). Conversation quanta are visually placed anywhere the user likes them to be on the landscape ($t = 1$). Here, we assume that the arbitrary arrangement of objects results in several spatial clues to remind the user of the nature of the located objects. The conversational knowledge is synthesized continuously (from $t = 0$ to $t = n$) on the landscape by connecting, rearranging and developing conversation quanta that result from the quantization spiral.

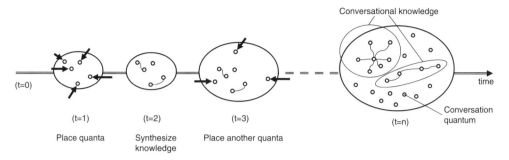

Figure 10.2   Spatio-temporal memory model

In our approach, the knowledge landscape on the 3D virtual globe gives shape to the spatio-temporal memory. The knowledge landscape consists of a zoomable globe and panoramic content presentation.

10.3.1 Zoomable Globe

Using the zoomable globe is our approach to manage a large archive. We adopt the zoomable globe as a memory space to make it easy to arrange conversational quanta arbitrarily. The globe is a sphere like a terrestrial globe with latitude and longitude lines, the landmarks for the north and south poles, and the equator (Figure 10.3). It has a zoomable feature which allows multi-scale representation like electronic atlases. A quantum is represented as a rectangle on the globe. A screenshot of the implemented knowledge landscape is shown in Figure 10.4. The details of the implementation are given in section 10.3.3.

The reason why we adopt a globe shape is our hypothesis about its good arbitrariness for contents arrangement. There could be many types of topologies in the space: a finite plane, an infinite plane, 2-dimensional torus, and so on. A finite plane appears to be unsuitable for a user to expand the contents on the edges of the plane. An infinite plane does not pose this problem; however, it is difficult for people to grasp infinite space. The 2-dimensional torus also has no edge; however, such a topology may be unfamiliar to people. A globe shape would be the appropriate solution where people can arrange contents more freely than in a finite plane because a globe has no edge. Moreover, a sphere is more familiar than a 2-dimensional torus since it is similar to a terrestrial globe.

A zooming graphical interface (Bederson and Hollan 1994) is effective to manage large content on a finite screen size. It enables a user to view the entire content by changing the scale, or to observe specific content by changing the focus. In general, zooming interfaces can be classified as linear and nonlinear. Linear zooming (Bederson and Hollan 1994) magnifies and shrinks the entire information in a manner similar to a multi-scale map. Nonlinear zooming (Furnas 1986; Lamping *et al.* 1995) distorts an arrangement of information to focus on specific information. We adopt the linear zooming interface to manage content because we aim to capitalize on the memory of the contents arrangement, which should not be distorted.

10.3.2 Panoramic Content Presentation

A panoramic content presentation is a topological model of contents arrangement in the knowledge landscape. The panorama includes tree structured content cards and stories, each of which consists of a sequence of viewpoints (Figure 10.5). A conversation quantum is represented by a content card in the knowledge landscape. The content card is an extended form of a knowledge card (Kubota *et al.* 2002a)

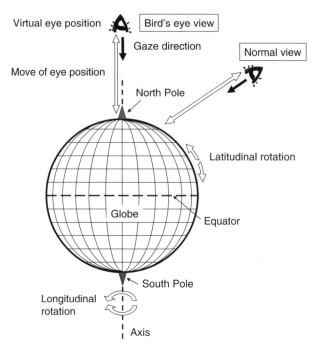

Figure 10.3   Zoomable globe

that is a reusable representation of content in conversational and editorial situations. The knowledge card allows only an image file or a movie clip, whereas a content card allows any files to make it more expressive.

A content card consists of three parts: an embedded file, a title of a card, and the annotation of the card. Figure 10.6 shows an XML description of a content card. The <card> element represents a unit of a content card. The <card> element contains three child elements: <title>, <url>, and <annotation>. The <title> element contains the title text of the card, the <url> element contains the URL of an embedded file (e.g., a document, an image, a movie clip, a slide, and so on), and the <annotation> element contains the annotation text of the card.

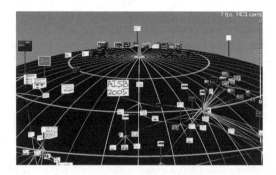

Figure 10.4   Screenshot of the implemented knowledge landscape

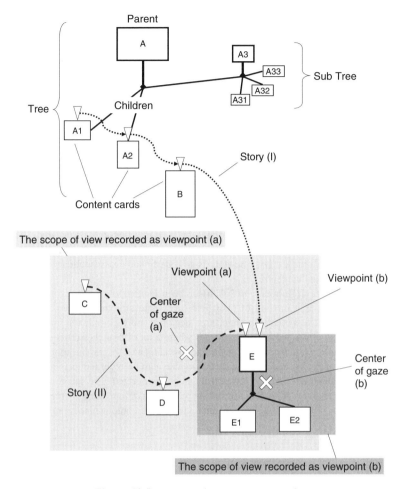

Figure 10.5   Panoramic content presentation

```
<?xml version="1.0" encoding="UTF-8"?>
<card version="1.0">
<title>Overview of EgoChat</title>
<url> files/egochatscreen.jpg</url>
<annotation>
We have developed a virtualized-ego system called "EgoChat". This slide shows overview of
EgoChat.
</annotation>
</card>
```

Figure 10.6   Description of a content card

A tree structure is a standard way of representing categories. By categorizing the content cards, a set of cards can be easily arranged and retrieved. Cards A1, A2, and A3 in Figure 10.5 are the children of card A, and A3 also represents a subcategory of card A. The relationship between a parent and a child is drawn by an edge. The knowledge landscape allows multiple trees on a virtual globe because the arrangement is limited if it allows only one tree. An independent card that does not belong to a tree is also allowed for the same reason.

Each content card is bound to one or more viewpoints. The viewpoint represents a scope of view in the panorama that is recorded as information about the eye distance from the globe and the center of gaze. The viewpoint is a very important factor that includes the context and background of the card. The scope of view recorded as viewpoint (a) in Figure 10.5 focuses the cards that compose story (II). This scope of view gives information about all the flow of story (II). The scope of view recorded as viewpoint (b) focuses the cards in tree E. This scope of view gives information about a specific category represented by cards E, E1, and E2.

Story representation is important to describe a sequence of contents. Here, a story is represented by a sequence of viewpoints that shows a panoramic storyline of the contents. The relationship between a previous viewpoint and the next viewpoint is drawn by an arrow that cuts across trees. The sequence from A1 to E is story (I) in Figure 10.5. The sequence from C to E is another story (II). A content card is allowed to belong to multiple stories. The story is so orderly and cross-boundary that the people can easily create cross-contextual contents.

Contour map representation (Kubota *et al.* 2007b) has been also proposed as an expansion of panoramic content presentation. This is a method of visualizing tree data structures arranged on a plane. The graphical representation of a spatially arranged tree becomes more complex as the number of nodes increases. Tree nodes get entangled in a web of trees where a lot of branches cross each other. Contour map representation simplifies such complex landscapes by using a chorographic approach. Contour lines show the arrangement of peaks, and besides, the enclosures can be regarded as hierarchical structures. For people, a contour map has a familiar design that can be memorized. The space efficiency of contour map representation is relatively high because the shape of the contour lines is freer than the shape of primitives like circles or rectangles. A 3-dimensional view of the contour map representation of SKG is shown in Figure 10.7.

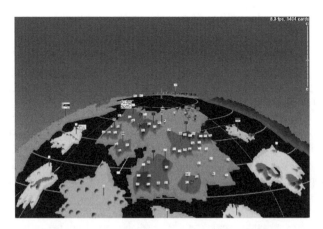

Figure 10.7 Contour map representation on SKG

10.3.3 Implementation of the Sustainable Knowledge Globe

The knowledge landscape is incorporated into SKG, which works on the Windows platform with .NET and Managed DirectX. The SKG landscape is a virtual sphere similar to a terrestrial globe, but not related to actual nations. Immediately after starting SKG, the user-defined home position is shown, and a user can then explore the landscape by using a wheel mouse. The sphere has three degrees of freedom (latitudinal direction, longitudinal direction, and depth direction) because a standard mouse device appears to be unsuitable for operating a sphere with many degrees of freedom. A user also changes the view angles of a sphere. A normal view is an oblique view that has depth, while a bird's eye view is an overhead view that has fewer occlusion problems (see Figure 10.3).

The quantization spiral revolves around SKG. An available quantum is any electronic file with natural language annotations. Files are imported as quanta to the SKG by dragging and dropping from the PC's desktop or downloading from an SKG network server. Quanta are also created by an inline editor in SKG. A quantum in SKG is attached to a thumbnail image. An annotation is automatically attached to a quantum if the source content has annotations (e.g., speaker's note on a PowerPoint® slide). An annotation can also be attached later.

Most quantization and construction operations are performed by humans. SKG can be used as an advanced whiteboard or for transcribing notes in meeting situations. Conversation quanta are created on the SKG by the participants of the conversation. Related handouts, photos, or web pages are also imported onto SKG. They are placed and connected on SKG along with the discussions.

The story structures of conversation quanta enable the tailoring process. A presentation agent is implemented in SKG by using story structures. The agent reads out a text annotation of a conversation quantum by using a commercial speech synthesis system. The order of the presentation is the same as that of a story structure of the conversation quanta. The agent walks on SKG to the target viewpoint, and then speaks in the target scope of view. Such an automated presentation is useful for concluding a meeting. People can create a summary of a meeting as an automated presentation for absentees. Story structures also help people to talk in a presentation. A thumbnail image on SKG is available as a presentation slide. Buttons for going to the previous or next slide along with a story structure are also implemented for easy presentation.

Gardening interface has been also developed as an expansion of SKG. It is necessary for people to arrange their contents according to their intentions. Fully-automated arrangements made by a computer system are not always desirable. This is analogous to a private room. A person tends to lose possessions in a private room if it has been tidied by a family member. The same is true for tidying by a fully-automated system. Semi-automated arrangements according to the user's intentions are proposed in the contents garden system (Kubota *et al.* 2007). This allows a person to arrange a large number of contents by simply indicating an arrangement policy to the system. The load for describing the policy is decreased by a perceptual interface such as gardening, for example, making crowded contents sparse, transforming broadly arranged contents into a narrow arrangement, creating a space in crowded contents in order to add new contents, or moving a group of contents by pushing and pulling them roughly.

10.4 Experiments

We have evaluated effectiveness of SKG through three experiments in practical conversational situations. The first two experiments are conducted to watch developments of group and personal contents. The last experiment is to investigate the ecology of reusing conversation quanta.

10.4.1 Building Proceedings of Meeting

We have made proceedings of a working group by using SKG. The participants discussed the study of an intelligent room. The screen of SKG was projected on a large screen during the meetings. One operator imported handouts, noted down the speeches, and located the transcripts on the surface of the globe. Speeches on the same topic were grouped into one or two cards that were grouped into a tree-like structure that was sorted on the basis of their dates. The operator also manipulated the globe according to the requests from participants to focus on specific content.

The meetings were held ten times from August to November in 2004. The total duration of the meetings was 20 hours. The average number of participants was six. From the experiment, we have acquired 151 contents that include 12 trees and three stories.

A summary story of ten meetings was created. The viewpoints in the story show the group of monthly proceedings and significant topics. The storyline is shown in Figure 10.8. The viewpoints are connected from viewpoints 1 to 9. The synopsis of the viewpoints and annotations is shown in Table 10.1.

The waking path of the presentation agent is shown in Figure 10.8(a). The walk from viewpoints 1 to 2 indicates a transition from the focused view to the overview. The walk from viewpoints 2 to 3 indicates a

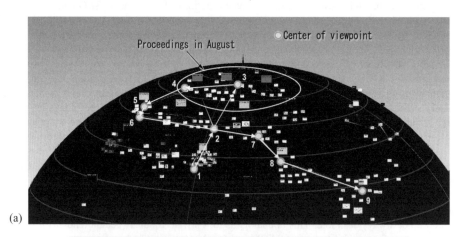

(a)

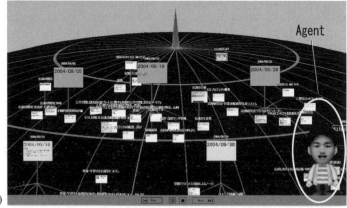

(b)

Figure 10.8 Examples of the story: (a) Storyline about a summary of proceedings (the number shows the order of the story); (b) Presentation agent that talks about a scope of view on SKG

Table 10.1   Synopsis of viewpoints and annotations

Viewpoint	Scope of view	Annotation
1	A close view of the card that describes a title of the story	I would like to talk to you about a meeting supported by SKG.
2	A distant view of the entire proceedings	151 contents are created in 3 months on SKG.
3	A distant view of the proceedings in August	These are the proceedings in August.
4	A close view of the cards that describe the main topics in the proceedings in August	We discussed the methods for evaluating knowledge productivity in a meeting.

transition from the overview to the focused view. The same transition is observed from viewpoints 3 to 4. These fluctuating transitions make the presentation quite panoramic. The speaking manner of the presentation agent is shown in Figure 10.8(b). The total playtime of the presentation is approximately 2 minutes. The presentation agent can summarize the presentation in order to create a compact overview.

We interviewed the participants about the effectiveness of SKG and received positive comments about the good overview and a reminder of the meetings. We also received a comment that it is difficult for the speaker to operate SKG by using a mouse device. The user should not be engaged with mouse operations in conversational situations because it disturbs communication using natural gestures. We are now developing a novel immersive browser (Kubota *et al.* 2007) that can improve the operativity of SKG by using physical interfaces such as a motion capturing system in a surrounding information space.

10.4.2 Building Personal Contents

We have also conducted a study to determine SKG usage by individuals for personal objectives (Kubota *et al.* 2007a). This study focused on investigating the diversity of the card arrangement that SKG offers. Three subjects (user (I), user (II), and user (III)) sustainably constructed their individual landscape on SKG in order to manage their contents. The contents were mainly research slides and movies, and also included photos, bookmarks, and memos. The content cards were created by using a built-in editor on SKG or converted from the contents on the user's PC. We discuss the landscapes created between 9 August 2004 and 28 January 2005. This was the period during which contents that could be quickly collected were imported and arranged on the SKG.

We observed highly personalized arrangements on the landscapes. Examples of how the stories were arranged are shown in Figure 10.9. The story in (A) (horizontal turn style) courses horizontally from back left to front right, while the story in (B) (vertical turn style) courses vertically from front left to back right. (C) Spiral and (D) clockwise are smooth arrangements without sharp curves. (C) is a space-saving compact spiral. (D) is a clockwise arrangement that depicts time flow by using a clock analogy.

Each user arbitrarily created individual arrangement styles and used them. Users (I) and (II) arranged their contents mainly by using the horizontal turn style. User (III) arranged his contents mainly by using the spiral style and the counter-clockwise style. Therefore, each user had a different preferred style.

The results of the user study revealed that the SKG provides general methods to arrange contents by using tree structures. In addition to using tree structures, most contents were also arranged according to various story arrangement styles. Such characteristic arrangements seem to help people remember the locations of their contents. It also seems that the SKG is a good platform for people to manifest the arbitrary structures that are stored in their brains. The landscapes are subjective rather than objective; they reflect a kind of cognitive map rather than an ontological map. Arranging the landscape on the SKG

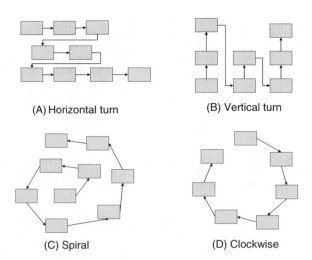

(A) Horizontal turn   (B) Vertical turn

(C) Spiral   (D) Clockwise

Figure 10.9   Examples of story arrangement styles

is akin to arranging objects in a room that provides a personalized environment for managing a large number of objects.

10.4.3 Reusing Conversation Quanta

Saito *et al.* (2006) used SKG as a system for creating new presentation content by reusing the conversation quanta. Figure 10.10(a) shows the SKG where conversation quanta extracted from conversation videos are arranged. A conversation quantum comprised a slide with a corresponding movie clip (Figure 10.10(b)). New presentation contents are created (Figure 10.10(c)) by reusing these contents. The conversation quanta of the recently created slides are connected to and placed near these slides (Figure 10.10(d)). A presenter can easily refer to the conversation quanta connected to the corresponding slides.

Figure 10.11 shows a simulation of the presentation contents shown in Figure 10.10. In situation A, a presenter refers to conversation quantum (a); a participant then comments on the conversation quantum. Conversation quantum (b) is quantized from situation A. In another presentation situation (situation B), the presenter refers to the conversation quanta that involves conversation quantum (b), which has been quantized from situation A. The subjects related to the conversation quanta, which are reused in the presentation, increase due to the repeated extraction. Valuable insights are obtained by simulating the presentation using conversation quanta. The reusable conversation quanta, which can be reused as effective presentation materials, contain the following information:

- nonverbal representation of participants such as gestures, facial expressions, etc.
- group setting or synchronized behaviour of people
- conversational rhythms such as rapid conversations, jokes, etc.
- real-world objects observed by the participants, knowledge of operating instruments, DIY, etc.
- personal memories such as subjective opinions, experiences, etc.

Conversation with rich context can be easily performed by reusing the conversation quanta involving some of this information. First, conversation quanta are useful in communicating nonverbal information. The nonverbal actions of participants such as gestures and facial expressions are included in extracted

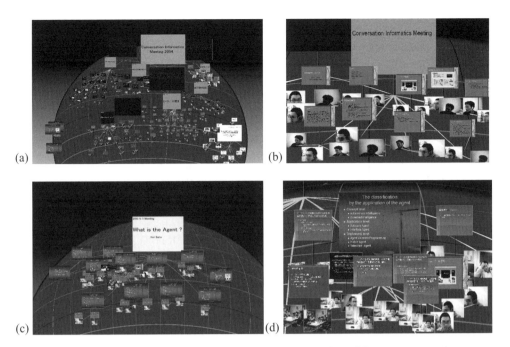

Figure 10.10 Examples of reusing conversation quanta: (a) overview of the past conversation quanta; (b) conversation quantum represented as a slide with the corresponding movie clips; (c) overview of the new presentation; (d) arrangement of conversation quanta near the new slides

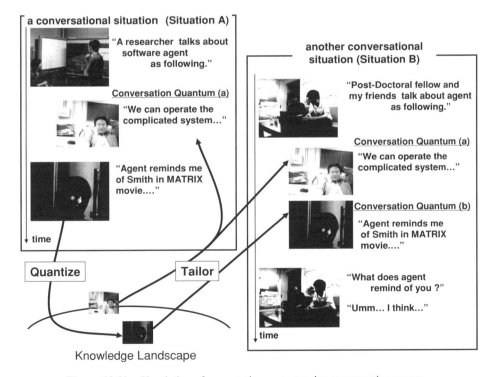

Figure 10.11 Simulation of presentation content using conversation quanta

conversation quanta. The conversation in a group setting and synchronized behaviors are also included in the extracted conversation quanta. They tend to be important in order to understand the interactions and considerations within the group members who share the same atmosphere. People can explain these contexts effectively by reusing the conversation quanta.

The conversation quanta involving conversational rhythms are useful for sharing the conversational presence and atmosphere. It is difficult to elaborate on conversational rhythm, presence, and atmosphere. For example, even if a speaker engaged in a rapid conversation, the listeners would be unable to comprehend the degree of its speed. Moreover, people can seldom reuse jokes because they depend on the context such as conversational rhythm, presence, and atmosphere. People can easily understand and share a previous conversation rhythm, presence, and atmosphere by showing the conversation quanta.

Furthermore, the conversation quanta involving real-world objects observed by the participants are useful in order to intuitively explain and understand the objects visually. It is difficult for a speaker to elaborate on his knowledge of operating instruments. On the other hand, it is difficult for listeners to understand the description of an unknown object. A speaker can easily describe the objects, while the listeners would intuitively and visually understand them by reusing the conversation quanta.

Finally, the conversation quanta involving personal memories, such as subjective opinions and individual experiences, are useful for creating a conversational agent called *virtualized ego*. (Kubota *et al.* 2002a) We can effectively communicate the characteristic conversational atmosphere of a person even in their absence by referring to the conversation quanta.

10.5 Discussion

We have discussed the methods for managing conversational materials and have confirmed that conversation quanta can be reused to build proceedings, personal contents, and presentation contents. It is easy to create these conversational contents by editing and reorganizing the conversation quanta. The conversation quanta comprising the nonverbal representation of participants, group settings and the synchronized behaviors of people, conversational rhythms, real-world objects observed by participants, and personal memory play an important and interesting role in their reuse. We conclude that conversation quanta involving such information are reusable and contribute toward an effective conversation.

The automatic capture of conversation quanta is also an important concern. Another conversation quantization system that comprises conversation capturing systems in poster presentation situations and a conversational agent system has been developed (Kumagai *et al.* 2006) using conversation quanta. This system detects microstructures in conversations by using physical references such as touches on a poster. Minoh *et al.* (2003) developed a robust method for automatically recognizing communicative events in a real classroom by integrating audio and visual information processing. Rutkowski *et al.* (2003) studied a method for monitoring and estimating the efficiency of interpersonal communication based on the recorded audio and video content by computing the correlation between the activities of the sender and responses of the receiver. Shibata *et al.* (2003) integrated robust natural language processing techniques and computer vision to automatically annotate videos with a closed caption.

Our studies in this chapter are focused on the management of conversational content and its reusability. The approach for capturing more casual chance conversations is strongly related to wearable and ubiquitous computing. The interaction corpus project (Sumi *et al.* 2004) uses wearable sets and ubiquitous sensors for capturing personalized experiences in events. Ubiquitous memory systems (Kawamura *et al.* 2004) capture personal experiences linked with real-world objects. Aizawa *et al.* (2004) capture a life log by using wearable cameras, microphones, GPS, gyrocompasses, and acceleration sensors to obtain contextual information about human life. Kern *et al.* (2005) study cues for interpreting human activities by analyzing information using only acceleration sensors or environmental audio. These automated approaches that are based on wearable and ubiquitous paradigms are emerging issues that should be examined from the viewpoint of evolving conversational knowledge.

10.6 Conclusions

This chapter has described a novel concept, its implementation, and its empirical evaluation for supporting knowledge-creating conversation. We proposed conversation quantization, which is a computational approach for approximating a continuous flow of conversation by using a series of conversation quanta. We also proposed the quantization spiral, which is a feasible model for utilizing conversational knowledge by reusing a large number of conversation quanta. The implementation of the system required to support the quantization spiral was described. This system is called the sustainable knowledge globe (SKG). SKG manages conversational quanta by using a zoomable globe and panoramic content presentation. Three experiments were conducted using SKG in practical conversational situations in order to evaluate its effectiveness. The proceedings of a working group and personal contents were built on SKG. The conversation quanta comprising the nonverbal actions of participants, group settings and the synchronized behaviors of people, conversational rhythms, real-world objects observed by participants, and personal memory played an important and interesting role in the reuse of conversation quanta.

References

Aizawa K., Hori T., Kawasaki S. and Ishikawa T. (2004) Capture and efficient retrieval of life log. In *Proceedings of the Pervasive Workshop on Memory and Sharing Experiences*, pp. 15–20.

Bederson B.B. and Hollan J.D. (1994) Pad++: a zooming graphical interface for exploring alternate interface physics. In *Proceedings of ACM UIST '94*.

Furnas G.W. (1986) Generalized fisheye views. In *Proceedings of Human Factors in Computing Systems Conference* (CHI 86), pp. 16–23.

Kawamura T., Ueoka T., Kono Y. and Kidode M. (2004) Relational analysis among experiences and real world objects in the ubiquitous memories environment. In *Proceedings of the Pervasive Workshop on Memory and Sharing of Experiences*, pp. 79–85.

Kern N., Schmidt A. and Schiele B. (2005) Recognizing context for annotating a live life recording. In *Personal and Ubiquitous Computing*.

Kubota H., Kurohashi S. and Nishida T. (2002a) Virtualized-egos using knowledge cards. In *Proceedings of Seventh Pacific Rim International Conference on Artificial Intelligence* (PRICAI-02), WS-5 International Workshop on Intelligent Media Technology for Communicative Reality (IMTCR2002), pp. 51–54.

Kubota H., Yamashita K. and Nishida T. (2002b) Conversational contents making a comment automatically. In *Proceedings of KES'2002*, pp. 1326–1330.

Kubota H., Sumi Y. and Nishida T. (2005) Sustainable knowledge globe: a system for supporting content-oriented conversation. In *Proceedings of AISB Symposium on Conversational Informatics for Supporting Social Intelligence & Interaction*, pp. 80–86.

Kubota H., Nomura S., Sumi Y. and Nishida T. (2007a) Sustainable memory system using global and conical spaces. Special Issue on Communicative Intelligence, *Journal of Universal Computer Science*, pp. 135–148.

Kubota H., Sumi Y. and Nishida T. (2007b) Visualization of contents archive by contour map representation. In T. Washio *et al.* (eds), *New Frontiers in Artificial Intelligence: Joint JSAI 2006 Workshop Post-proceedings*, vol. 4384, Springer.

Kumagai K., Sumi Y., Mase K. and Nishida T. (2006) Detecting microstructures of conversations by using physical references: case study of poster presentations. In T. Washio *et al.* (eds), *Proceedings of JSAI 2005 Workshops*, (LNAI 4012), pp. 377–388.

Lamping J., Rao R. and Pirolli P. (1995) A focus+context technique based on hyperbolic geometry for visualizing large hierarchies. In *Proceedings of ACM CHI'95 Conference on Human Factors in Computing Systems*.

Minoh M. and Nishiguchi S. (2003) Environmental media, in the case of lecture archiving system. In *Proceedings of International Conference on Knowledge-based Intelligent Information and Engineering Systems* (KES2003), vol. 2, pp. 1070–1076.

Nagao K., Kaji K., Yamamoto D. and Tomobe H. (2004) Discussion mining: annotation-based knowledge discovery from real world activities. In *Proceedings of the Fifth Pacific-Rim Conference on Multimedia* (PCM 2004).

Nakano Y., Murayama T., Kawahara D., Kurohashi S. and Nishida T. (2003). Embodied conversational agents for presenting intellectual multimedia contents. In *Proceedings of the Seventh International Conference on Knowledge-based Intelligent Information and Engineering Systems* (KES'2003), pp. 1030–1036.

Nishida T. (2004) Social intelligence design and communicative intelligence for knowledgeable community (Invited Talk). *International Symposium on Digital Libraries and Knowledge Communities in Networked Information Society* (DLKC'04), Kasuga Campus, University of Tsukuba.

Nishida T. (2005) Conversation quantization for conversational knowledge process (Special Invited Talk). In *Proceedings Fourth International Workshop on Databases in Networked Information Systems* (LNCS 3433), Springer, pp. 15–33.

Nonaka I. and Takeuchi H. (1995) *The Knowledge-Creating Company*. Oxford University Press, New York.

Rutkowski T.M., Seki S., Yamakata Y., Kakusho K. and Minoh M. (2003) Toward the human communication efficiency monitoring from captured audio and video media in real environment. In *Proceedings of International Conference on Knowledge-based Intelligent Information and Engineering Systems* (KES2003), vol. 2, pp. 1093–1100.

Saito K., Kubota H., Sumi, Y. and Nishida T. (2006) Support for content creation using conversation quanta. In T. Washio *et al.* (eds), *New Frontiers in Artificial Intelligence: Joint JSAI 2005 Workshop Post-proceedings*, vol. 4012, pp. 29–40.

Shibata T., Kawahara D., Okamoto M., Kurohashi S. and Nishida T. (2003) Structural analysis of instruction utterances. In *Proceedings of International Conference on Knowledge-based Intelligent Information and Engineering Systems* (KES2003), pp. 1054–1061.

Sumi Y., Mase K., Mueller C., Iwasawa S., Ito S., Takahashi M., Kumagai K. and Otaka Y. (2004) Collage of video and sound for raising the awareness of situated conversations. In *Proceedings of International Workshop on Intelligent Media Technology for Communicative Intelligence* (IMTCI2004), pp. 167–172.

11

Automatic Text Presentation for the Conversational Knowledge Process

Sadao Kurohashi, Daisuke Kawahara, Nobuhiro Kaji, and Tomohide Shibata

11.1 Introduction

Humans exchange, modify, and create knowledge most effectively through conversation. Using conversation, we can exchange our opinions, stimulate each other, and accelerate the creation of new ideas. We can call the process in which people construct mutual understanding and knowledge through conversation the *conversational knowledge process* (Nishida 2003).

Supporting the conversational knowledge process, therefore, can lead to great contributions to social problem-solving. For example, let us consider the typical social problem of earthquake disaster mitigation. Although several types of warning information are provided by central and local governments, ordinary people cannot always catch and assimilate this information and as a result cannot always behave in a reasonable way. As a solution to this problem, local residents and people who have an interest in the same geographical area constitute a community, which can in turn function as a knowledge mutual-aid system used to solve social problems. In order to facilitate such a system, information technology supporting *conversation* can play an important role. Based on such an investigation, this chapter proposes a support system for the conversational knowledge process, in particular, an automatic text presentation system.

It is obvious that conversation is based on spoken language, as people in general talk to each other using spoken language. On the other hand, the arranged and established knowledge of a particular group of people or organizations are generally recorded as texts, such as technical papers and reports. Indeed, most human wisdom has been stored in this form. Furthermore, the rapidly growing development of the World Wide Web has allowed it to become a kind of knowledge cyberspace. Texts are written, of course, using written language.

Conversation and text each have their own advantages, but their styles are very different, and heretofore it has been difficult to mix and exploit their advantages. That is, in order to have a deep discussion on some subject, discussants must read related documents beforehand. It is usually impossible to participate in a discussion while reading related documents.

Recent drastic advances in, and the spread of, computers and computer networks, however, have promoted the development of language, speech, and agent technology, making it possible to introduce

Conversational Informatics: An Engineering Approach Edited by Toyoaki Nishida
© 2007 John Wiley & Sons, Ltd

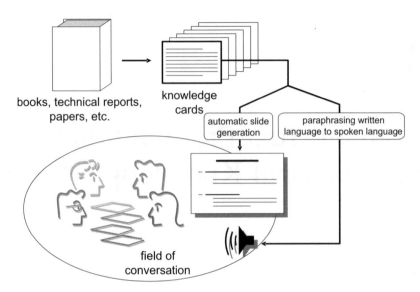

Figure 11.1   Automatic text presentation for the conversational knowledge process. Reproduced by permission of ELRA

texts into the field of conversation. For example, it is now possible to, in a conversation, retrieve a portion of a text that is suitable to the conversation or that can be an answer to a question of one of the participants. Further, this text can be presented by a 3D-CG agent along with the relevant gesture and facial expression. The original written text can thereby be transformed into natural spoken-language expressions and synthesized into speech, while a summarization slide is displayed at the same time (Kubota *et al.* 2002).

As a step for such a future intelligent conversation support system, this chapter proposes an automatic text presentation system that consists of two methods of presentation: a written-language to spoken-language conversion method and an automatic abstraction method that can create an abstract slide from a text (Figure 11.1), targeting Japanese texts. We first introduce the recent advances in natural language processing, or NLP, that facilitate automatic text presentation.

11.2 Current State of Natural Language Processing

With the advent of electronic computers in the 1950s, their use in studies of handling natural languages began. However, it is not very easy for computers to deal with natural languages. One reason is that since the complexity of languages reflects the complexity of the world, world knowledge should first be entered into computers in order to precisely handle languages, but this cannot be achieved easily. Another reason is the existence in languages of various inherent ambiguities, which computers have difficulty in handling but people can disambiguate intuitively using wide-ranging contextual information.

However, by virtue of recent progress in computers and network environments, a large amount of online text has become available, and this situation has allowed NLP studies to develop significantly. Some kinds of world knowledge can be learned from huge amounts of data, and the optimum parameters of disambiguation can be learned from hand-crafted language data with correct annotation using machine learning techniques.

In the remainder of this section, we briefly introduce the current state of fundamental NLP resources and our recent research outcomes.

11.2.1 Fundamental Resources of NLP

The first steps of NLP are morphological analysis and syntactic analysis. Morphological analysis clarifies word boundaries, parts of speech, inflection forms, and so forth in given sentences. Syntactic analysis identifies the relationships between words, such as modifier–head relations and predicate–argument relations. These two types of analyses have been improved through the elaboration of word dictionaries and basic grammar. For instance, a morphological analyzer achieves approximately 99% accuracy on newspaper articles, and a syntactic analyzer performs with approximately 90% accuracy for phrase-wise dependency.

Thesauri have also been developed as a clue to handling word senses. A thesaurus is a dictionary in which words are classified according to their senses, and it usually has a tree structure in which words with similar senses are grouped together in a node. For example, the Japanese thesaurus developed by Nippon Telegraph & Telephone Company contains about 300 000 entries stored in a tree consisting of eight layers (Ikehara *et al.* 1997). A similarity between two words can be computed according to the distance between them in the tree. The tree is also utilized for word clustering.

11.2.2 Automatic Construction of Large-scale Case Frames and Analysis of predicate–Argument Structure

What is represented in natural language text is basically expressed in predicate–argument structures, in which a predicate is associated with its related arguments via case-marking postpositions such as "*ga*", "*wo*", and "*ni*". These structures cannot be completely clarified by syntactic analysis alone. For example, a case marker is often hidden by the topic-marking postposition "*wa*" or a clausal modifiee. Furthermore, arguments are often omitted in Japanese text.

To identify predicate–argument structures in text, it is necessary to first store in a computer what actions or events occur frequently, and then to match the text with this stored knowledge. Such knowledge is called *case frames*. We have developed a gradual method for automatically constructing wide-coverage case frames from a large amount of text (Kawahara and Kurohashi 2002). This method is summarized as follows.

First, a large raw corpus is automatically parsed, and modifier–head examples that have no syntactic ambiguity are extracted from the resulting parses. The biggest problem in case frame construction is verb sense ambiguity. Verbs with different meanings require different case frames, but it is hard to precisely disambiguate verb senses. To deal with this problem, the modifier–head examples are distinguished by single coupling a verb and its closest case component. For instance, examples are not distinguished by verbs such as "*naru*" (make/become) or "*tsumu*" (load/accumulate), but by couples such as "*tomodachi-ni naru*" (make a friend), "*byouki-ni naru*" (become sick), "*nimotsu-wo tsumu*" (load baggage), and "*keiken-wo tsumu*" (accumulate experience).

As a result of the above process, we can obtain the initial basic case frames from the parsing results, and therefore case structure analysis based on these case frames becomes possible. In the second stage, case structure analysis is applied to the corpora. The case structure analysis results provide richer knowledge, such as double nominative sentences, non-gapping relations, and case changes.

We constructed case frames from 470M Japanese sentences extracted from the Web (Kawahara and Kurohashi 2006). The result consists of approximately 90 000 predicates, and the average number of case frames for a predicate is 14.5. Some examples of constructed case frames are shown in Table 11.1.

Table 11.1 Case frame examples (written in English). The number following each example indicates its frequency.). (This table is reproduced from (Kawahara and Kurohashi 2006) by permission of ELRA.)

	CS	examples
yaku (1) (broil)	ga wo de	I:18, person:15, craftsman:10, ··· bread:2484, meat:1521, cake:1283, ··· oven:1630, frying pan:1311, ···
yaku (2) (have difficulty)	ga wo ni	teacher:3, government:3, person:3, ··· fingers:2950 attack:18, action:15, son:15, ···
yaku (3) (burn)	ga wo ni	maker:1, distributor:1 data:178, file:107, copy:9, ··· R:1583, CD:664, CDR:3, ···
⋮	⋮	⋮
oyogu (1) (swim)	ga wo de	dolphin:142, student:50, fish:28, ··· sea:1188, underwater:281, ··· crawl:86, breaststroke:49, stroke:24, ···
⋮	⋮	⋮
migaku (1) (brush)	ga wo de	I:4, man:4, person:4, ··· tooth:5959, molar:27, foretooth:12 brush:38, salt:13, powder:12, ···
⋮	⋮	⋮

Using the case frames enables us to obtain denser predicate–argument structures than those obtained from mere syntactic analysis. Hidden case markers of topic-marked phrases and clausal modifiees are clarified by matching them with the appropriate case slot (CS) in a case frame. As an example, let us describe this process using the following simple sentence that includes a topic-marked phrase:

> *bentou-wa* *taberu*
> lunchbox-TM eat
> (eat lunchbox)

In this sentence, the case marker of "*bentou-wa*" is hidden by the topic marker "*wa*". The analyzer matches "*bentou*" (lunchbox) with the most suitable case slot in the following case frame of "*taberu*" (eat):

	CS	examples
taberu	ga wo	person, child, boy, ··· lunch, lunchbox, dinner, ···

Since *"bentou"* (lunchbox) is included in the *"wo"* examples, its case is analyzed as *"wo"*. As a result, we obtain the predicate–argument structure *"φ:ga bentou:wo taberu"*, which means that the *"ga"* (nominative) argument is omitted and the *"wo"* (accusative) argument is *"bentou"* (lunchbox).

As there are many case frames for each predicate, the above process also includes the selecting of an appropriate case frame. This procedure is described in detail in Kawahara and Kurohashi (2002). In addition, the resolution of zero anaphora (omitted arguments) can be achieved by using these case frames. A detailed description of zero anaphora resolution can be found in Kawahara and Kurohashi (2004).

11.2.3 Discourse Structure Analysis

A method for analyzing relations between sentences in a text is called discourse structure analysis. This type of analysis is important for capturing the flow of discussion and extracting important sentences/phrases in a text.

We can suppose a tree structure as a model of discourse structure, in which a sentence is linked to related sentences through coherence relations. We take into account the following coherence relations between sentences: list, contrast, topic chaining, topic-dominant chaining, elaboration, reason, cause, and example. An example of discourse structure analysis is shown in Figure 11.2. The words sandwiched by parentheses indicate the coherence relation to the parent sentence.

We start with the starting node and build a discourse tree in an incremental fashion (Kurohashi and Nagao 1994). As a new sentence comes in, by checking surface information we find a connected sentence and the coherence relation between them by calculating the confidence scores for all possible relations. Table 11.2 shows examples of rules for calculating confidence scores. Each rule specifies a condition for a pair of a new sentence and a possible connected sentence: the range of possible connected sentences (how far from the new sentence) and patterns for the two sentences. The last row in Table 11.2, for example, captures the tendency that a sentence that starts with *"nazenara (because)"* often indicates a reason of the preceding sentence. If a pair meets the condition, the relation and score in the rule are given to it.

Finally, we choose the connected sentence and coherence relation that have the highest score.

11.3 Unit of Conversation: the Knowledge Card

The main issue in introducing a chunk of text to conversation is the unit, or size, of the chunk. Since the dynamic exchange of information is the nature of communication, the reading out of very long text

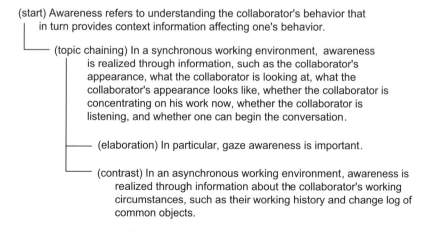

Figure 11.2 Example of discourse structure analysis

Table 11.2 Examples of rules for discourse structure analysis

Coherence relation	Score	Applicable range	Patterns for a connected sentence	Patterns for a new sentence
Start	10	*	*	*sate* (then)···
List	15	1	*	*mata* (and)···
List	30	1	*daiichini* (first)···	*dainini* (second)···
Contrast	30	1	*	*mushiro* (rather than)···
Contrast	40	*	X···	X'($\simeq$ X)···
Elaboration	30	1	*	*tokuni* (especially)···
Reason	30	1	*	*nazenara* (because)···

interminably is not appropriate. On the other hand, it is difficult for current technology to utter very short, appropriate comments according to the situation of communication.

In this study, as the unit of conversation, we consider a chunk of text that has semantic coherence can be useful information when standing alone, and is as short as possible. We call such a unit a *knowledge card*, which typically consists of a dozen or so sentences. An entry in an encyclopedia or a newspaper article can be considered a knowledge card just as it is. A section or a subheading of coherent text can also be a knowledge card. However, if the text is too long or if the addition of relations to the preceding or following context is necessary, slight editing of the knowledge card by hand may be required. In this study, we use units explicitly given in original texts as knowledge cards.

11.4 Paraphrasing Written Language to Spoken Language

A set of knowledge cards is stored in a computer, and the card that most relates to the conversation at a particular moment or the card that answers a question from a participant is chosen and presented. This can be realized by techniques of information retrieval and speech synthesis.

One problem with this process is that of how to present the knowledge card. One possibility is to read aloud the content of the card as it is by a synthesized voice. However, the knowledge card, which is originally a part of a document, is represented in written language, and as such it is difficult to naturally bring the card into conversation by using only speech synthesis.

We propose a method of first paraphrasing a written language knowledge card into spoken language, and then using speech synthesis to read it aloud. This section describes this paraphrasing technique.

11.4.1 Expressions to be Paraphrased

On the basis of generating natural spoken expressions from written language, the target expressions of styles of written language can be classified as follows:

1. Expressions peculiar to written language
   (a) Non-polite style (style without interpersonal expressions)

不本意　　だ が　···
disinclination although

(b) Literary style (formal and oldish style)

会議室 にて　　　ミーティングを 行う
at the conference room meeting　　　conduct

(c) Difficult words (unusual difficult-to-understand expressions)

規則に　違反 する
regulation contravene

2. Expressions with complicated structures
   (a) Nominalized predicates (compound nouns including nominalized predicates)

内閣の 大幅な 改造 を 行う
Cabinet large　　reform perform

(b) Long sentences including embedded sentences, subordinate clauses, etc.

11.4.2 Outline of the Paraphrasing System

We have developed a system of paraphrasing difficult words and compound nouns into natural spoken language out of the expressions introduced in the previous subsection. This system performs the following steps in order.

0. *Preprocessing.* Morphological, syntactic, and predicate–argument analysis are applied to the input sentence.
1. *Paraphrasing of nominalized predicates.* Predicate–argument analysis is applied to the nominalized predicate in a compound noun, and case information is complemented.
2. *Paraphrasing of difficult words.* Difficult words that are not suitable for speech are paraphrased into easier expressions.
3. *Paraphrasing of expressions in non-polite and literary style.* Expressions in non-polite and literary style are paraphrased into colloquial expressions.

11.4.3 Paraphrase of Nominalized Predicates

Predicate–argument analysis based on the case frames is applied to the nominalized predicates, and then they are paraphrased based on the results.

- noun + deverbative noun + "する"
  An appropriate case-marking postposition is added to the noun according to the analysis result of the deverbative noun.

構造　改革 する ⇒ 構造 を 改革 する
structure reform do

- deverbative noun + "を" + "行う, する, 図る"

This expression is transformed into 'deverbative noun + "する"', and then modifying words are changed properly according to the analysis result of the deverbative noun.

内閣の 大幅な 改造を 行う ⇒ 内閣 を 大幅 に 改造 する
Cabinet large   reform perform

- ... deverbative noun + noun
  "する" is added to the deverbative noun, if the relation between the last noun and the deverbative noun is nominative, accusative or non-gapping. Appropriate case-marking postpositions are added to the nouns preceding the deverbative noun according to the analysis result of the deverbative noun.

機種   依存   文字 ⇒ 機種 に 依存 する 文字
machine dependence character

11.4.4 Paraphrasing of Difficult Words

Difficult words that are peculiar to written language are paraphrased into easier expressions using the relationship between the head words and their definition sentences in an ordinary dictionary. For instance, the following paraphrases are possible:

方針を 決定する ⇒ 方針を はっきり決める
policy   determine   policy clearly decide

規則に   違反する ⇒ 規則 を   やぶる
regulation contravene   regulation break

To realize such paraphrases, it is necessary to perform accurate analyses, word sense disambiguation (in case of polysemous words) and case marker transformation (the second example above). We have developed a method that grasps several senses of both a headword and its def-heads (the head of the definition) as a form of case frame. These case frames are then aligned, resulting in the acquisition of word sense disambiguation rules and the detection of the appropriate equivalents and case marker transformations (Kaji *et al.* 2002).

An outline of the method is depicted in Figure 11.3. The headword *keitou* and the def-heads *necchuu* and *shitau* have several case frames depending on their several usages. For each case frame of *keitou*, we find the most similar case frame among the def-head case frames. At this matching stage, case component correspondences are also found. For example, the case frame

{she, I ...} *ga* {president} *ni*

of *keitou* is matched to the case frame

{student, he ...} *ga* {teacher, professor ...} *wo*

of *shitau*, and their case components are aligned as shown in the lower part of Figure 11.3. From this correspondence, we could find that this usage of *keitou* has the *shitau* meaning, that the *ni* case should be transformed into the *wo* case, and that the equivalent is *shitau*.

In addition, it is necessary to judge what words should be paraphrased. We have previously proposed a method for automatically acquiring difficult words to be paraphrased from written-like text and spoken-like text collected from the Web. This procedure is detailed in Kaji *et al.* (2004).

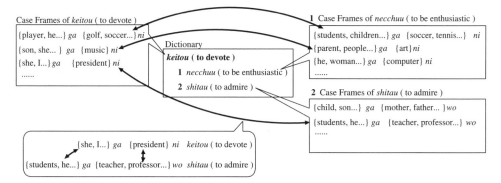

Figure 11.3 Verb paraphrase based on case frame alignment. Reproduced by permission of ACL, Kaji *et al.*, 2002

11.4.5 *Paraphrasing of Expressions in Non-polite and Literary Style*

Expressions in non-polite and literary style are represented by suffixes and function words, and thus their variations are limited. Therefore, they can be paraphrased into expressions in colloquial style by hand-crafted rules.

The non-polite style lacks interpersonal expressions and is rarely used in spoken language in normal, public situations. Expressions in non-polite style are paraphrased into those in polite style, which are generally used in social utterances. For instance, the non-polite-style word "だ" is paraphrased into the polite-style word "です" as follows:

不本意　　だ が　　 … ⇒ 不本意 です が　 …
disinclination although

Similarly, expressions in literary style are formal and oldish, and as such they are not suitable for speech. They are paraphrased into their equivalent expressions normally used in daily life. For example, the literary word "にて" is paraphrased into the usual word "で" as follows:

会議室 にて　　　　ミーティングを 行う ⇒ 会議室 で ミーティングを 行います
at the conference room meeting　　　　　　conduct

These paraphrasing rules are described flexibly by referring to morphological and syntactic information, in particular parts of speech, inflection forms and dependencies.

11.5 Automatic Slide Generation

As described in the previous section, the use of a synthesized voice to read out a knowledge card that has been paraphrased from written language to spoken language enables written texts to be brought into the field of conversation. The next problem related to this task is that of what visual information is presented along with the audio information. If chart, image, or video is attached to a knowledge card, we can present it along with the knowledge card. Another solution would be having a 3D-CG agent present the spoken language with facial expressions and gestures in synchronization with the contents of the card. In this study, we adopt a method of automatically summarizing a knowledge card and presenting summary slides, which are commonly used in lectures. Since current speech synthesis systems often produce speech that is unnatural or hard to listen to, the presenting of summary slides is important for compensating audio information.

In this section, we describe a method for automatically generating a summary slide (Shibata and Kuro-hashi 2005). Automatic slide generation consists of three steps: discourse structure analysis (described in section 11.2.3), the extraction of topic parts and reduction of non-topic parts, and the displaying of them in a slide. We describe the last two steps in detail in the following sections.

11.5.1 Extraction of Topic Parts / Reduction of Non-topic Parts

We extract a topic part and non-topic part from a sentence, which are then placed in a slide. Since, in general, non-topic parts are relatively long, we reduce the non-topic parts to make the produced slide easier to read.

Extraction of Topic Parts

A phrase whose head word is marked with the topic marker "*wa*" is extracted as a topic. A topic can also be extracted by several cue phrases. When there are multiple cue phrases in a sentence, a topic is extracted only by the first cue phrase. Some of these are shown below:

- *inner city deha* ... (At an inner city, ...)
- *syukkagenin-ga-hanmei-shita-kasai ni-oite* ... (In the fire whose cause was revealed, ...)
- *3-sen-no-futsuu ni-yori* ... (Due to the interruption of the three train services, ...)

Reduction of Non-topic Parts

In order to make a slide easier to read, it is important to reduce a text while preserving the original meaning. We thus reduce non-topic parts by (1) pruning extraneous words/phrases based on syntactic/case structure, and (2) deleting extraneous parts of the main predicate according to some rules.

1. *Pruning extraneous words/phrases*. Extraneous words/phrases are pruned based on syntactic/case structure. The following materials are pruned:
   - conjunctions
   - adverbs
   - adverb phrases
     Example: *computer-teishi-no-tame data-hikitsugi-mo-konnandatta.* (Because of the computer going down, it is difficult to turn over the data.)
   - appositive phrases and examples
     Example: *nourinsuisansyou, kokudotyou-nado kuni-no-kakukikan.* (Agencies, such as the Agriculture, Forestry and Fisheries Ministry and National Land Agency)
2. *Removing extraneous parts of a main predicate*. Extraneous parts of a main predicate are removed by the following rules:
   - deverbative noun-*suru/sareta* : *suru/sareta* is deleted
     Example: *jisshi-sareta* $\Rightarrow$ *jisshi* (carried out)
   - deverbative noun-*ga-okonawareta* : *ga-okonawareta* is deleted
     Example: *yusou-ga-okonawareta* $\Rightarrow$ *yusou* (transport)
   - noun-copula : copula is deleted
     Example: *genin-da* $\Rightarrow$ *genin* (cause)

Note that extraneous parts of the main predicate are not removed if the predicate has a negation expression.

11.5.2 Slide Generation

Topic/non-topic parts that are extracted in the previous section are placed in a slide based on the discourse structure that is detected in section 11.2.3. The heuristic rules for generating slides are the following:

- If the text has a title, it is adopted as the title of the slide; otherwise, let the first topic in the text be the title of the slide.
- If there is a topic in a sentence, the topic is displayed first, and in the next line the non-topic parts are displayed in the subsequent indent level. If there is not a topic in a sentence, only the non-topic parts are displayed.
- The indent of each sentence is determined according to the coherence relation to its parent sentence:
  - *start:* the level is set to 0 because a new topic starts.
  - *contrast/list:* the same level.
  - *topic chaining:* if a topic is equal to that of the parent sentence, the level is set to the same level, and the topic is not displayed; otherwise, the level is decreased by one and the topic and non-topic part are displayed.
  - *otherwise:* the level is increased by one.

Figure 11.4 shows an example of a slide automatically generated based on the discourse structure in Figure 11.2. In this example, "awareness", which is a topic in the first sentence, is adopted as the title of the slide, and the non-topic part of this first sentence is displayed in the first line of the main part in the slide. Then, based on the contrast relation between the second and the forth sentence, "synchronous working environment" and "asynchronous working environment", which are the topics in the second and forth sentences, respectively, are displayed at the same indent level, while non-topic parts are displayed in the next lines following each topic part. The non-topic part extracted from the third sentence is displayed at an indent level increased by one against that of the second sentence.

As a number of researchers have pointed out (Marcu 1999; Ono *et al.* 1994), discourse structure can be a clue to summarization: units found closer to the root of a discourse structure tree are considered to be more important than those found at lower levels in the tree. Concurrent with this idea, it can be

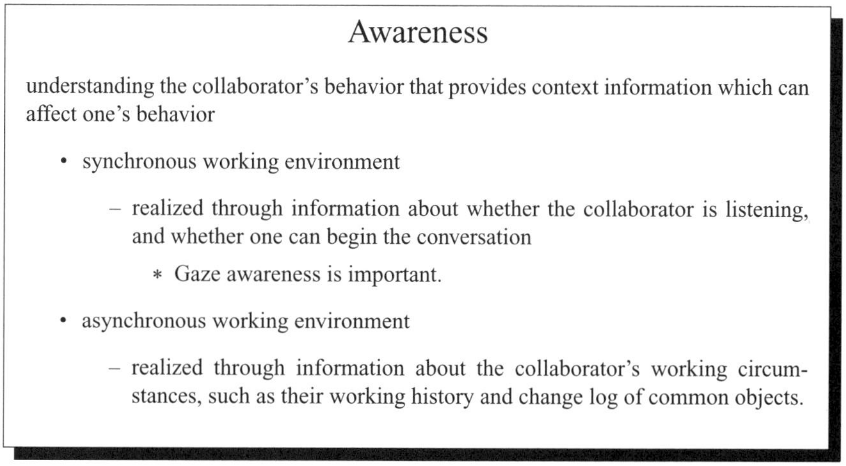

Awareness

understanding the collaborator's behavior that provides context information which can affect one's behavior

- synchronous working environment
  - realized through information about whether the collaborator is listening, and whether one can begin the conversation
    * Gaze awareness is important.
- asynchronous working environment
  - realized through information about the collaborator's working circumstances, such as their working history and change log of common objects.

Figure 11.4   Example of generated slide (translated into English)

> **Category:** emergency response in the devastated area
> **Subcategory:** volunteer
> **Subsubcategory:** category/activity of volunteers
> **Lesson:** 1.8 million people from various parts of the country rushed in to the devastated area as volunteers and gave courage and hope to many citizens. The roles of the volunteers were at first the provision of medical care, food/supply distribution, the confirmation of the safety of the aged, and refuge organization, but these later changed to supply allocation, house-moving/repair, and care for the aged and disabled persons, as time passed. Since April, the number of volunteers has been decreasing dramatically because many outside volunteers, consisting mainly of students, have withdrawn. Those who came from outside of the prefecture accounted for more than 60% and many volunteers acted individually.

Figure 11.5 Example of a knowledge card (translated into English). (Reproduced by permission of Cabinet Office, Government of Japan and Hyogo Earthquake Memorial 21st Century Research Institute)

considered that topic/non-topic parts in lower units (deep in a discourse structure) are not placed in a slide. However, in cases in which automatically generated slides are presented along with speech synthesis, the above treatment is not applied because speech without any corresponding description in a slide is not natural.

As for a knowledge card like the one used in the above example, which is logical and has sufficient cue phrases, automatically generated slides can grasp the main points of such cards. In particular, a slide in which a contrast/list structure is detected is far easier to read than an original text.

11.6 Experiments and Discussion

To demonstrate the effectiveness of our proposed method, we performed experiments. According to our motive of wanting to utilize such technology for solving social problems, we adopted the *Hanshin–Awaji Daishinsai Kyoukunn Siryousyuu* (Great Hanshin–Awaji Earthquake Research Paper; cabinet office 2000) as a collection of text. This material has a hierarchical structure: "Category", "Subcategory", "Subsubcategory", and "Lesson". We considered one "Lesson" as comprising one knowledge card. An example of a knowledge card is shown in Figure 11.5. The total number of knowledge cards are 400, each of which contained an average of 3.7 sentences with an average length of 50 Japanese characters.

The system runs on a Web browser, as shown in Figure 11.6 (the translated presentation is shown in Figure 11.7). The system first retrieves the knowledge card that is most similar to the user's query with the text retrieval technique (Kiyota *et al.* 2002). Then, the system converts the written text into spoken language, and feeds it into a speech synthesis engine while presenting automatically generated summary slides. We adopted HitVoice 3.0 developed by Hitachi Ltd[1] as a speech synthesis engine.

Twenty queries, such as "role of volunteers", "what is the cause of the fire?", and "damage caused by gas", were given to the system, and we evaluated the results. First, as for the paraphrases from written language to spoken language, expressions in non-polite and literary style were transformed into those in polite style, resulting in the synthesized voice sounding more natural. Thus, this type of paraphrasing was fully successful.

---

[1] See http://www.hitachi.co.jp/

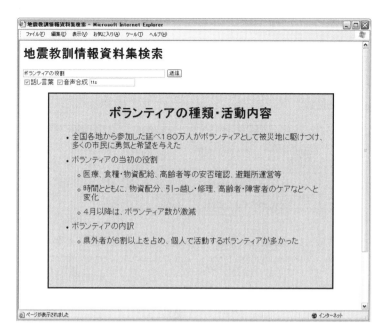

Figure 11.6 Screenshot of automatic presentation system. (Reproduced by permission of Cabinet Office, Government of Japan and Hyogo Earthquake Memorial 21st Century Research Institute)

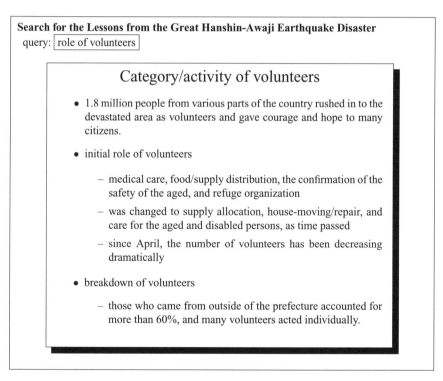

Figure 11.7 Automatic presentation system (translated into English). (Reproduced by permission of Cabinet Office, Government of Japan and Hyogo Earthquake Memorial 21st Century Research Institute)

There were 159 paraphrases of other types. We classified these paraphrased results into three classes: 141 appropriate paraphrases, 11 unnatural paraphrases, and 7 inappropriate paraphrases. Some examples of each class are shown below.

Appropriate paraphrases

液状化が 生じず ⇒ 液状化 しないで
liquefaction (not) engender　liquefaction (not) do

甚大な 被害 ⇒ 大きな 被害
tremendous damage　large　damage

被災者の 避難 場所 ⇒ 被災した 人の 避難する 場所
stricken-victims evacuation place　stricken people　evacuation place

Unnatural paraphrases

携帯電話 ⇒ 携帯の 電話
cellular phone　portable phone

供給に 支障を 及ぼす ⇒ 供給に さし さわりを 及ぼす
supply impediment cause　supply offense　cause

最大の 要因 ⇒ 最大の 主な 原因
biggest determinant　biggest main cause

Inappropriate paraphrases

協力 会社では ⇒ 協力した 会社では
cooperative dcompany　cooperated company

自治体の 要請を 受けて ⇒ 自治体が 要請されて
municipal requirement receive　municipality required

The major causes of the inappropriate paraphrases were the insufficiency of the handling of compound nouns and the failure of predicate–argument analysis.

As for the automatically generated slides, there were no crucially inappropriate slides (i.e., cases in which the produced slide prevented the user from understanding the content), although there is still some room for improvement in topic extraction and indentation setting. As illustrated in Figure 11.6, by extracting topics and setting the indentations, automatically generated slides are far easier to read in comparison with knowledge cards in their original state. When the quantity of displayed text exceeds a certain threshold, multiple slides are generated by splitting the slide as illustrated in Figure 11.8 (the translated presentation is shown in Figure 11.9). In this case, the system switches the slide to the next slide in synchronization with the speech synthesis. Since this material has an appropriate title at "Subcategory", it is adopted as the title of the slide instead of the topic of the first sentence.

As for points in need of improvement as shown in Figures 11.6 and 11.7, for example, it is preferable that "initial" be removed from the second topic and that "initial", "as time passed", and "since April" be

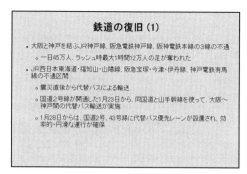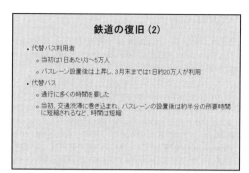

Figure 11.8 Example of generated multiple slides. (Reproduced by permission of Cabinet Office, Government of Japan and Hyogo Earthquake Memorial 21st Century Research Institute)

placed at the same indentation level in order to indicate that these items are contrasted. Thus, by detecting the contrast structure in a whole text and extracting the topic of each sentence precisely as mentioned above, better slides can be produced.

We plan to perform additional evaluation experiments to demonstrate that the whole system can produce natural presentation and contribute to the improvement of user understanding.

11.7 Conclusions

Blockades in knowledge circulation are considered crucial factors in the occurrence of social problems and stagnation. The introduction of software agents offers one solution to such problem, but their current insufficiencies are not acceptable because people are very sensitive to the appropriateness/inappropriateness of language usage. This study proposed a methodology to address this problem using NLP technologies. In the future, we will conduct a field trial in which we introduce this system fused with a software agent to the field of conversation with the aim of verifying the effectiveness of the proposed framework.

Railway Recovery (1)

- interruption of the three train services, JR Kobe-line, Hankyu Express Kobe-line and Hanshin Electric Railway
  - 450,000 people per day, 120,000 people per hour at the peak of rush, had no transportation
- interruption sections in West Japan Railway Toukaidou Line, Sannyou Line, Hankyu Takarazuka, Imazu and Itami Line and Kobe-Electric Arima-line
  - after the earthquake occurred, transportation by alternate-bus was provided
  - from January 23rd, when National Route 2 was opened, transportation by alternate-bus between Osaka and Kobe was provided
  - from January 28th, the alternate-bus priority lane was set up, and the smooth transportation was maintained

Railway Recovery (2)

- alternate-bus user
  - at first, 30,000 to 50,000 people per day
  - the number of users increased after the alternate-bus lane was set up, and around 200,000 people per day used until the end of March
- alternate-bus
  - it took a lot of time to transit
  - at first, an alternate bus was caught in a traffic jam, and after the alternate-bus lane was set up, time required was cut in half

Figure 11.9 Example of generated multiple slides (translated into English). (Reproduced by permission of Cabinet Office, Government of Japan and Hyogo Earthquake Memorial 21st Century Research Institute)

References

Cabinet Office Government of Japan, Hyogo Earthquake Memorial 21st Century Research Institute (2000) Great Hanshin–Awaji Earthquake Research Paper, http://www.iijnet.or.jp/kyoukun/index.html.

Japanese Lexicon.

Kaji N., Kawahara D., Kurohashi S. and Sato S. (2002) Verb paraphrase based on case frame alignment. *Proceedings of the 40th Annual Meeting of the Association for Computational Linguistics*, pp. 215–222.

Kaji N., Okamoto M. and Kurohashi S. (2004) Paraphrasing predicates from written language to spoken language using the web. *Proceedings of the Human Language Technology Conference*, pp. 241–248.

Kawahara D. and Kurohashi S. (2002) Fertilization of case frame dictionary for robust Japanese case analysis. *Proceedings of 19th COLING* (COLING02), pp. 425–431.

Kawahara D. and Kurohashi S. (2004) Zero pronoun resolution based on automatically constructed case frames and structural preference of antecedents. *Proceedings of the 1st International Joint Conference on Natural Language Processing*, pp. 334–341.

Kawahara D. and Kurohashi S. (2006) Case frame compilation from the web using high-performance computing. *Proceedings of the 5th International Conference on Language Resources and Evaluation.*

Kiyota Y., Kurohashi S. and Kido F. (2002) Dialog navigator: a question answering system based on large text knowledge base. *Proceedings of 19th COLING*, pp. 460–466.

Kubota H., Kurohashi S. and Nishida T. (2002) Virtualized-egos using knowledge cards *Seventh Pacific Rim International Conference on Artificial Intelligence* (PRICAI-02), WS-5 International Workshop on Intelligent Media Technology for Communicative Reality (IMTCR2002), pp. 51–54.

Kurohashi S. and Nagao M. (1994) Automatic detection of discourse structure by checking surface information in sentences. *Proceedings of 15th COLING*, vol. 2, pp. 1123–1127.

Marcu D. (1999) Discourse trees are good indicators of importance in text. In I. Mani and M. Maybury (eds), *Advances in Automatic Text Summarization*. MIT Press, pp. 123–136.

Nishida T. (2003) Conversational knowledge process support technologies as advanced communication infrastructure for social technologies. *Social Technology Research Transactions* **1**, 48–58 (in Japanese).

Ono K., Sumita K. and Miike S. (1994) Abstract generation based on rhetorical structure extraction. *Proceedings of the 15th COLING*, pp. 344–348.

Shibata T. and Kurohashi S. (2005) Automatic slide generation based on discourse structure analysis. *Proceedings of Second International Joint Conference on Natural Language Processing* (IJCNLP-05), pp. 754–766.

12

Video Content Acquisition and Editing for Conversation Scenes

Yuichi Nakamura

12.1 Introduction

In this chapter, we introduce our approach for capturing and editing conversations for providing useful archives as multimedia contents. The purposes are to capture and archive meeting scenes, and to provide images or videos that can be used for interactive or conversational video archives. The basic problems are video capturing, video editing, and multimedia presentation. To resolve these two problems, we propose a novel system for automated video capture, and a novel method for automatic editing. For multimedia presentation, there are interesting works such as POC by Nishida *et al.* (1999), and we are planning tight cooperation.

As the background of this research, there are considerable demands for meeting recording and archiving. Providing the minutes of a meeting by an automated system is a promising application even if it requires human assistance to a certain extent. Videos have, moreover, much richer information than a text minute. They can provide not only "who spoke what" but also "how it was spoken", "how was the attitude of the listeners", "feeling or emotional aspect of a meeting", etc. These kinds of information are essential for correctly or precisely understanding what happened in a meeting.

We can also consider a conversation as a collection of informative knowledge resources. Capturing conversations and use of appropriate portions of recorded conversations can be a good communication tool. Conversation records are also good data for cognitive science and physiology. Researchers in those fields need detailed and precise records for observing phenomena that happen in a conversation.

As related works, there have been not a few works proposed for recording meetings (Lee *et al.*, 2002; Erol *et al.*, 2003), lectures (Murakami *et al.*, 2003), etc. For example, Rui's work (Rui *et al.*, 2001) realized a simple and portable meeting recording system, in which participants' faces are always recorded with an omni-directional video recording system. Playback with clipping out of a speaker's figure is an efficient method for obtaining multimedia presentation of a meeting record. They capture minimum data that enable us to roughly grasp what happened in meetings at a low cost.

Our research focuses on a different point: we aim to capture and provide videos with good picture compositions and good editing as seen in movies and TV programs. We need more detailed capturing

Conversational Informatics: An Engineering Approach Edited by Toyoaki Nishida
© 2007 John Wiley & Sons, Ltd

of a meeing than the previous works, and need more effective ways of replaying the captured scenes. In this sense, we need more sophisticated techniques on camerawork and editing. We need at least several cameras located at different points, and editing to produce comprehensible and not boring videos. For this purpose, we can see many examples and rules collected and explained in textbooks on film studies, such as Bordwell and Thompson (1997) and Arijon (1976), since a conversation scene is one of the most important elements in movies. We put a great deal of emphasis on realizing a system that simulates these techniques.

As the first step toward sophisticated meeting capturing and browsing, we designed a system for meetings among two to four people, and applied the system to actual meetings. Although scalability for more attendees is one of the most important issues, differences between our idea and the previous works and its effectiveness can be shown in experiments even with small meetings. Moreover, those meetings often require support for recording, since we often have such a meeting without an agenda, also without a secretary or a person who takes notes.

12.2 Obtaining Conversation Contents

Figure 12.1 shows the overview of our framework. The purposes of this research are production of rich archives and generation of virtual conversations using the archives. This implies that what, how, or when we want to present to the users has tight relations to camerawork, indices, and editing. For examples, three or more kinds of shots are necessary for producing attractive video contents as shown in movies and TV programs. The design is based on the idea that we need to capture most of the targets that are potentially necessary to be presented, and need to provide editing patterns suitable for most of potential usage.

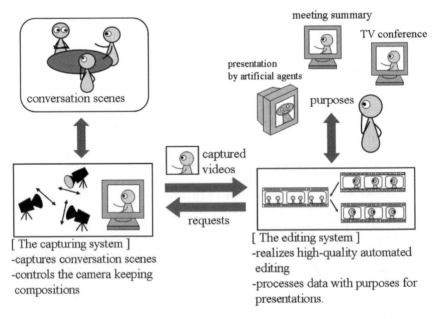

Figure 12.1 Scheme of contents production. Reproduced by permission of © IEICE

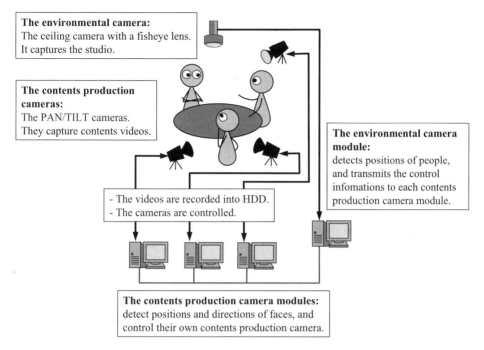

The environmental camera:
The ceiling camera with a fisheye lens.
It captures the studio.

The contents production cameras:
The PAN/TILT cameras.
They capture contents videos.

The environmental camera module:
detects positions of people, and transmits the control infomations to each contents production camera module.

- The videos are recorded into HDD.
- The cameras are controlled.

The contents production camera modules:
detect positions and directions of faces, and control their own contents production camera.

Figure 12.2   Overview of the capturing system. Reproduced by permission of © IEICE

Our contents acquisition scheme mainly consists of contents capturing and editing. Contents capturing consists of filming[1] and index acquisition. Filming is to obtain attractive shots as seen in movies or TV programs without disturbing natural conversations. We constructed a system as shown in Figure 12.2 that observes the positions of talking people, and automatically controls cameras for filming them by attractive shots. Index acquisition potentially includes recognition of a variety of phenomena such as speech, tone of voices, nonverbal actions, facial or body expressions, etc. Currently, we are using amplitude and utterance of speech, nodding actions, and deictic movement as important features.

Editing has two requirements, that is, quality and appropriateness: resultant videos need to be of high quality and suitable for the situation where it is used. To handle this problem, we need a mechanism that produces different editing results for different requirements. For this purpose, we developed a computational editing model based on optimization with constraints satisfaction (Ogata *et al.*, 2003). We can examine a variety of editing patterns by adjusting evaluation functions and the constraints in this model, or just inserting/leaving-out some of them.

12.3 Capturing Conversation Scenes

12.3.1 Overview of Capturing System

As shown in Figure 12.2, this system is composed of two modules: the "environmental camera module" detects and tracks people; a "contents capturing camera module" captures each shot while keeping an appropriate picture composition. This arrangement is necessary, since realizing both observation and

---

[1] We use "filming" as the activity of shooting a scene by cameras without necessarily using physical films.

Figure 12.3   Contents capturing camera module: (a) close-up shots of talker A; (b) close-up shots of talker B; (c) over-the-shoulder shots of talker A; (d) long shots capturing two talkers

capturing by a single camera is difficult. A wide-angled camera is required for tracking people that arbitrarily move in a room, and narrow-angled cameras are required for capturing close-up shots, and they cannot be pan/tilted frequently or rapidly.[2]

Each contents capturing camera module transmits its ID and capturing state to the environmental camera module at regular intervals. The environmental camera module assigns each contents capturing camera module to a shot: close-up, etc. Then, the environmental camera module estimates the pan/tilt angles of the contents capturing camera modules based on the positions of people and the positions of the contents capturing camera modules.

12.3.2 Picture Compositions for Conversation Scenes

Figure 12.3 shows the shots captured by contents capturing modules. As shown in this example, this system controls cameras to keep typical *picture compositions* that are commonly used in movies and TV programs.

- *Close-up/bust shot.* The captured face occupies a dominant portion of a screen as shown in Figures 12.3(a) and (b). The center of the face is placed at around 1/3 from the top of the screen. Usually, if a face is full frontal, the face is placed at the center of a screen. If a face looks oblique or in profile, the face is placed so that there is larger space in the front side of the face than in the back side.
- *Over-the-shoulder shot.* This captures a face with the shoulder of the opposite person as shown in Figure 12.3(c). A face is captured as there is enough space between the face and the shoulder. The center of a face is placed at around the 1/4 from the top of the screen.
- *Long shot.* A "long shot" in this research means the shot that captures people with a wide-angled frame as shown in Figure 12.3(d). In our research, we usually use a shot that captures the side view of two people who sit face to face.

12.3.3 Contents Capturing Camera Module

To continue capturing a face while keeping a good picture composition, a contents capturing camera module needs to continuously detect and track a face.

---

[2] Rapid or frequent pan/tilt causes shaky and annoying videos. We need to avoid unnecessary camera movements as much as possible.

Face Detection and Tracking

First, this module applies the face detection method of Rowley *et al.* (1998) to a captured image, and this detection is repeated. Although the detection accuracy of this method is satisfactory, it requires considerable computation time (perhaps one second), and we need another method to check the face position at every frame. For this purpose, the camera module tracks the face at every frame by using template matching, where the template is the detected face region at the previous frame. When the template matching is failed, the camera module calls the above face detection. The camera module also calls the face detection at regular intervals and renews the template, since template matching is sometimes misdirected by noise or a target's appearance change.

Estimation of Face Orientation

Face orientation is also essential information for keeping good picture composition. Our system estimates face orientation by using the center of the face, the center of the left eye, and the center of the right eye.

The detection of an eye is based on the method of Oda *et al.* (1998). Figure 12.4 shows the process of eye detection. First, *eye candidate regions* are detected each of which has a color far from the skin-color and lies inside of the face region. Then, the left and right eye positions are estimated by checking every pair of eye candidate regions.

The camera module also tracks eyes by template matching by using the eye candidate region at the previous frame. The camera module computes the weighted sum of the positions obtained by the above two methods, and estimates the current eye positions by the Kalman filter.

Finally, the camera module estimates the face orientation by the position of the eyes and the position of the face.

Camera Control

To keep an appropriate picture composition, the best position where a face should be located on a screen is estimated at every frame, and the camera's pan/tilt angle is controlled based on this position. For this purpose, a contents capturing camera module has three states, and it changes the current state by face tracking results.

- *Talking state.* While the face stays in a certain area on a screen, the camera module is in the talking state. In this state, the camera module computes the average position of the face position at regular intervals. If the distance between the average position and the target position where the face should be located is larger than the threshold, the camera module changes the pan/tilt angle. At that time, the speed of the pan/tilt is set slow.

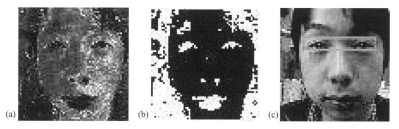

Figure 12.4 Eye detection: (a) a skin-color distance; (b) binarized; (c) eye candidate region. Reproduced by permission of © IEICE

Figure 12.5   Experiment: the environmental camera module

- *Moving state.* When the face is around the edge of the screen and moving toward the outside of the screen, the state is changed to the moving state. In this state, the module changes pan/tilt angle quickly so that the face does not go out of the screen. At that time, the speed of pan/tilt is set fast.
- *Sleeping state.* When the camera module fails to detect a face, the state is changed to the sleeping state. The camera module waits until the face appears on the screen again or the face position is reported by the environmental camera module.

12.3.4 Environmental Camera Module

To support the contents capturing camera modules, we use an environmental camera with a fisheye lens, which detects people in a meeting room. Figure 12.5 shows an example of human detection by the environmental camera module. The ellipses show the positions of people.

The environmental camera module computes inter-frame subtraction, and labels each region whose pixel has a difference value larger than the threshold. The camera module applies clustering to the detected regions, and obtains the number of humans and their rough positions.

The system also uses template matching in combination with the inter-frame subtraction, since the difference is not well detected if a person moves slowly or stays still. The template is renewed only when the displacement of a person is large and clear differences are detected by the inter-frame subtraction.

12.3.5 Communication among Camera Modules

First, the environmental camera module assigns each contents capturing camera module to a shot (close-up, etc.). Then, the environmental camera module estimates the pan/tilt angles for each contents capturing camera module based on the positions of people and the position of the camera module.

The contents capturing modules and the environmental camera module communicate the following data at regular intervals:

- A contents capturing camera module sends the module ID and its state as mentioned above.
- The environment module sends a contents capturing camera module the detected position of the person whom the camera module is tracking.

This enables cooperative tracking of people.

12.4 Editing Conversation Scenes

Editing is essential for video production and effective utilization of video. As the first step toward automated editing, we investigated the problem of "camera switching" for a scene taken by multiple cameras as discussed in the previous sections.

For the camera switching problem, so far, event-driven algorithms that switch views at the time of typical event occurrences have been proposed (Atarashi *et al.*, 2002; Onishi *et al.*, 2001). However, such event-driven methods are sensitive to slight changes of event occurrence time, and do not have sufficient functions for systematically investigating a variety of editing patterns.

We propose a novel editing method based on optimization with constraint satisfaction. This model searches for the best editing patterns out of all possible shot combinations that satisfy given constraints. This model, as a result, has greater flexibility in integrating various editing rules simultaneously.

12.4.1 Computational Editing Model

First, we consider a video as a sequence of short video segments, each of which has a length of 0.5 or 1 second. Next, we define video editing as the problem of assigning an appropriate shot to each video segment.

The model is formally composed of five elements:

$$Editing = \{S, E, C, O, V\} \tag{12.1}$$

The explanation of the above terms is given below:

- *Shots.* S is a set of shots: $S = \{s_0, \ldots, s_n\}$, where s_i is a shot, such as "a bust shot of person A", "a long shot of person B", etc.
- *Video.* V is a sequence of video segment units: $V = \{v_0, \ldots, v_{tmax}\}$, each of which has a length of 0.5 or 1 second. An appropriate shot (s_i) is assigned to each video segment (v_j).
- *Events.* E is a collection of events (e_i), each of which occurs in the scene. An event is something important to be watched or to be a trigger for view switching. If e_i occurs at time t with the certainty of 0.9, we denote it as $e_i(t) = 0.9$.
- *Evaluation.* O is a set of evaluation (objective) functions o_i, each of which gives the appropriateness of each assignment of a shot (or shots) to a video segment (or segments).
- *Constraints.* C is a set of constraints c_i. Since combinatorial explosion will occur if we allow for all the possible editing patterns, the number of candidates must be limited by constraints before they are thoroughly evaluated.

The objective of this model is optimization of G that is the sum of the above evaluation function values:

$$G = \sum_{t=0}^{t_{max}} \sum_{i=0}^{N_o} o_i(t) \tag{12.2}$$

Figure 12.6 shows the flow of computation. The flow is mainly composed of three steps: *pre-scoring*, *candidate searching*, and *post-scoring and selection*. In the pre-scoring step, the relevance of each shot is evaluated for each video segment. In other words, each shot at each time is scored based on events occurring by evaluation functions.

In the candidate-searching step, based on the given scores and constraints, the possible editing candidates are searched. Then, in the post-scoring step, each candidate, or shot sequence, is scored by evaluation functions. Finally, the editing pattern(s), i.e., a shot sequence or sequences that received the highest score, is/are chosen.

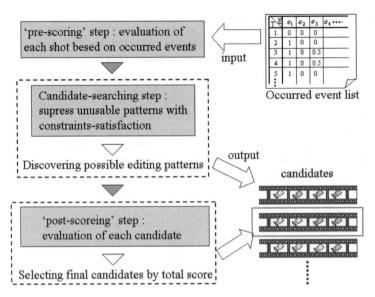

Figure 12.6   Flow of computation for editing. Reproduced by permission of © IEICE

12.4.2 Events, Evaluation Functions, and Constraints Definition

We need to consider a variety of events that might have relevance to video editing, such as speeches, various kinds of motion, facial expressions, object movements, and so on. Table 12.1 shows typical events. They are often focused on in videos as well as they are often used for triggers for view switching. Table 12.2 shows the shots and events that we used in our experiments.

For making good videos, we need to consider a variety of criteria:

- *Focus.* The video should lead the audience to the focus(es) of attention, though there can be two or more focus candidates. This is the most common request for video editing.
- *Not misleading.* A shot or a combination of shots often suggests something that is not presented on a screen. "Montage" is a famous term for explaining this effect. Although this is powerful and fundamental function of video editing, we must be careful to avoid misleading composition. Table 12.3 shows some examples that would cause misunderstanding.
- *Perceptual load.* A good video gives the audience an appropriate perceptual load. Too monotonous a composition is boring, while too many stimuli cause too much perceptual strain. A good illustration of perceptual load is the issue of shot length. A succession of short-length shots (e.g., less than 1 second) strains audience perception, while a long-length shot (e.g., 30 seconds) can easily become boring. Table 12.4 shows examples that causes inadequate perceptual load.

As mentioned above, there is no clear distinction between an evaluation function and a constraint. The computational feasibility of our editing model, however, relies on the constraints' role in reducing the number of possible editing patterns. Table 12.5 shows examples of constraints. For convenience, evaluation functions are used in two ways. One is pre-scoring, which evaluates every shot at each time. The other is post-scoring, which evaluates a combination of shots. Table 12.6 shows examples of evaluation functions. The upper part shows examples of pre-scoring evaluation. For example, o_1 gives scores to shot s_1 ("bust shot of person A") at time t, if person A speaks at time t. The lower part of Table 12.6 shows

Table 12.1 Events used in our experiments

Speech	The fact of speaking, starting speech, or ending speech, is one of the most important clues. The spoken words are also good clues that specify where the focus of attention is.
Gestures	Movements, especially deictic movements and illustrations,[a] strongly suggest where the focus of attention is.
Hand manipulations	Hand manipulations of objects are also good clues for focus detection.
Facial expression and head motion	Facial expressions and head motions, such as nodding and turning the head, express listeners' attitudes, and therefore strongly suggest where to look.
Touching between persons	Body touching, such as tapping, touching, hand shaking, etc. are also key events to be watched.

[a] These categories are given in Ekman and Friesen (1969).

Table 12.2 Shot and event setting for experiments: shots, events, constraints, and evaluation functions

shots (S)

s_1	long shot of two people
s_2	close shot of table-top
s_3	bust shot of person A
s_4	bust shot of person B
s_5	over-the-shoulder shot of person A
s_5	over-the-shoulder shot of person B

events (E)

e_1	person-A is speaking
e_2	person-B is speaking
e_3	a keyword is spoken that refers an object or a work on a table
e_4	person-A is nodding
e_5	person-B is nodding

Table 12.3 Examples of negative effects caused by inadequate editing (misunderstanding) ·

Violation of the 180 degree system	This leads to misunderstanding of spatial arrangements or motion directions.
Inappropriate shot changes	Shot changes suggest changes of the focus of attention. An inappropriate shot change directs the audience's attention to a wrong portion.
Connecting shots with similar angles	Jump cuts between similar angles create the illusion of apparent motions.

Table 12.4　Examples of negative effects caused by inadequate editing (perceptual load)

Too few shot changes	This usually causes a boring video.
Too many shot changes	This requires too much perceptual strain.
Too many close shots	This requires too much perceptual strain. perception.
Inappropriate shot changes	This disturbs the natural understanding of a story and causes considerable strain.

Table 12.5　Examples of constraints

c_1	Prohibits shots equal to or shorter than 2 seconds
c_2	Prohibits shots that have scores equal to or less than 0 and continue more than 3 seconds
c_3	Stipulates that the establishing shot (s_1) must be contained in the first 10 seconds

examples of post-scoring evaluation, where o_6 gives the preference for avoiding negative effects shown as the third example in Table 12.3.

12.4.3 Parameter Setting

The editing result changes depending on constraints, evaluation functions, and the parameters of them. Tables 12.6 and 12.5 show parameter examples in our experiments. In the current configuration, the scores given by each evaluation function range between 0 and 100, and scores over 50 points are used in cases of strong preference. The parameter values in the evaluation functions and constraints are empirically determined.

Table 12.6　Examples of pre-scoring functions (upper) and post-scoring functions (lower): a vector in each figure shows a collection of the values for each shot at each time

	Scoring
$o_1(t)$	$[0, 0, 10e_1(t), 0, 0, 0]$
$o_2(t)$	$[0, 0, 0, 10e_2(t), 0, 0]$
$o_{3A}(t)$	$[0, 10e_3(t), 0, 0, 0, 0]$
$o_{3B}(t)$	$[0, 30e_3(t), 0, 0, 0, 0]$
$o_4(t)$	$[0, 0, 0, 50(e_2(t) \wedge e_4(t)), 0, 0]$
$o_5(t)$	$[0, 0, 0, 0, 0, 50(e_1(t) \wedge e_5(t))]$
$o_6(t)$	Decrease the score by 15 points, if s_3 and s_5, or s_4 and s_6 are connected
$o_7(t)$	Increase the score by 15 points, if bust shots of two persons have the same or similar length

12.5 Example of Capture and Editing

12.5.1 Examples of Capture

As an experiment of video capturing, we applied our system to the scene where two people sat down around a table and talked faced to face. In this scene, two people walked toward a table, sat down on the chairs near the desk facing each other, and started conversation.

The captured videos are shown in Figures 12.5 and 12.3. The result of human detection by the environmental camera module is shown in Figure 12.5, and the videos captured by contents capturing camera modules are shown in Figure 12.3. We can see the environmental camera module successfully detected people while they walked and sat down around the desk. Then, contents capturing camera modules successfully detected the faces and started tracking and capturing faces while keeping appropriate picture compositions and suppressing shaky movements.

This experiment shows good performance of our system, although it needs further improvements on some points, especially on the accuracy of face tracking.

12.5.2 Examples of Editing

For an editing experiment, we filmed a conversation scene between two people for about 2 minutes using multiple cameras. We generated various editing patterns by including/excluding evaluation functions or constraints, and thereby verified that our computational model really works to produce a variety of editing results.

Table 12.7 shows conditions for editing. These S, E, C, O are presented in section 12.4. Figure 12.7 shows editing patterns, each of which obtained the best score under its respective condition. Edit1 and edit2 is the parameter value of evaluation function o_3. By making the score for the table shot (s_2) greater, edit2 employs the shot from 8 seconds through 11 seconds.

Edit3 is edited adding c_2, the number of editing candidates was drastically reduced from 1 002 156 to 596, while the same editing pattern was selected for both. In edit4, evaluation functions o_4 and o_5 were added in order to show, by over-the-shoulder shots, the listener's attitudes as well as the speaker's face. It is, however, undesirable since connecting similar shots such as s_3 and s_5 may cause apparent motion. To prevent such transitions, evaluation function o_6 is added in edit5. As a result, shot s_4 from 9 seconds through 13 seconds is replaced by shot s_3. In edit7, evaluation is done referring to an event that occurred 1 second in the future. In edit8, instead of o_{1A} and o_{2A}, the system used o_{1B} and o_{2B} that give medium score to long shot s_1 as well as bust shots s_3 and s_4. This setup meant that long shot s_1 was often selected when both of two persons were simultaneously speaking.

Table 12.7 Conditions for editing and the number of candidates

	Applied C, O	Number of candidates
edit1	$o_{1A}, o_{2A}, o_{5A}, c_1$	1,002,156
edit2	$o_{1A}, o_{2A}, o_{5B}, c_1$	1,002,156
edit3	$o_{1A}, o_{2A}, o_{5A}, c_1, c_2$	596
edit4	$o_{1A}, o_{2A}, o_{5B}, o_6, o_7, c_1, c_2$	1,752
edit5	$o_{1A}, o_{2A}, o_{5B}, o_6, o_7, c_8, c_1, c_2$	1,752
edit6	$o_{1A}, o_{2A}, o_{5B}, c_1, c_3$	459,816
edit7	$o_3, o_4, o_{5B}, c_1, c_2$	610
edit8	$o_{1B}, o_{2B}, o_{5B}, c_1, c_3$	459,816
edit9	o_{1B}, o_{2B}, o_{5B}	3×10^{15}

Figure 12.7 Editing results with nine different conditions

Edit9 is the editing result with the same evaluation functions of edit8, but no constraints are used. This setting causes frequent shot changes at almost all seconds, since the shot that obtains the best score changes frequently. Consequently, the editing result is choppy and the movie clip is virtually unwatchable.

As shown in the experiments, the model has good ability to generate a variety of editing patterns by including/excluding editing rules.

12.6 Performance Evaluation

12.6.1 Evaluation of Camerawork

To evaluate camera controls in capturing phase, we conducted subjective evaluation. In this experiment, 30 subjects watched the short video clips recorded in the capturing phase, and scored the following four factors with values from 1 to 5:

A.1 "How was the rotation speed of the camera ?" (1:too slow – 5:too fast)
A.2 "How was the frequency of the camera control ?" (1:too low – 5:too high)
A.3 "Did the short video clips have good picture composition?" (1:poor – 5:good)
A.4 "Was the camera control adequate for the situation ?" (1:inadequate – 5:adequate)

Note that these factors did not require the subjects to discuss the effects of editing of the produced video clips. Figure 12.8 shows the average scores for each factor. In A.1 and A.2, the best score was 3, while in A.3 and A.4 the best score was 5. In the figure, vertical/horizontal lines on the bars indicate deviation. As shown in A.1 and A.2 in Figure 12.8, our method had an overall score of 3, indicating that our method realized almost the best control of speed and frequency of moving the pan/tilt/zoom cameras. In contrast,

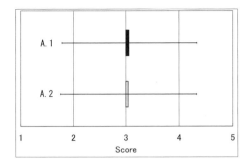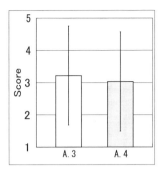

Figure 12.8   Evaluations of automated capturing

there was room for improvement of picture composition and adequate camera control, as shown in A.3 and A.4 in Figure 12.8. These scores of A.3 and A.4 were also supported by comments returned by the subjects; some noted that the accuracy of picture composition requires improvement. It is necessary to improve this factor in our future studies.

12.6.2 Evaluation of Recognition Rate

We first report the recognition rate of our capture system and then discuss the relationship between the recognition rate and the quality of video editing.

In this experiment, we discuss four conversation states: nodding, utterances of demonstrative pronouns, utterances of conjunctions, and occurrences of utterances. We conducted four experiments to evaluate the recognition rate of our capture system. The first three used 15 video clips of about 15 minutes in total length to evaluate nodding, 47 video clips of about 38 minutes in total length to evaluate utterances of demonstrative pronouns, and 47 video clips of about 35 minutes in total length to evaluate utterances of conjunctions. The results are shown in Table 12.8 with precision and recall rates; the precision and recall rates were mostly more than 90%. To evaluate occurrences of utterances, another experiment was conducted on 50 recorded videos, each of which is about 120 seconds in length and contains an average of about 100 seconds of utterances (Table 12.9).

12.6.3 Evaluation of Resultant Videos

To analyze the influence of recognition rate on the resultant video quality, we conducted a subjective evaluation experiment. We created five types of automatically edited videos of the same scene by changing recognition rates in five ways, and compared the resultant videos. Figure 12.9 shows examples of the five edit types that subjects watched, and Table 12.10 shows the recognition rates for each edit type.

Type 1 is regarded as an ideal situation, where all the conversation states are perfectly recognized. That means there are no recognition errors in type 1. Type 2 corresponded to our actual capturing system.

Table 12.8   Recognition ratio 1: nodding and utterances of keywords

	Clips	Length (min)	Number	Detected	Error	Failure	Precision	Recall
Nodding	15	15	80	62	6	18	90%	78%
Demonstratives	47	38	54	50	5	4	90%	93%
Conjunctions	47	35	52	48	0	4	100%	92%

Table 12.9   Recognition ratio 2: utterance period

Number of video clips	Average of precisions	Average of recalls
50	88 %	81 %

Type 3 assumes that utterance states are completely recognized, while nodding was not recognized. On the other hand, type 4 assumes no recognition of utterance state, while nodding is recognized completely. Finally, in type 5 no conversation state is recognized.

By applying these five editing types to four conversation scenes, we created 20 resultant videos. Subjects watched the videos and scored their impressions on the following six factors (Figure 12.10).

B.1 "Did you understand the statuses of the speakers?" (1:no – 5:yes)
B.2 "Did you understand the statuses of the listeners?" (1:no – 5:yes)
B.3 "Did you recognize the locations of all persons?" (1:no – 5:yes)
B.4 "Did you feel the atmosphere of the conversation?" (1:no – 5:yes)
B.5 "Was view switching good ?" (1:no – 5:yes)
B.6 "How did you feel about the frequency of view switching?" (1:boring – 5:busy)

In B.1, evaluations for types 1, 2, and 3 were high. This was because the editing system tended to use speaker shots according to the recognition rates given by types 1, 2, and 3. In B.2, all types had almost the same score. This indicated that subjects could understand aspects of listeners if only long shots were inserted. Thus, the insertion of listener shots did not improve the impressions of B.2. In B.3, all types again showed almost the same score. We had envisioned that evaluations for types 4 and 5 would be much higher than the other types because we initially felt that a long shot would allow subjects to recognize the locations of all participants, and long shots were inserted frequently for these types that had a few valid states. However, the results indicated that almost half of the subjects could recognize the locations even if a long shot was inserted only a few times.

In B.4, evaluations of types 1, 2, and 3 that used speaker shots were high. We can see that occasional insertion of speaker-related video clips improved understanding of the atmosphere of the conversation. In B.5, types 1, 2, and 3 scored around 3, though their scores were significantly better than those of types 4 and 5. We need further improvement on the timing of switching. In B.6 (Figure 12.11), we found that the frequency of view switching was appropriate for types 1, 2, and 3.

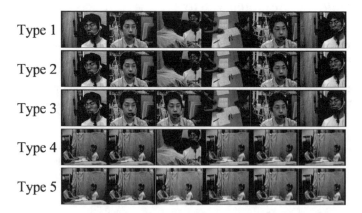

Figure 12.9   Example snapshots of edit types

Table 12.10   Five types of different recognition ratios

	Times of nodding		Demonstratives		Conjunctions		Utterance periods	
	Precision	Recall	Precision	Recall	Precision	Recall	Precision	Recall
Type 1	100	100	100	100	100	100	100	100
Type 2	91	78	91	93	100	92	88	81
Type 3	0	0	100	100	100	100	100	100
Type 4	100	100	0	0	0	0	0	0
Type 5	0	0	0	0	0	0	0	0

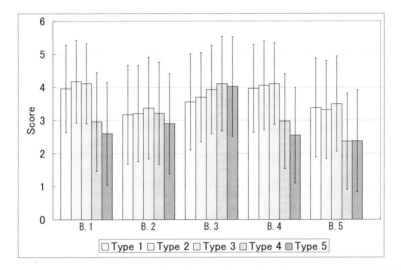

Figure 12.10   Evaluations of the video clips (1). Reproduced by permission of © IEICE

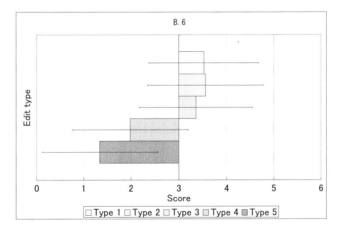

Figure 12.11   Evaluations of the video clips (2)

12.7 Conclusion

This chapter has presented a scheme for automated video capturing and editing for conversation scenes. The purposes are to archive meeting scenes, and to produce comprehensible video clips by automated editing. It has been demonstrated that conversation scenes are well captured by automatic camera control, and the videos are well edited by using various editing rules.

Subjective evaluation showed that the scheme for camerawork is satisfactory. The resultant videos were scored almost the same as the ideal case with perfect recognition. This implies that the recognition rate of our capturing system is sufficient.

However, there are areas for future improvement. One is the scalability for meetings with more people. We need a design for making bigger systems at reasonable cost. We believe this problem will be solved in the near future, as cameras are getting cheaper, and computers are getting faster. We also need to improve the accuracy of face orientation recognition in the capturing subsystem, and improve the timing of view switching in the editing subsystem, etc. It is also necessary to explore and evaluate recognition of other conversation states.

References

Arijon D. (1976) *Grammar of the Film Language*. Focal Press.

Atarashi Y., Kameda Y., Mukunoki M., Kakusho K., Minoh M. and Ikeda K. (2002) Controlling a camera with minimized camera motion changes under the constraint of a planned camera-work. In *Workshop on Pattern Recognition and Understanding for Visual Information Media, in Cooperation with ACCV*, pp. 9–14.

Bordwell D. and Thompson K. (1997) *Film Art: an introduction*, 5th edn. McGraw-Hill.

Ekman P. and Friesen W. (1969) The repertoire of nonverbal behavior: categories, origins, usage, and coding. *Semiotica* 1, 49–98.

Erol B., Lee D. and Hull J.J. (2003) Multimodal summarization of meeting recordings. In *Proceedings of the IEEE International Conference on Multimedia and Expo*, vol. 3, pp. 25–28.

Lee D., Erol B., Graham J., Hull J.J. and Murata N. (2002) Portable meeting recorder. In *Proceedings of ACM Multimedia 2002*, pp. 493–502.

Murakami M., Nishiguchi S., Kameda Y. and Minoh M. (2003) Effect on lecturer and students by multimedia lecture archive system. In *4th International Conference on Information Technology Based Higher Education and Training* (ITHET2003), pp. 377–380.

Nishida T., Fujihara N., Azechi S., Sumi K. and Yano H. (1999) Public opinion channel for communities in the information age. *New Generation Computing* 17 (4), 417–427.

Nishizaki T., Ogata R., Nakamura Y. and Ohta Y. (2004) Video contents acquisition and editing for conversation scenes. *Knowledge-Based Intelligent Information & Engineering Systems*, 2004

Oda T., Mori K. and Suenaga Y. (1998) A method for estimating eye regions from facial images. IEICE Technical Report, PRMU98-197, pp. 39–46.

Ogata R., Nakamura Y. and Ohta Y. (2003) Computational video editing model based on optimization with constraint-satisfaction. In *Proceedings Fourth Pacific-Rim Conference on Multimedia*, pp. CD-ROM, 2A1-2 (6 pages).

Onishi M., Kagebayashi T. and Fukunaga K. (2001) Production of video images by computer controlled cameras and its application to tv conference system. In *Proceedings of IEEE Conference on Computer Vision and Pattern Recognition*, vol. 2, pp. 131–137.

Ozeki M., Nakamura Y. and Ohta Y. (2002) Human behavior recognition for an intelligent video production system. In *IEEE Proc. Pacific-Rim Conference on Multimedia*, pp. 1153–1160.

Rowley H., Baluja S. and Kanade T. (1998) Neural network-based face detection. *IEEE Transactions on Pattern Analysis and Machine Intelligence* 20 (1), 23–38.

Rui Y., Gupta A. and Cadiz J.J. (2001) Viewing meetings captured by an omni-directional camera. In *Proceedings of ACM's Special Interest Group on Computer–Human Interaction*, pp. 450–457.

13

Personalization of Video Contents

Noboru Babaguchi

13.1 Introduction

We are facing an information flood from huge multimedia spaces such as WWW and TV/cable broadcasts in our daily life. A process of iteratively interacting with the multimedia spaces and filtering out the information which is really necessary can be viewed as a form of communication or conversation between the information spaces and humans. It is of great importance to develop a means for facilitating communication with these huge spaces. In this chapter, we focus on personalization and summarization as promising means to this end. *Summarization* aims to give only the main information and not its details. In recent years, some approaches to text, speech or video summarization have been attempted. In addition, *personalization* or personal content adaptation is indispensable for making communication more smooth. Personalization is a task to make contents adaptable to a user. In what follows, we proceed to discuss personalization and summarization of video contents.

In video applications, personalization is sometimes considered in the context of personal TV systems (Merialdo *et al.* 1999; Smyth and Cotter 2000; Jasinschi *et al.* 2001). We think that an alternative embodiment of personalization is to tailor video contents for a particular user (Maglio and Barrett 2000). For example, we Japanese may be more interested in the batting result of Matsui, who is a player of Yankees, than in the game result. Thus we may prefer to see a digest including all scenes of Matsui's at-bats rather than see a digest including highlight scenes of the game. To realize this, the user's preference has to be represented in an appropriate form, to be automatically acquired, and to be used in various applications. In this chapter, we aim at personalized tailoring in summarization of sports videos, taking account of the user's preference.

Video summarization is defined as creating shorter video clips or video posters from an original video stream (Aigrain *et al.* 1996). In recent years, it has become one of the most demanding video applications. The scheme of video summarization is divided into two classes. The first is to create a concise video clip by temporally compressing the amount of video data. This is sometimes referred to as video skimming and its examples are movie trails and sports digests. Recent researches (Smith and Kanade 1997; Lienhart *et al.* 1997; He *et al.* 1999; Oh and Hua 2000; Lee *et al.* 2003; Ekin *et al.* 2003; Babaguchi *et al.* 2004; Ngo *et al.* 2005) are in this class. The second is to provide image keyframe layouts representing the whole video contents on a computer display like a storyboard. This is suitable for at-a-glance presentation by means of spatial visualization. Such systems were reported in Yeung and Yeo (1997), Uchihashi *et al.* (1999), Chang and Sundaram (2000), Toklu *et al.* (2000), Ferman and Tekalp (2003), and Chiu *et al.* (2004). The first and second classes are commonly called time compression and spatial expansion, respectively.

In this chapter, we present a method of generating two types of sports video summaries, specifically broadcasted TV programs of baseball games, taking personalization into consideration (Babaguchi *et al.* 2004; Takahashi *et al.* 2004, 2005). It is assumed that metadata described with MPEG-7 is accompanied by the video data. This method can be viewed as video summarization based on highlights that are closely related to semantical video contents. As representation forms, we consider video clips and video posters which correspond to time compression and spatial expansion, respectively.

As described before, we concentrate on personalization (Riecken 2000; Maglio and Barrett 2000) in making video summaries. It has been extensively attempted in a variety of application fields such as web mining and user interfaces. Also in the field related to video processing, the idea of personalization has been recently introduced to personal TV applications (Merialdo *et al.* 1999; Smyth and Cotter 2000; Jasinschi *et al.* 2001). Our aim is to apply it to video summarization. We emphasize that because the significance of scenes could vary according to personal preferences and interests, the resultant video summaries should be individual. For example, one who favors the San Diego Chargers wants to see more scenes about them than scenes about their opponent. To this end, a profile is provided to collect personal preferences such as his/her favorite team, player, and play. It is expected to act as a bias in making video summaries.

13.2 Related Work

As stated earlier, video summarization is divided into two classes: *time compression* and *spatial expansion*. We first describe existing methods for the time compression type. Smith and Kanade (1997) proposed a method of great interest, what is called video skimming. They extracted significant information from video such as audio keywords, specific objects, camera motions, and scene breaks with integrating language, audio, and image analyses. They reported the compression ratio was about 1/20 although the essential content was kept. Lienhart *et al.* (1997) tried to assemble and edit scenes of significant events in action movies, focusing on actor/actress's close-up, text, and sounds of gunfire and explosions. These two methods are based on surface features of the video rather than on its semantical contents. Oh and Hua (2000) developed a method of summarizing video using its interesting scenes. Their method is able to automatically uncover the remaining interesting scenes in the video by choosing some interesting scenes. They reported successful results for news video. Babaguchi *et al.* (2004) discussed video summarization based on its semantical content in the sports domain. To select highlights of a game, an impact factor for a significant event in two-team sports was proposed. He *et al.* (1999) proposed a method to create summaries for online audio–video presentations. They used pitch and pause in audio signals, slide transition points in the presentation, and users' access patterns. Their compression rate ranged from 1/5 to 1/4. They also considered the desirable properties for an ideal summary, the 4 Cs: conciseness, coverage, context, and coherence. They evaluated the generated summaries from the 4 Cs viewpoint.

Let us next mention existing methods of the spatial expansion type. Their goal is to visualize the whole contents of the video. Because of the keyframe layouts, they are suitable for at-a-glance browsing. Yeung and Yeo (1997) proposed a method to automatically create a set of video posters (keyframe layouts) by the dominance value for each shot. Uchihashi *et al.* (1999) presented a similar method of making video posters whose size can be changed according to the importance measure. Chang and Sundaram (2000) made shot-level summaries of time-ordered shot sequences or hierarchical keyframe clusters, as well as program-level summaries.

Several systems have been proposed to provide users with personalized video summaries by taking into account their preferences. The systems can be divided into two classes. The first class used metadata like MPEG-7. Jaimes *et al.* (2002) proposed a system of extracting semantic features from MPEG-7 metadata to train a highlight classifier and generating personalized video digests for soccer videos. Tseng *et al.* (2004) developed a system that fully exploited not only MPEG-7 but also MPEG-21 to produce personalized video summaries according to his/her preference and environments. Unfortunately, they indicated no way to learn user's preferences.

The second class did not use MPEG-7-like metadata. Personal Broadcast News Navigator, P-BNN (Maybury *et al.* 2004), is capable of user modeling and tailored presentation of news programs. P-BNN stores a set of search keywords given by a user to retrieve news stories with respect to a topic of interest, and uses it as a profile for the topic. The user can explicitly specify news sources, time periods, etc. as his/her preference. P-BNN also implicitly infers the user's interests during its keyword expansion process. Syeda-Mahmood and Ponceleon (2001) modeled video browsing behaviors with the Hidden Markov Model while using Media Player. They applied their method to making interesting video previews, which is a kind of video summarization. Yu *et al.* (2003) proposed a method of measuring interestingness and importance of a shot based on the browsing log of a user, as well as generating video summaries.

13.3 Metadata

This section describes metadata used in this method. We deal with sports videos accompanied by their metadata which describes their semantic structure and contents (Ogura *et al.* 2002). As mentioned before, broadcast baseball videos are considered in this case study.

This metadata is described with MPEG-7 (Martinez 2001), which is a scheme for describing multimedia contents based on XML. Its purpose is to standardize descriptors (tags), in order to efficiently retrieve and summarize the multimedia contents. We focus on broadcast sports videos whose metadata is concerned with structure of the video and semantic contents of video segments. Figure 13.1 shows the format of

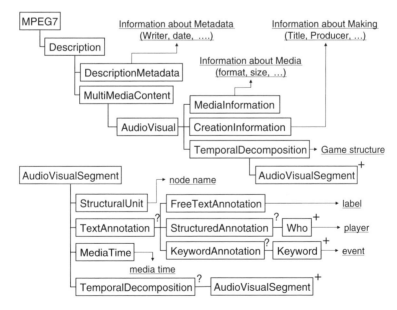

Figure 13.1 Format of metadata: the symbol + denotes one or more occurrences, and the symbol? denotes zero or one occurrence

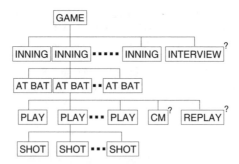

Figure 13.2   Game tree of baseball: the symbol ? denotes zero or one occurrence and CM is an abbreviation for Commercial Message

the metadata. In its header, information on metadata description, on media format, and on production is described.

The semantic structure of a video recording a broadcast sports game can be represented as a tree, called a *game tree*, which is hierarchical. For example, as shown in Figure 13.2, the game tree of baseball is composed of the nodes of the 1st through 9th innings, and moreover each inning node is composed of the at-bat nodes, and so on. The at-bat nodes have their child nodes of play. The play is a meaningful unit in the sports contents. In baseball video, we regard an interval of every pitch as the play. The play scene is decomposed into shots, which mean consecutive image frames given from a single camera. In this chapter, a temporal and partial interval of a video corresponding to the node in the game tree shown in Figure 13.2 is called a *video segment* or simply a *scene*. Note that the video segments can be defined at any node in the tree. Namely, the video segment may represent a scene at an at-bat node or at a play node depending on the situation. It starts with the head of a shot and ends with the tail of either the same shot or another shot. Of course, some shots can be included in it.

Each node is described in terms of **AudioVisualSegment** tags. The tag contains the following five items: node name, label, player, event, and media time. We realize the hierarchical structure using several **AudioVisualSegment** tags via **TemporalDecomposition** tags. Figure 13.3 shows an example of the MPEG-7 description. The description is as follows: node name (at-bat), label (Arias), player (Arias, Kudou), event (SoloHomeRun, OpenTheScoring), media time (...). The media time is omitted in the example. The keywords used for player's items are player's names, while those for event items are defined based on the terminology in baseball.

In the metadata, the contents description is given from the game node, which is a root of the tree, to the play nodes. Because a shot is often too short to describe meaningful contents, no contents description is given to the shot nodes. Ideally metadata should be produced automatically. To this end, we have to achieve automated analysis of the semantic contents of videos. However, since highly reliable analysis of such contents is very difficult at present, we assume the metadata is given in this case.

13.4 Profile

To reflect on user's preference, we introduce a *profile*, which is a formal description for a particular user. A fundamental description $<Key, v_{Key}>$ is a pair of a keyword *Key* and its weight v_{Key}, called a *preference degree*. The weight v_{Key} expresses the degree of the user's preference for the keyword *Key*, ranging between -1 and 1. A set of fundamental descriptions forms the profile.

It is noted that there are two kinds of profiles: the profile for people and that for events, which represent the people and events in which the user is interested, respectively. The keywords in the profile for people

```
<AudioVisualSegment>
  <StructuralUnit>
    <Name>at-bat</Name>
  </StructuralUnit>
  <TextAnnotation>
    <FreeTextAnnotation>Arias</FreeTextAnnotation>
    <StructuredAnnotation>
      <Who>
        <Name>Arias</Name>
        <Name>Kudou</Name>
      </Who>
    </StructuredAnnotation>
    <KeyWordAnnotation>
      <Keyword>SoloHomeRun</Keyword>
      <Keyword>OpenTheScoring</Keyword>
    </KeyWordAnnotation>
  </TextAnnotation>
  <MediaTime>•••••••••••</MediaTime>
  <TemporalDecomposition>
    <AudioVisualSegment>•••••••••••</AudioVisualSegment>
    <AudioVisualSegment>•••••••••••</AudioVisualSegment>
                                    • •
  </TemporalDecomposition>
</AudioVisualSegment>
```

Figure 13.3   Example of a part of the metadata

are the player's names, while those in the profile for events are the play names such as SingleHit and Homerun. We assume the description of the keywords is identical to that of metadata. These two profiles are utilized for generating video summaries.

13.5 Definition of Video Summarization

One of the strong demands for sports videos is to understand their content in short. Video summarization is a solution for this demand.

Definition 13.5.1 (Video Summarization) *Video summarization is defined as creating a concise description that represents the whole contents from an original video.*   □

If the description should be a video clip, such summarization scheme is referred to as time compression; if it should be a video poster or a storyboard, its scheme is referred to as spatial expansion. Based on (He *et al.* 1999), the quality of video summaries should be evaluated in terms of conciseness, coverage, context and coherence. We now proceed to define two types of video summarization.

13.5.1 Time Compression

We first define the time compression type. Let L denote the summary length (the time specified by the user) and N the number of all play scenes. Let $p_i, i = 1, \ldots, N$, be a play scene, and $s(p_i)$ and $l(p_i)$ be the significance degree and temporal length of the scene p_i, respectively. A function φ is a mapping from $l(p_i)$ to $[0, l(p_i)]$, which changes the length of a play scene p_i. The formulation is as follows.

Definition 13.5.2 (Time Compression) *The time compression problem is to select a subset, called a highlight set, such that $P' = \{p_j \mid j = 1, 2, \ldots, k\}$ $(k < N)$ from a set $P = \{p_1, p_2, \ldots, p_N\}$, subject to*

$$\sum_{p_j \in P'} s(p_j) \longrightarrow \max \quad and \quad \sum_{p_j \in P'} \varphi(l(p_j)) \le L$$

☐

Thus, we can define this problem as a combinational optimization problem under constrained conditions. Note that this is equivalent to the well-known *knapsack problem*. The resultant video clip through time compression summarization is called a *video digest* in what follows. In time compression, an hour-length video can be compressed into a minute-length digest. Actual examples of this type are movie trails and sports digests in TV news programs.

13.5.2 Spatial Expansion

Let A denote the area of a 2-dimensional plane where keyframes should be displayed. Note that a keyframe should be a representative image frame that expresses contents of a scene. It is an important but open problem how appropriate keyframes should be selected from all the image frames constituting the scene. The beginning frame is frequently selected for simplicity. Let N be the number of all play scenes which is equal to that of all keyframes. Let p_i, $i = 1, \ldots, N$, be a play scene, and $s(p_i)$ and $a(p_i)$ be the significance degree and keyframe area of the scene p_i, respectively. A function ψ is a mapping from $a(p_i)$ to $[0, A]$, which changes the keyframe area of a play scene p_i. The formulation is as follows.

Definition 13.5.3 (Spatial Expansion) *The spatial expansion problem is to select a subset such that $P' = \{p_j \mid j = 1, 2, \ldots, k\}$ $(k < N)$ from a set $P = \{p_1, p_2, \ldots, p_N\}$, subject to*

$$\sum_{p_j \in P'} s(p_j) \longrightarrow \max \quad and \quad \sum_{p_j \in P'} \psi(a(p_j)) \le A$$

☐

Similar to the earlier problem, this problem is also formulated as a combinational optimization problem under constrained conditions. The result through spatial expansion summarization is called a *video poster* in what follows.

As stated in the above definitions, the most critical issue in video summarization is how significance of scenes should be determined. Because it depends on the problem domain, it is very difficult to determine it in a general way. The next section discusses a case of two-team sports such as baseball.

13.6 Scene Significance

It is evident that significance of scenes is closely related to video contents. In addition, important scenes may change depending on the user's preferences or interests. Therefore, each play scene should be ranked according to not only its semantic significance but also the user's preference degree. In this section, we propose a method of evaluating the significance of play scenes using metadata.

Each play scene is evaluated in terms of the following four components: the play ranks, the play occurrence time, the number of replays, and the user's preference.

Play Ranks

We here assume that a game is played between two teams, A and B, and that one team's goal is to get more scores than its opponent. Under this assumption, there are three states of the game situation: two-team tie, team A's lead, and team B's lead. If a play can change the current state into a different state, we call it a State Change Play (SCP). It is evident that SCPs are candidates of the highlights.

The ranks of various play scenes are defined as follows:

- Rank 1: SCPs.
- Rank 2: score plays except SCPs.
- Rank 3: plays closely related to score plays.
- Rank 4: plays with score chance.
- Rank 5: fine plays and a game-ending play.
- Rank 6: all other plays that are not in Rank 1 to 5.

Note that all of the ranks can be obtained from the metadata description.

Now, s_r ($0 \leq s_r \leq 1$), the rank-based significance degree of a play scene p_i, is defined as

$$s_r(p_i) = 1 - \alpha \cdot \frac{r_i - 1}{5} \tag{13.1}$$

where r_i denotes the rank of the ith play scene p_i, and α ($0 \leq \alpha \leq 1$) is the coefficient to consider how much the difference of the rank affects the significance of play scenes.

Play Occurrence Time

The score play scenes which are close to the end of the game largely affect the game's outcome. Thus, such play scenes are of great significance. We define the occurrence-time-based significance degree of a play scene, s_t ($0 \leq s_t \leq 1$), as

$$s_t(p_i) = 1 - \beta \cdot \frac{N - i}{N - 1} \tag{13.2}$$

where N is the number of all play scenes, and β ($0 \leq \beta \leq 1$) is the coefficient to consider how much the occurrence time affects the significance of play scenes.

Number of Replays

An important play scene has many replays and more important play scenes tend to have more replays than others in an actual TV broadcast. Therefore, a play scene which has many replays is important. We define the number-of-replays-based significance degree of a play scene, s_n ($0 \leq s_n \leq 1$), as

$$s_n(p_i) = 1 - \gamma \cdot \frac{n_{\max} - n_i}{n_{\max}} \tag{13.3}$$

where n_i denotes the number of replays of the play scene p_i, and $n_{\max}$ is the maximum number of n_i. Also γ ($0 \leq \gamma \leq 1$) is the coefficient to consider how much the number of replays affects the significance of the play scene.

The foregoing three components are, in a sense, heuristics obtained from actual broadcast digests of baseball and American football. Because they are heuristics, not all digests may follow them.

User's Preference

We strongly claim that a video summary has to be personalized because the significance of play scenes is changeable for each user. For this purpose, we provide a profile to describe the sets of the keyword which shows user's preferences or interests and its preference degree. Its items are as follows: (a) favorite team, favorite players, and events; (b) user's preference degree of each favorite team, player, and event. Recall the fundamental description of the profile is $<Key, v_{Key}>$. The user's preference-based significance degree of a play scene, $s_p(p_i)$, is calculated as

$$s_p(p_i) = \theta^{v_{Key}} \tag{13.4}$$

where v_{Key} ($-1 \leq v_{Key} \leq 1$) denotes the user's preference degree of Key, and θ ($\theta \geq 1$) is the coefficient to consider how much user's preference affects the significance of the play scene. When v_{Key} is close to 1, the user wants to see the scene corresponding to this keyword. When v_{Key} is, in contrast, close to -1, the user will not want to see the scene corresponding to this keyword. $\theta = 1$ means that user's profile has no effect on the significance of the play scenes.

As a consequence, the significance degree of a play scene p_i is given by

$$s(p_i) = s_r(p_i) \cdot s_t(p_i) \cdot s_n(p_i) \cdot s_p(p_i) \tag{13.5}$$

Changing the parameters α, β, γ, and θ enables us to control the composition of a video summary. For example, larger α can emphasize the significance of the play ranks. The other parameters behave in a similar manner.

13.7 Generation of Video Digest

As described in Section 13.5, the problem of video summarization can be reduced to selection of highlight scenes so that the sum of their significance degrees can be maximized. Given the digest length, the algorithm for generating video digests will select the highlight scenes from all the scenes with different length under the condition that the total length of the selected scenes should be within the digest length. The algorithm is capable of making a variety of digests with the arbitrary length which the user wants.

We first sort the play scenes in descending order of their significance, forming a scene list. Next, we select a highlight scene with highest significance from the list, until the sum of the length of the selected play scenes exceeds the digest length. In this case, the following function (cf. Definition 13.5.2) that operates on the scene length is considered in order to put bounds to the length of the play scene.

$$\varphi(l(p_i)) = \min\left[l(p_i),\ l_{th} + \delta \cdot L'\right] \tag{13.6}$$

where l_{th} denotes a threshold of the minimum time required for the user to grasp the content of a play scene, L' denotes the current remaining time after subtracting the total time of the selected play scenes from L, and δ is the coefficient to consider how much L' affects the length of play scenes. This function enables us to cut short the play scenes dynamically considering the digest length. The bigger the value of l_{th} or δ is, the longer the length of each play scene becomes; consequently the number of the play scenes contained in the video summary decreases. When l_{th} or δ is smaller, although the number of the play scenes increases, each play scene segment may not fully represent the content. These parameters should be determined so that the generated summary can satisfy both requirements.

The algorithm for generating video digests is described by the following pseudo code:

Algorithm Video Digest Generation
Input L : the time specified by the user (digest length)

N : the number of all play scenes

$s(p_1), s(p_2), \ldots, s(p_N)$: p_i's significance

$l(p_1), l(p_2), \ldots, l(p_N)$: p_i's time length

l_{th}, δ :   parameters

Output   *DIGEST*;

$LIST \leftarrow empty$

$L' \leftarrow L$

Sort out $p_1, p_2, \ldots, p_N$ to be $s(p_1) \geq s(p_2) \geq \cdots \geq s(p_N)$

for $i = 1, 2, \ldots, N$ **do**

$l(p_i) \leftarrow \min [l(p_i), \ l_{th} + \delta \cdot L']$

if $(l(p_i) \leq L')$ **then**

PUT p_i into *LIST*;

$L' \leftarrow L' - l(p_i)$;

end;

end;

Make *DIGEST* by connecting scenes in *LIST* in original temporal order;

end;

Note that the highlight scenes in the video digest flow chronologically as in TV sports digests.

13.8 Generation of Video Poster

We realize spatial-expansion-based video posters on a PC display, which provide keyframes each of which represents a play scene. In this case, the keyframe is the first frame of each scene. Figure 13.4 shows the interface of our system. Let us describe key the functions.

- *Hierarchic presentation according to the game tree.* The system can display logical structure of videos based on the game tree, as illustrated in Figure 13.2 on its interface for quick access to the video segment corresponding to each node in the tree. From the user's side, this presentation function is used in browsing the video contents. For baseball videos, the keyframes are hierarchically displayed at the game, inning, at-bat, and play levels by traversing the game tree.
- *Keyframe number control.* With the video poster, a user can directly specify the number of keyframes to be displayed. He/she can directly specify either N_{new}, which denotes the number of keyframes to be newly displayed, or the decrease ratio ζ $(0 \leq \zeta \leq 1)$ such that $N_{new} = \zeta \cdot N_f$, where N_f denotes the number of keyframes presently displayed. Additionally, only the keyframes which match with the keywords such as players and events can be displayed.
- *Retrieval by text.* We think that the most intuitive and effective way to access videos is retrieval by text. In our system, users can retrieve players and plays using player's names or play events as queries. It searches the metadata description which matches with the text query, and if found, presents the keyframes which are colored.
- *Playback of each play scene.* A user can view only the play scenes which he/she wants to view by clicking the corresponding keyframes.
- *Annotations of video content.* Since it is difficult to understand the semantic content only from the keyframes, the system also shows annotations about players and plays by the corresponding keyframes.

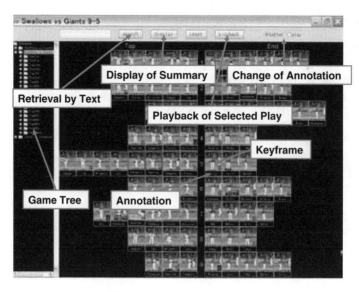

Figure 13.4   System interface

13.9 Experimental Results

In this section, we show some experimental results. To evaluate the quality of generated video digests, we first compare them with video digests produced by professional editors. Then, we investigate the questionnaires to the users to evaluate the effectiveness of our system including video posters. Lastly, we evaluate the effect of user's preferences on video summarization.

We used professional baseball videos which were on-air as examples of sports videos. The play ranks for baseball were defined as shown in Table 13.1. We prepared five baseball videos broadcasted by three different TV stations with their corresponding metadata. The time length of each baseball video was about three and a half hours. We also prepared five video digests broadcasted as highlight videos by the same TV stations as broadcasting the original videos. We call these digests man-made digests in this case.

Table 13.1   Play ranks for baseball

Rank	Play
Rank 1	"OpentheScoring", "TietheGame", "TurntheGameAround", "GoAhead"
Rank 2	"AddaPoint"
Rank 3	All plays except out-plays ("StrikeOut", "SingleOut", and "DoublePlay") before score plays (Rank 1 and 2) in the same inning
Rank 4	The fifth or its subsequent plays in the inning without score plays, The plays after "SacrificeHit", "TwoBaseHit", or "ThreeBaseHit"
Rank 5	"FinePlay", "CaughtoffBase", "TriplePlay", "DoublePlay", "StrikeOut"
Rank 6	All other plays

Table 13.2   Recall and precision

Game (date)	No. of plays in both	No. of plays on TV	No. of plays in our digest	Recall (%)	Precision (%)
Swallows/Tigers (9/11)	5	6	6	83	83
Swallows/Tigers (9/12)	2	3	3	67	67
Giants/Tigers (9/13)	7	11	7	64	100
Tigers/Giants (10/5)	4	7	6	57	67
Giants/Swallows (10/10)	7	12	7	58	100
Average	5.0	7.8	5.8	66	83

We assume that the play scenes in the man-made digests are the correct answer set for video summarization and evaluate the results with respect to the recall and the precision defined as follows:

$$Recall = \frac{number\ of\ play\ scenes\ included\ in\ both\ man\text{-}made\ and\ our\ digest}{number\ of\ play\ scenes\ included\ in\ man\text{-}made\ digest}$$

$$Precision = \frac{number\ of\ play\ scenes\ included\ in\ both\ man\text{-}made\ and\ our\ digest}{number\ of\ play\ scenes\ included\ in\ our\ digest}$$

The digests generated with our method were compared with the man-made digests. The value of the parameters were experimentally determined as $\alpha = 0.8$, $\beta = 0.1$, $\gamma = 0.3$, $l_{th} = 14$, and $\delta = 0.02$. Since the man-made digests are for general people and do not reflect user's preferences, we let $\theta = 1$. The comparative results between our and man-made digests are shown in Table 13.2. In this case, each digest length was the same length of each man-made digest. The detailed composition of our and man-made digests for the game of Swallows vs Tigers (9/11) are shown in Figure 13.5.

The experimental results demonstrated that our method was able to produce favorable digests. It is advantageous for our method to contain only the important play scenes by excluding the redundant shots. Such an editing function for shots is different from our previous method (Babaguchi *et al.* 2004).

We implemented the algorithm of generating video posters, which was an embodiment of spatial expansion summary, in our sports video browsing system (Takahashi *et al.* 2005). To examine the quality of the generated video posters, a user study was performed. We gave nine users the following questions:

Q.1: Is the system's operability good?
Q.2: Is each function convenient?
Q.3: Are there any advantages or disadvantages in the system?

Figure 13.5   Example of baseball video of Tigers vs Swallows

Table 13.3 Questionnaire results

	1	2	3	4	5	Average evaluation
Operability	0%	0%	11%	78%	11%	4.0
Keyframe number control	0%	11%	11%	44%	33%	4.0
Retrieval by text	0%	0%	11%	44%	44%	4.3
Playback of each play scene	0%	0%	0%	22%	78%	4.8
Annotations of video content	0%	0%	44%	22%	33%	3.9

The results for Q.1 and Q.2 are shown in Table 13.3. The users responded on a scale of 1–5, with 1 being very bad and 5 being very good. According to the responses to Q.3, the following issues have been derived:

- The keyframe which better expresses the content of a play scene should be selected.
- More complex queries should be handled.
- Retrieval by similar scenes should be introduced.

These issues have to be considered in the future to improve the system.

Next, we think about the effect of the user's preferences. Changing θ and v_{Key}, we examined how the rank of the play scene corresponding to keywords included in the profile was changing. We focused on a keyword "Nioka", who is a famous baseball player of Giants, assuming the user's profile includes <Nioka, v_{Nioka}>. When there were five play scenes including the keyword "Nioka" in the metadata for the game of Giants vs Swallows (10/10), we examined where Nioka's play scenes ranked as the user's preference degree v_{Nioka} was changing. Figures 13.6 and 13.7 show the relationship between v_{Nioka} and the number of Nioka's play scenes which ranked in the top-fives/tens, respectively. Even if $v_{Nioka} = 1$ under $\theta = 2$, the numbers of Nioka's play scenes which ranked in the top-fives/tens were only three and four, respectively. Since the user's preferences did not affect the results so much for $\theta = 2$, we think it appropriate θ is between 5 and 10.

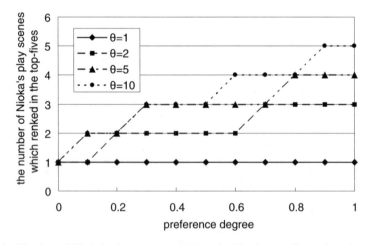

Figure 13.6 Number of Nioka's play scenes which ranked in the top-fives when the user's profile includes the keyword "Nioka"

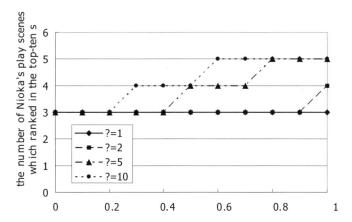

Figure 13.7 Number of Nioka's play scenes which ranked in the top-tens when the user's profile includes the keyword "Nioka"

In addition, we investigated the case where the user's preference degrees were negative. We used the profile which included <*Nioka*, 0.5> and <*Homerun*, $v_{Homerun}$>, and $v_{Homerun}$ was changed from 0 to -1 by 0.1. Figure 13.8 shows the change of the rank of the Nioka's homerun scene using this profile for the game of Giants vs Swallows (10/10). The horizontal axis shows the user's preference degree $v_{Homerun}$, and the vertical axis shows the rank of Nioka's homerun scene. The rank of Nioka's homerun scene was top at first because we used the profile that included <*Nioka*, 0.5>. The rank of Nioka's homerun scene fell down as the user's preference degree $v_{Homerun}$ got smaller; as θ got larger, the rank dropped faster.

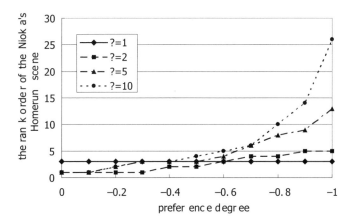

Figure 13.8 Change of the rank order of Nioka's homerun scene

13.10 Discussion

Generality of this Method

The current system offers only baseball video summaries. However, its framework is applicable to other types of sports videos with similar game structures. The idea of the significance degree for each scene, as stated in section 13.6, can be useful for different two-team sports such as football and basketball. Moreover, the system would deal with videos such as news programs that have logical structure and contain important topics and events. For wide applicability, the domain knowledge could be indispensable. Verification of its applicability through experimental considerations is our future work.

Metadata

Although we used metadata described with MPEG-7, there is still a problem how it is made less costly. At present, we have to make it manually but the task is very laborsome. To solve this, we should introduce a more flexible method to automatically detect the significant scenes with image/video analysis. As a promising approach, we have tried intermodal collaboration (Babaguchi *et al.* 2002), which has been extensively developed in recent years.

Profile

The notion of the profile is essentially derived from the research on information filtering. Currently, we must write down all the descriptions in the profile keywords and their preference degrees. It is of great importance to acquire the profile automatically by some means. Learning to personalize from operations for our video browsing system has been explored (Ohara *et al.* 2003; Babaguchi *et al.* 2003).

Evaluation of Video Summaries

The task of video summarization is considerably subjective. Accordingly, it is very difficult to evaluate the quality of the video summaries objectively. It seems that there is no correct answer to video summarization. Our solution is the comparison between man-made and machine-made summaries. However, quantitative evaluation in terms of difference or similarity between them is an open problem. We have to continue to pursue an evaluation scheme taking account of human factors and subjective aspects.

13.11 Conclusions

In this chapter, we have proposed a method of automatically creating two types of summaries, video digests and video posters from broadcasted sports videos considering the semantic contents. Our method explores the metadata to take account of the semantic contents, such as which team is winnings or which player has scored. The significance of play scenes can be determined based on the play ranks, the play occurrence time, and the number of replays derived from its metadata. In addition, our method is capable of reflecting on user's preferences or interests to generate a personalized video summary. Video digests whose length is controllable with a user can be generated, including the highlights of the sports game. In video posters, a user was able to gradually access target scenes from the whole game to its detail by tracing the tree structure of the game. As a result of experiments with baseball videos, we obtained significant play scenes with recall of 66% and precision of 83% compared with the actual digests broadcasted on TV. The remaining work is (1) acquisition of the use's preference, (2) improvement of the video browsing system, and (3) transplanting into other devices such as mobile phones.

References

Aigrain P., Zhang H.J. and Petkovic D. (1996) Content-based representation and retrieval of visual media: a state-of-the-art review. *Multimedia Tools and Applications* **3**, 179–202.

Babaguchi N., Kawai Y. and Kitahashi T. (2002) Event based indexing of broadcasted sports video by intermodal collaboration. *IEEE Trans. Multimedia* **4**, 68–75.

Babaguchi N., Ohara K. and Ogura T. (2003) Effect of personalization on retrieval and summarization of sports video. In *Proceedings of Fourth IEEE Pacific-Rim Conference on Multimedia* (PCM2003), Singapore, 2B1.3(2003–12).

Babaguchi N., Kawai Y., Ogura T. and Kitahashi T. (2004) Personalized abstraction of broadcasted american football video by highlight selection. *IEEE Trans. Multimedia* **6**, 575–586.

Chang S.-F. and Sundaram H. (2000) Structural and semantic analysis of video. In *Proceedings of IEEE ICME*.

Chiu P., Girgensohn A. and Liu Q. (2004) Stained-glass visualization for highly condensed video summaries. *Proc. IEEE ICME 2004*.

Ekin A., Tekalp A.M. and Mehrotra R. (2003) Automatic soccer video analysis and summarization. *IEEE Trans. Image Processing* **12**, 796–807.

Ferman A.M. and Tekalp A.M. (2003) Two-stage hierarchical video summary extraction to match low-level user browsing preferences. *IEEE Trans. Multimedia*, 244–256.

He L., Sanocki E., Gupta A. and Grudin J. (1999) Auto-summarization of audio-video presentations. In *Proc. ACM Multimedia*, pp. 489–498.

Jaimes A., Echigo T., Teraguchi M. and Satoh F. (2002) Learning personalized video highlights from detailed MPEG-7 event metadata. *Proc. of ICIP 2002*, vol.1, pp. 133–136.

Jasinschi R., Dimitrova N., McGee T., Agnihotri L. and Zimmerman J. (2001) Video scouting: an architecture and system for the integration of multimedia information in personal tv applications. In *Proc. IEEE ICASSP*, pp. 1405–1408.

Lee J.-H., Lee G.-G. and Kim W.-Y. (2003) Automatic video summarizing tool using MPEG-7 descriptors for personal video recorder. *IEEE Trans. Consumer Electronics* **49**, 742–749.

Lienhart R., Pfeiffer S. and Effelsberg W. (1997) Video abstracting, *Commun. ACM* **40** (12), 55–62.

Maglio P. and Barrett R. (2000) Intermediaries personalize information streams. *Commun. ACM* **43** (8), 96–101.

Martinez J.M. (2001) Overview of the MPEG-7 standard (version 6.0), ISO/IEC JTC1/SC29/WG11 N4509.

Maybury M., Greiff W., Boykin S., Ponte J., McHenry C. and Ferro L. (2004) Personalcasting: tailored broadcast news. *User Modeling and User-Adapted Interaction* **14**, 119–144.

Merialdo B., Lee K.T., Luparello D. and Roudaire J. (1999) Automatic construction of personalized tv news programs. *Proc. ACM Multimedia*, pp. 323–331.

Ngo C.-W., Ma Y.-F. and Zhang H.-J. (2005) Video summarization and scene detection by graph modeling. *IEEE Trans. Circuits and Systems for Video Technology*, **15**, 296–305.

Ogura T., Babaguchi N. and Kitahashi T. (2002) Video portal for a media space of structured video streams. *Proc. IEEE ICME 2002*, pp. 309–312.

Oh J.H. and Hua K.A. (2000) An efficient technique for summarizing videos using visual contents. In *Proc. IEEE ICME*.

Ohara K., Ogura T. and Babaguchi N. (2003) On personalizing video portal system with metadata. *Proc. KES2003*.

Riecken D. (2000) Personalized views of personalization. *Commun. ACM* **43** (8), 27–28.

Smith M.A. and Kanade T. (1996) Video skimming and characterization through the combination of image and language understanding techniques. In *Proc. IEEE CVPR*, pp. 775–781.

Smyth B. and Cotter P. (200) A personalized television listings service. *Commun. ACM* **43** (8), 107–111.

Syeda-Mahmood T. and Ponceleon D. (2001) Learning video browsing behavior and its application in the generation of video previews. *Proc. ACM Multimedia 2001*, pp. 119–128.

Takahashi Y., Nitta N. and Babaguchi N. (2004) Automatic video summarization of sports videos using metadata. *Proceedings of Fifth IEEE Pacific-Rim Conference on Multimedia* (PCM2004).

Takahashi Y., Nitta N. and Babaguchi N. (2005) Video summarization for large sports video archives. *Proceedings of IEEE International Conference on Multimedia and Expo* (ICME2005).

Toklu C., Liou S.-P. and Das M. (2000) Videoabstract: a hybrid approach to generate semantically meaningful video summaries. In *Proc. IEEE ICME*.

Tseng B.L., Lin C.-Y. and Smith J.R. (2004) Video personalization and summarization system for usage environment. *Journal of Visual Communication & Image Representation* **15**, 370–392.

Uchihashi S., Foote J., Girgensohn A. and Boreczky J. (1999) Video manga: generating semantically meaningful video summaries. In *Proc. ACM Multimedia*, pp. 383–392.

Yeung M. M. and Yeo B.-L. (1997) Video visualization for compact presentation and fast browsing of pictorial content. *IEEE Trans. Circuits and Systems for Video Technology*, **7**, 771–785.

Yu B., Ma W.-Y., Nahrstedt K. and Zhang H.-J. (2003) Video summarization based on user log enhanced link analysis. *Proc. ACM Multimedia*, pp. 382–391.

Part III

Conversational Environment Design

14

Conversational Content Acquisition by Ubiquitous Sensors

Yasuyuki Sumi, Kenji Mase, and Toyoaki Nishida

14.1 Introduction

The spread of computer networks and media technologies has enabled spatially and temporally distributed people to share knowledge. The knowledge exchanged through the media, however, is basically externalized and verbalized by humans manually. Therefore, the available knowledge tends to be limited to formalized, not tacit knowledge.

Conversation is one of the most popular and powerful ways to convey and create knowledge among people. Our daily lives are filled with conversations. We have various conversations in meetings, lectures, offices, and corridors. Through such conversations, we exchange and share tacit knowledge (awareness, common sense, know-how, nebulous ideas, atmosphere, etc.) with others, which is difficult to convey through today's media such as textbooks, e-mail, and the Web.

Recent advances in ubiquitous computing and augmented reality technologies are expected to capture our daily conversations and help us to access the accumulated data. Our aim is to build technologies for effectively recording and helping us to reuse our conversations.

Important issues in creating such a medium where we can capture and reuse conversations are the extraction and semantic indexing of tractable-sized conversational scenes from continuous conversations. These are the issues addressed in this chapter. Ideally, we would like to use speech recognition technology to verbally understand conversations (i.e., the contents and subject matter of the conversations), but current technologies are not yet mature enough for daily usage. This chapter focuses on nonverbal information that emerges along with the conversations, such as the existence of utterances, gazing, gesture, position, etc.

This chapter shows our attempt to build a communicative medium where we can capture and share conversations that occur in our daily lives. Throughout this chapter, we use the term "conversation" widely to describe various kinds of interactions between people, not only speaking to each other but also gazing at a particular object together, staying together for a particular purpose, and so on.

This chapter consists of three parts. The first part presents a system for capturing human interactions by ubiquious/wearable sensors. The system consists of multiple sensor sets ubiquitously set up around

Conversational Informatics: An Engineering Approach   Edited by Toyoaki Nishida
© 2007 John Wiley & Sons, Ltd

the room as well as wearable ones. The feature of our system is that we can obtain multiple viewpoint videos that capture a particular scene. The first goal for creating a medium to deal with the captured conversations is to extract conversation scenes of a meaningful and tractable size, as conversation quanta (Nishida 2004). We show a method for inferring interaction semantics, such as "talking with someone", "staying somewhere", and "gazing at something", among users and the environment by collaboratively processing data of those who interacted with each other. This can be performed without time-consuming audio and image processing by using an infrared ID system to determine the position and identity of objects in each video's field of view.

The second part of this chapter describes systems that help us to access the extracted conversations. Our approach is to make a "collage" of video and sound fragments associated with the extracted conversations. We investigated three kinds of "collage" systems:

- *Video summary*. Chronological collage of multiple-viewpoint videos.
- *Spatiotemporal video collage*. Visualization of conversation scenes in a 3D virtual space.
- *Ambient sound collage*. Acoustic visualization of past conversations in the real space.

The third part focuses on poster presentations in order to discuss how to detect more detailed structures of captured conversation scenes by using data about how speakers touched the poster (poster touch data). Every day, we converse with others using peripheral objects such as papers, notebooks, and whiteboards, etc. and often refer to them while speaking. Also at exhibition sites, many conversations between exhibitors and visitors are conducted around posters and people often point to the posters with their fingers. We will show a method of inferring the topics of the extracted conversations by associating segmented conversations with areas of the poster.

14.2 Capturing Conversation Scenes by Multiple Sensors

We prototyped a system for recording natural interactions among multiple presenters and visitors in an exhibition room (Sumi *et al.* 2004). The prototype was installed and tested in one of the exhibition rooms used for our research laboratories' open house.

Our approach is characterized by the integration of many ubiquitous sensors (video cameras, trackers, and microphones) set up around the room, and wearable sensors (video cameras, trackers, microphones, and physiological sensors) to monitor humans as subjects of interactions. Our system incorporates ID tags with an infrared LED (LED tags) and an infrared signal tracking device (IR tracker) to record position context along with audio/video data. The tracking device is a parallel distributed camera array where any camera can determine the position and identity of any tag in its field of view. When a user wears a tracking camera, his or her gaze can be determined. This approach assumes that gazing can be used as a good index for human interactions (Stiefelhagen *et al.* 1999).

Figure 14.1 shows a snapshot of the exhibition room set up for recording interactions among visitors and exhibitors. There were five booths in the exhibition room. Each booth had two sets of ubiquitous sensors, which included video cameras with IR trackers and microphones. LED tags were attached to possible focal points for social interactions, such as on posters and displays. Each presenter at a booth had a set of wearable sensors, including a video camera with an IR tracker, a microphone, an LED tag, and physiological sensors (heart rate, skin conductance, and temperature). A visitor could choose to carry the same wearable system as the presenters, just an LED tag, or nothing at all. One booth had a humanoid robot for its demonstration; this was also used as an actor to interact with visitors and record the interactions using the same wearable system as the human presenters.

Eighty users participated during the two-day open house providing about 300 hours of video data and 380 000 tracker data along with associated biometric data.

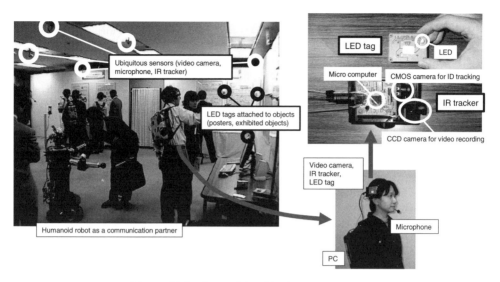

Figure 14.1   Setup of the ubiquitous sensor room

14.3 Segmentation and Interpretation of Scenes

We developed a method of segmenting interaction scenes from the IR tracker data. We defined interaction primitives, or "events", as significant intervals or moments of activities. For example, a video clip that has a particular object (such as a poster or user) in it constitutes an event. Since the location of all objects is known from the IR tracker and LED tags, it is easy to determine these events. We then interpret the meaning of events by considering the combination of objects appearing in the events. The basic events that we considered are illustrated in Figure 14.2.

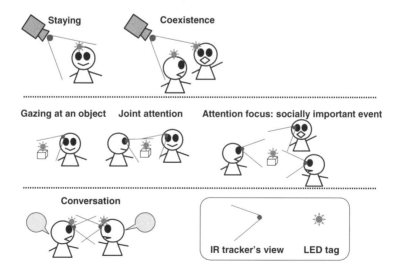

Figure 14.2   Social events that we considered as scene primitives

- *Stay.* A fixed IR tracker at a booth captures an LED tag attached to a user: the user *stays* at the booth.
- *Coexist.* A single IR tracker camera captures LED tags attached to two or more users at some moment: these users *coexist* in the same area.
- *Gaze.* An IR tracker worn by a user captures an LED tag attached to someone/something: the user *gazes* at someone/something.
- *Attention.* An LED tag attached to an object is simultaneously captured by IR trackers worn by two users: these users jointly pay *attention* to the object. When many users pay attention to the object, we infer that the object is playing a socially important role at that moment.
- *Facing.* Two users' IR trackers detect each others' LED tag: they are facing each other.

In order to make the captured interaction data more adaptable to various applications, we formalized the layered model shown in Figure 14.3 to interpret human interactions using a bottom-up approach (Takahashi *et al.* 2004). In this model, interpretations of human interactions are gradually abstracted, and each layer has unique databases for storing these interpretations as machine-readable indices.

In this layered model, each layer has to be composed independently from the others so that application designers can select appropriate layers for dealing with human contexts according to their needs, without learning any details of other layers or protocols of individual sensors. To achieve this, we set up four such layers in a bottom-up approach: the RawData Layer, the Segmentation Layer, the Primitive Layer, and the Composite Layer.

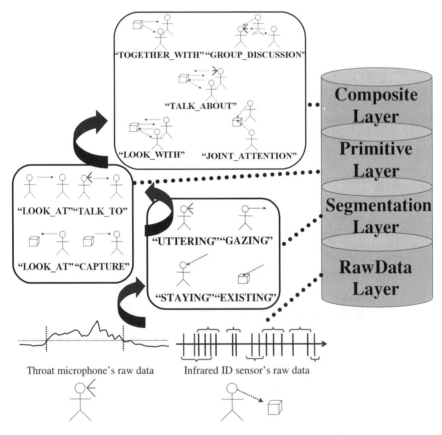

Figure 14.3 Layered model of interaction interpretations

The raw data acquired by individual sensors is stored in the first layer, the RawData Layer. This layer is composed of symbolic data in sensor-dependent formats including a lot of noise. For example, data from an infrared ID sensor is recorded in a discrete and intermittent format as the data pair (ID value, time-stamp), while that from a throat microphone is a continuous sequence of the volume of the user's voice.

In the second layer, the Segmentation Layer, the raw data is divided into meaningful clusters to provide the information that is necessary for interpreting human interactions for the upper layers. This layer eliminates noise and produces meaningful clusters in a sensor-independent format from the data stored in the RawData Layer. We can form these clusters by using two or more values that are close to each other to form one cluster for the intermittent data, such as the data from the infrared ID sensors. The clusters are also obtained by processing the thresholds for continuous data, such as the data from the throat microphones.

In the third layer, the Primitive Layer, basic elements of human interactions such as "LOOK_AT", "TALK_TO", and "CAPTURE", so-called interaction primitives, are extracted from the clusters provided by the Segmentation Layer. The situation where the infrared ID sensor worn by UserA captures UserB or ObjectX can be interpreted as "LOOK_AT", and if the additional information that UserA is uttering at this time is obtained, the situation is interpreted as "TALK_TO". The situation where the infrared ID sensor attached to ObjectX captures UserA can be interpreted as "CAPTURE", which means that UserA is staying near ObjectX.

In the fourth layer, the Composite Layer, interaction primitives are spatiotemporally connected to each other to enable the interpretation of complicated human interactions, such as social interactions. We have prepared five interpretations of the interactions in this layer.

- "GROUP_DISCUSSION" occurs when UserA talks to UserB and UserB talks to or looks at UserC in a close time-zone. In other words, a scene where two or more persons have a discussion at the same place.
- "JOINT_ATTENTION" occurs when UserA and UserB look at ObjectX at the same time. In other words, a scene where a socially significant event attracts many persons' attention.
- "TOGETHER_WITH" occurs when UserA and UserB are captured by ObjectX at the same time. In other words, a scene where two or more persons coexist at the same place. Additionally, if all of the persons who stay together look at ObjectX at the same time, the situation can be interpreted as "LOOK_WITH" or "TALK_ABOUT", scenes where they are looking at the exhibit together or talking about the exhibit.

These interpretations can be significant indices for the captured data in a poster exhibition site, enabling us, for example, to search for a particular scene from a huge amount of audio-visual data and develop various applications using human contexts. This model provides application designers with a common interface that suits the expected needs in developing applications by systematically assigning layers to application semantics by using a bottom-up approach from directly acquired raw data from individual sensors. The abstraction level of the index is higher in higher layers, but it takes more time to interpret human interactions in the upper layer. Therefore, the real-time performance is higher in lower layers. Thus, it is possible to query a wide range of information by using this layered model; that is, we can get information with greater abstraction in the upper layers and get information in real time in the lower layers. Moreover, we can use additional sensors to get in-depth contexts regardless of the form or format of the sensor outputs because their differences can be solved in the upper layers.

14.4 Video Summary: Chronological Collage of Multiple-viewpoint Videos

We were able to extract appropriate "scenes" from the viewpoints of individual users by clustering events having spatial and temporal relationships.

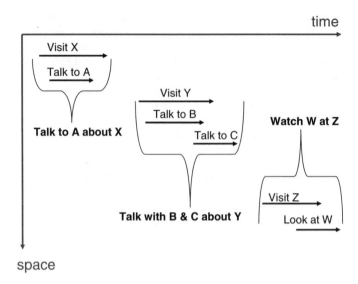

Figure 14.4 Interpreting events as scenes by grouping spatiotemporal co-occurences

A scene is made up of several basic interaction events and is defined based on time. Because of the setup of the exhibition room, in which five separate booths had a high concentration of sensors, scenes were location-dependent to some extent as well. Precisely, all the events overlapped were considered to be part of the same scene (Figure 14.4).

Scene videos were created in a linear time fashion using only one source of video at a time. In order to decide which video source to use to make up the scene video, we established a priority list. In creating the priority list, we made a few assumptions. One of these assumptions was that the video source of a user associated with a captured event of UserA shows a close-up view of UserA. Another assumption was that all the components of the interactions occurring in BoothA were captured by the ubiquitous cameras set up for BoothA.

The actual priority list used was based on the following basic rules. When someone is speaking (the volume of the audio is greater than 0.1 out of 1.0), a video source that shows the close-up view of the speaker is used. If no one involved in the event is speaking, the ubiquitous video camera source is used.

An example of video summarization for a user is shown in Figure 14.5 The summary page was created by chronologically listing scene videos, which were automatically extracted based on events. We used thumbnails of the scene videos and coordinated their shading based on the videos' durations for quick visual cues. The system provided each scene with annotations for time, description, and duration. The descriptions were automatically determined according to the interpretation of extracted interactions by using templates, such as *I talked with [someone]*; *I was with [someone]*; and *I looked at [something]*.

We also provided summary video for a quick overview of the events that the users experienced. To generate the summary video we used a simple format in which at most 15 seconds of each relevant scene was put together chronologically with fading effects between the scenes.

The event clips used to make up a scene were not restricted to ones captured by a single resource (video camera and microphone). For example, for a summary of a conversation "talked with" scene, the video clips used were recorded by: the camera worn by the user him/herself, the camera of the conversation partner, and a fixed camera on the ceiling that captured both users. Our system selected which video clips to use by checking the volume levels of users' individual voices. Remember that the

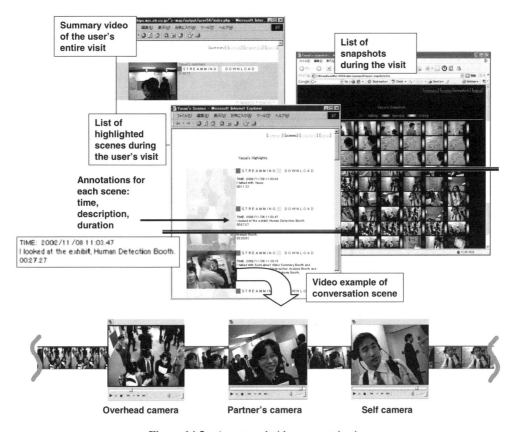

Figure 14.5 Automated video summarization

worn LED tag is assumed to indicate that the user's face is in the video clip if the associated IR tracker detects it.

14.5 Building 3D Virtual Space by Spatiotemporal Video Collage

Our ubiquitous sensor system records a huge amount of video data capturing conversation scenes. We are prototyping a system to build a 3D virtual space as a medium for re-experiencing the captured conversations. The important feature of our system here is that we have multiple-viewpoint videos that capture a particular scene. Suppose that the multiple-viewpoint videos that represent the scene are projected in a 3D virtual space according to the spatial point and direction of the individual video cameras (Figure 14.6(a)). In such a space with spatially mapped multiple videos, the particular object (a person or exhibit) that attracts a lot of attention at a particular time is expected to reveal its appearance at the intersection of videos. Conversely, areas that do not atract anybody's attention will not be visually rendered at all. We call this method "spatiotemporal (video) collage".

Most existing attempts to build 3D virtual spaces are based on uniform modeling from the so-called "God's view", so they are not good for representing a live atmosphere created by people residing in the space. On the other hand, a space produced by the spatiotemporal collage method is expected to reveal the social attention of participants in the real scenes, although the produced space may be geometrically

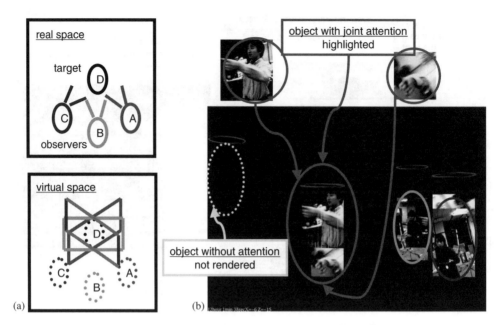

Figure 14.6 Building 3D virtual space by video collage: (a) collage of multiple viewpoint videos; (b) example snapshot of produced virtual space

inconsistent. The visualized social attention (i.e., intersection of multiple videos) will guide a visitor in the 3D virtual space a guidance during a walk-through (i.e., re-experience) of the 3D space.

For the video collage method, we have to tackle the following two issues at least:

1. Observer's information such as the spatial point and direction of cameras are necessary.
2. If we try to align videos in a strict point temporally and spatially, there will be insufficient video resources to render each scene.

Regarding the first issue, we are prototyping an LPS (local positioning system) to track every participant's location and gazing direction by putting the previously mentioned IR tracker on the top of the head of each participant and regularly attaching many LED tags on the ceiling. Regarding the second issue, we are examining a method of increasing the visual resources that capture a particular object by extending the temporal window.

An example snapshot of a 3D virtual space reproduced by the video collage method is shown in Figure 14.6(b). Only three objects that drew the attention of participants in the scene are displayed while four objects were detected in this view by LPS. Such automatic selection naturally reveals the viewpoints of participants in the scene and helps a user to re-experience the scene.

14.6 The Ambient Sound Shower: Sound Collage for Revealing Situated Conversations

The Ambient Sound Shower system (Müller *et al.* 2004) provides a user who is touring an exhibition with ambient sound via an earphone (e.g., a mixture of the conversations by past visitors), in order to intuitively

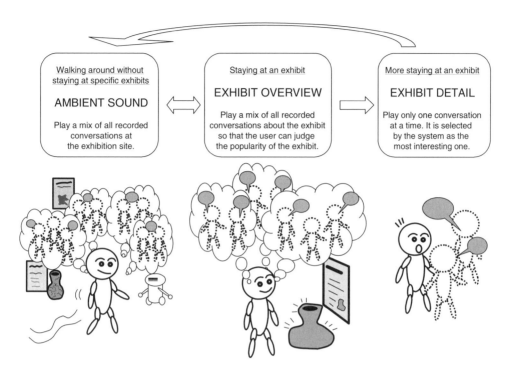

Figure 14.7   Usage scenario of the Ambient Sound Shower system

provide him/her with the atmosphere of the exhibition site on the fly. The system uses ubiquious sensors to infer the user's current situation and changes sound modes according to his/her situation.

The system automatically changes between three different playback modes, as illustrated in Figure 14.7. If the user is at an exhibition with few visitors, the system will first try to establish a stimulating ambient atmosphere by playing back on the user's earphone a mix of all conversations that were held at the exhibits by other participants so far. If the user then shows interest in a particular exhibit by focusing on it, the system will switch to play back a mix of all conversations that were held only about (actually, beside) this particular exhibit. The number of conversations that the user can hear indicates the popularity of the exhibit. If he/she keeps the exhibit in focus, the system assumes that he/she is still interested in it and starts to play back conversation one. The presented conversation is assumed to be of particular interest for the user. It is selected by taking into account several pieces of context information of the user and his/her environment.

The most important part of the system performs matchmaking between the user and the previous participants of the conversation by looking at their context history. To detect the user's current situation and decide if providing additional information would be useful or disturbing, the system takes into account the user's context as sensed by wearable sensors such as a throat microphone or IR tracker. The system infers from the sensory data contextual information such as the user's

- conversation status (whether or not involved in a conversation)
- accompaniment status (whether accompanied by people)
- interest in particular exhibits (whether focusing on a particular exhibit).

One of our premises is to avoid the use of automatic speech recognition and natural language understanding systems since they are still difficult to use in real enviromnents and are error prone.

14.7 Inferring Semantic Information about Detected Conversation Scenes by Nonverbal Information

So far, we have shown a method of extracting conversation scenes and giving situational information (i.e., talked "when", "to whom", and "about what") to them by analyzing gaze and utterance data. This section shows another of our attempts to detect more fine-grained structures of conversation scenes by focusing on other nonverbal data, the speaker's actions for refering to peripheral objects. It describes a method of giving more detailed structures to the extracted conversations from poster presentation scenes by using data about how the speaker touches the poster.

14.7.1 Conversational Content Acquisition from Poster Presentation Scenes

We devised a poster touch capturing method and investigated the correlation between how people touched the poster and conversation structure; that is, the topic of conversation is related to a sub-theme in the poster that an exhibitor touched at the time when it occurred. By analyzing the relationship between conversations and poster touch data, we can detect the transition of topics and divide a long conversation composed of many topics into some microstructured conversations consisting of a single topic or context.

We utilized a large touch panel display to capture the exhibitor's poster touch data. Exhibitors provided explanations to visitors with reference to an electronic poster on a touch panel display (Figure 14.8), and poster touch data was accumulated when the poster was touched. An example of accumulated poster touch data, represented by dots, is shown in Figure 14.9.

We suppose that posters have several topics corresponding to their layout; that is, one topic is related to one area, and an exhibitor provides an explanation by selecting one of them. Analyses of preliminary experiments revealed that exhibitors often touched more than one area within a short span of time. In such a situation, the exhibitors tend to speak about the topics of both areas that they touch alternately (Figure 14.10); therefore, we suppose that the touched areas of the poster are semantically related.

We used a tree model to represent the semantic structure of posters. The image in the left panel of Figure 14.11 represents a typical poster having three main parts: the title, element A, and element B. Both elements have areas of explanatory text and images. Based on the semantic position of each area in the poster, we can derive a tree model as the hierarchical structure of the poster layout; this is depicted

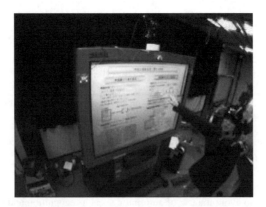

Figure 14.8 Explanations with reference to an electronic poster

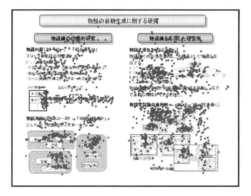

Figure 14.9   Example of accumulated poster touch data

Figure 14.10   Scene of alternate touches between different topics

in the right panel of Figure 14.11. In the model, sentences of element A correspond to sentence A and images of element A correspond to image A (the same is the case with element B). The title area usually corresponds to All. If sentence A has some keywords, we may be able to divide an area of sentence A into some areas of keyword A; however, this aspect requires further discussion because it is more difficult to touch smaller areas accurately.

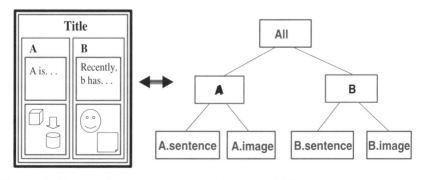

Hierarchical layout of a poster   **A tree model for a poster**

Figure 14.11   Tree model based on the poster structure

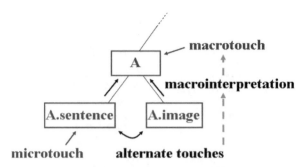

Figure 14.12   *Macrointerpretation* of alternate touches

In the scene depicted in Figure 14.10, the exhibitor touched several areas alternately in a span of approximately thirty seconds, while speaking continuously about the topics in those areas. According to our method, we can interpret such a scene by calculating a common parent of those areas and considering the exhibitor's alternate touches on the poster as *macrotouches* in the tree model (we term this process *macrointerpretation*). Thus, using the tree model, we can deal with alternate touches as *macrotouches*, in addition to *microtouches* (Figure 14.12).

14.7.2 Relationship between Conversation Topics and Poster Touch Data

Now we explain the relationship between poster conversations and poster touch data. In general, it can be considered that an exhibitor explicitly indicates the current topic by touching a particular area of the poster; however, poster touch data does not necessarily correspond exactly to the conversations between the exhibitor and the visitors.

For example, the exhibitor may chat with visitors for a period of time ignoring the poster and without touching it. Furthermore, visitors often change topics suddenly, and an exhibitor may begin to explain with touches after they have finished asking their questions. It is also possible that the topic of conversation could change if the conversation continues for a long time without a poster touch. Therefore, the problem is how to relate conversations to poster touch data. In order to resolve this, we establish the following hypothesis:

> *Conversations made almost synchronously with poster touch actions are closely related to the topics indicated by the touch.*

We now discuss specific procedures for relating conversations to poster touch data. First, we make segments from ON-OFF data of utterances obtained. The segments have start and end times. Second, we process segments according to the following three cases (Figure 14.13):

- *Case 1.* If there is poster touch data pertaining to one node of a poster between the start and end of a segment, we relate the segment to it.
- *Case 2.* If there is some poster touch data pertaining to some nodes of a poster between the start and end of a segment, we divide the segment into sub-segments based on the time of the poster touch data and relate segments to corresponding poster touch data.
- *Case 3.* If there is no poster touch data between the start and end of a segment, we relate the segment to the nearest poster touch data.

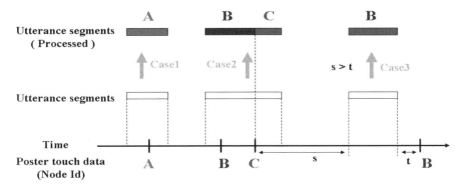

Figure 14.13   Examples of processing segments

Case 2 includes scenes in which an exhibitor provides an explanation. In this case, we make a segment with a long time span; however, we divide it into sub-segments with shorter time spans by using poster touch data and relate each sub-segment to corresponding poster touch data.

Case 3 includes scenes in which either the topic of the conversation changes to one that is not related to the poster, or the topic suddenly changes from topic C discussed for a long time to topic B. In this case, we relate a segment to the time nearest to it; however, it is not always obvious whether we should consider the weights of both "s" and "t" to be equal. Further discussion on this aspect is required.

14.7.3 Experiment and Analyses

To examine our method we conducted an experiment with four pairs of an exhibitor and a visitor who engaged in a conversation about a poster at a given time. The poster and the tree model of it that we used in the experiment are shown in Figure 14.14. The tree model was constructed manually from the poster by the exhibitor beforehand.

The poster has four main parts, and each part has some micro-parts. We created the tree model manually according to sentences and images. In this example, each section (A, B, C, and D) was divided into a large number of parts, in which case the second-layer nodes in the tree model would have more

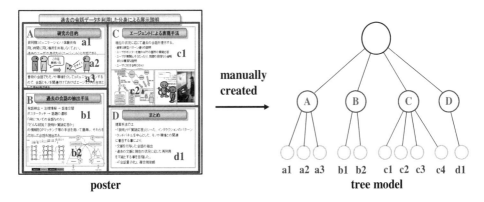

Figure 14.14   Example of a poster and its tree model

bottom-layer nodes. In such a case, we can distinguish one touch from another more accurately but experience difficulties in finding groups of related touches. From the experiment, we obtained three findings.

- New topics were often started when the area touched by the exhibitor changed and visitors began making utterances at that time.
- The dispersion of touches differed between the situation in which the exhibitor provided explanations to visitors and that in which he engaged in discussion with them.
- There was a connection between the touches of visitors and the beginning of a discussion on a new topic.

With regard to the first finding, the utterances of visitors had an effect on the change of the area touched by the exhibitor. The reason for this is as follows. Initially, the exhibitor was speaking while touching one area. Next, visitors asked questions or spoke some opinions, causing the exhibitor to touch another area in order to indicate his thoughts in relation to the poster. This situation mainly occurred when the exhibitor and visitors engaged in discussion.

While providing an explanation, the exhibitor spoke continuously and touched some areas. The visitors rarely spoke. In this case, not all of the utterance data of visitors could be used, but a typical pattern of touches was detected: the touches accompanying explanations were inclined to be dispersed within one part of the poster, while the accompanying discussions tended to be collected in micro-areas in this one part of the poster (Figure 14.15). This indicates that when the exhibitor provided explanations, he/she spoke while touching almost all of one part, but when he/she engaged in discussion, he/she touched only a few micro-areas of one part because visitors often asked questions based on that part.

Tree models with touches for both situations are shown in Figure 14.16. In this case, the touches in the explanation situation were interpreted by macro-interpretation as belonging to the upper node, and those in the discussion situation were interpreted as belonging to the lower node. This interpreted touch data reminded us of the change in the area that the exhibitor touched during a long time span and enabled us to detect topic transition in the explanation situation more accurately.

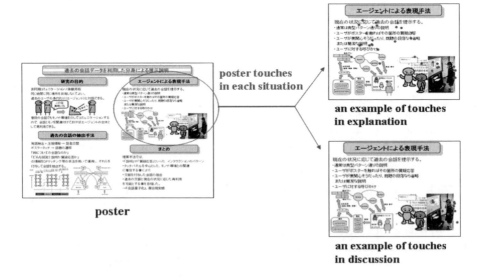

Figure 14.15 Example of touches in both situations: explanation and discussion

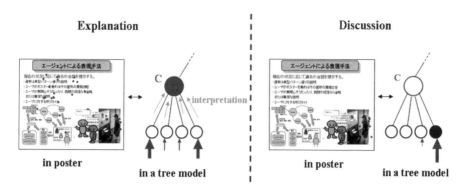

Figure 14.16   Tree model with touches in both situations

14.8 Related Work

There have been many studies on smart environments for supporting humans in a room by using video cameras set around the room; examples are: the Smart room (Pentland 1996), Intelligent room (Brooks *et al.* 1997), AwareHome (Kidd *et al.* 1999), Kidsroom (Bobick *et al.* 1999), and EasyLiving (Brumitt *et al.* 2000). They all shared the same goal of recognizing human behavior using computer vision techniques and an understanding of human intentions. On the other hand, our interest is to capture not only an individual human's behavior but also interactions among multiple humans (networking of their behaviors). Therefore, we focused on understanding and utilizing human interactions by using an infrared ID system to simply identify human presence.

There have also been studies on wearable systems for collecting personal daily activities by recording video data; e.g., Mann (1998) and Kawamura *et al.* (2002). Their aim was to build an intelligent recording system used by single users. We, however, aim to build a system used collaboratively by multiple users to capture their shared experiences and promote their further creative collaborations. Such a system enables our experiences to be recorded from multiple viewpoints, and individual viewpoints will become obvious.

This chapter presented a system that automatically generates video summaries for individual users as an application of the interaction data captured by our system. In relation to this system, some systems for extracting important scenes of a meeting from its video data have been proposed; e.g., Chiu *et al.* (1999). These systems extract scenes according to changes in the physical quantity of video data captured by fixed cameras. On the other hand, our interest is not to detect changes in visual quantities but to segment human interactions (perhaps derived by the humans' intentions and interests) and then extract scene highlights from a meeting naturally.

There is a wide range of projects related to the Ambient Sound Shower system in the area of nomadic information systems; some of them use ubiquitous or wearable sensors. In the HIPS project (Benelli *et al.* 1999) a hand-held electronic museum guide was developed. The Museum Wearable (Sparacino 2002) uses a similar wearable and ubiquitous sensor system to ours. In both projects the presented information already exists in the sense that it is stored in and retrieved from a static database. The sensory data is only used to decide which information should be selected or in which way it should be arranged, but it is not captured or presented to other visitors as additional information about the exhibits. In our approach, on the other hand, we try to enable experience sharing among visitors, which was also a concept of the C-MAP system described in Sumi and Mase (2002). With the Ambient Sound Shower, visitors and exhibitors can contribute to a dynamic repository of data that is used to provide additional information about exhibits.

With systems like the Audio Aura (Mynatt *et al.* 1998) or the system described by Rekimoto *et al.* (1998), people can attach digital messages to objects and these are automatically or manually retrieved and presented if other persons gaze at the object or enter a certain location. This message retrieval is comparable to the Ambient Sound Shower system in that it is not personalized, which means that all messages attached to an object or location are presented to the user regardless of whether they are of interest to him or not.

14.9 Conclusions

This chapter described our attempt to build a communicative medium for capturing and re-experiencing conversations taking place in real space. We described a system that captures and interprets conversation scenes by means of ubiquitous sensors. Based on this system, we presented three approaches to visualizing and helping users access the extracted conversation scenes: chronological summarization of videos, spatio-temporal collage of videos, and ambient sound display. Finally, we showed a method of detecting more detailed structures of captured conversation scenes in poster presentation settings by using poster touch data of speakers.

Acknowledgments
We thank our colleagues at Kyoto University and ATR for their contributions to the development of the systems and their help in the experiments described in this chapter. The research presented here is supported by the National Institute of Information and Communications Technology and Grant-in-Aid for Scientific Research.

References

Benelli G., Bianchi A., Marti P., Not E. and Sennati D. (1999) HIPS: hyper-interaction within physical space. In *Proceedings of IEEE International Conference on Multimedia Computing and Systems*, Vol. 2, pp. 1075–1078.

Bobick A.F., Intille S.S., Davis J.W., Baird F., Pinhanez C.S., Campbell L.W., Ivanov Y.A., Schütte A. and Wilson A. (1999) The KidsRoom: a perceptually-based interactive and immersive story environment. *Presence* **8** (4), 369–393.

Brooks R.A., Coen M., Dang D., Bonet J.D., Kramer J., Lozano-Pérez T., Mellor J., Pook P., Stauffer C., Stein L., Torrance M. and Wessler M. (1997) The intelligent room project. In *Proceedings of the Second International Cognitive Technology Conference* (CT'97), pp. 271–278.

Brumitt B., Meyers B., Krumm J., Kern A. and Shafer S. (2000) EasyLiving: technologies for intelligent environments. In *Proceedings of HUC 2000* (Springer LNCS1927), pp. 12–29.

Chiu P., Kapuskar A., Reitmeier S. and Wilcox L. (1999) Meeting capture in a media enriched conference room. In *Proceedings of CoBuild'99* (Springer LNCS1670), pp. 79–88.

Kawamura T., Kono Y. and Kidode M. (2002) Wearable interfaces for a video diary: towards memory retrieval, exchange, and transportation. In *Proceedings of 6th International Symposium on Wearable Computers* (ISWC2002), pp. 31–38.

Kidd C.D., Orr R., Abowd G.D., Atkeson C.G., Essa I.A., MacIntyre B., Mynatt E., Startner T.E. and Newstetter W. (1999) The aware home: a living laboratory for ubiquitous computing research. In *Proceedings of CoBuild'99* (Springer LNCS1670), pp. 190–197.

Mann S. (1998) Humanistic computing: "WearComp" as a new framework for intelligence signal processing. *Proceedings of the IEEE* **86** (11), 2123–2151.

Müller C., Sumi Y., Mase K. and Tsuchikawa M. (2004) Experience sharing by retrieving captured conversations using non-verbal features. In *Proceedings of First ACM Workshop on Continuous Archival and Retrieval of Personal Experiences* (CARPE 2004), pp. 93–98.

Mynatt E.D., Back M., Want R., Baer M. and Ellis J.B. (1998) Designing audio aura. In *Proceedings of CHI'98*, ACM, pp. 566–573.

Nishida T. (2004) Towards intelligent media technology for communicative intelligence. In *Proceedings of International Workshop on Intelligent Media Technology for Communicative Intelligence* (IMTCI 2004).

Pentland A. (1996) Smart rooms. *Scientific American* **274** (4), 68–76.

Rekimoto J., Ayatsuka Y. and Hayashi K. (1998) Augment-able reality: situated communication through physical and digital spaces. In *Proceedings Second International Symposium on Wearable Computers* (ISWC'98), pp. 68–75.

Sparacino F. (2002) The museum wearable: real-time sensor-driven understanding of visitors' interests for personalized visually-augmented museum experiences. *Proceedings of Museums and the Web.*

Stiefelhagen R., Yang J. and Waibel A. (1999) Modeling focus of attention for meeting indexing. In *Proceedings of ACM Multimedia '99*, pp. 3–10.

Sumi Y. and Mase K. (2002) Supporting the awareness of shared interests and experiences in communities. *International Journal of Human-Computer Studies* **56** (1), 127–146.

Sumi Y., Ito S., Matsuguchi T., Fels S. and Mase K. (2004) Collaborative capturing and interpretation of interactions. In *Proceedings of Pervasive 2004 Workshop on Memory and Sharing of Experiences*, pp. 1–7. http://www.ii.ist.i.kyoto-u.ac.jp/~sumi/pervasive04.

Takahashi M., Ito S., Sumi Y., Tsuchikawa M., Kogure K., Mase K. and Nishida T. (2004) A layered interpretation of human interaction captured by ubiquitous sensors. In *Proceedings of First ACM Workshop on Continuous Archival and Retrieval of Personal Experiences* (CARPE 2004), pp. 32–38.

15

Real-time Human Proxy

Rin-ichiro Taniguchi and Daisaku Arita

15.1 Introduction

A lot of network-based human interaction systems have been developed and used. These systems handle video and sound streams, which are captured by cameras and microphones, transmitted via networks, and presented by video displays and speakers. Although such simple audio-visual systems are easy to use, we admit they have the following disadvantages:

- *Inconsistency of spatial relations.* Since all visual information is presented on a 2D display, positional relations among participants are not consistent. This means that, for example, each participant cannot understand where other participants look at and point to.
- *Limitation of the number of participants.* Since video images of all participants are arranged on a 2D display, the number of participants is limited by the size and the resolution of the display. When there are many participants, we have to have a mechanism to dynamically choose an appropriate subset of participants. Of course, the network bandwidth becomes another restriction on the number of participants.
- *Privacy.* Audio-visual interaction systems sometimes convey personal information which the participants do not want to convey, such as their actual faces, their clothes, and their rooms. Blurring the images is one typical solution to this problem, but it causes unnatural feeling in communication.

To solve these problems of video-based interaction systems, there are several researches on virtual environments for human interaction such as NICE (Narrative Immersive Constructionist/Collaborative Environments) (Roussos *et al.* 1997) and Nessie World (Jeffrey and McGrath 2000). In these researches, a 3D virtual space is reconstructed, in which each participant is represented as an avatar generated by computer graphics. Through a reconstructed virtual space, each participant virtually sees and hears other participants' activities from the position where his/her avatar is represented.

Since a 3D virtual space is reconstructed, spatial relations among the participants can be consistent. This means that each participant can understand where other participants look at and point to, can see where he/she wants to see, and can move in the virtual space. In addition, since a display presents not video images of participants but a single virtual space, the limitation of the number of participants caused by the size of 2D display can be relaxed. The network bandwidth problem is also solved because no

Conversational Informatics: An Engineering Approach   Edited by Toyoaki Nishida
© 2007 John Wiley & Sons, Ltd

video image should be transmitted. Instead, only some parametric information to move the avatars is transmitted.

However, it is difficult for avatar-based interaction systems to acquire and present all of participants' information, especially nonverbal information. In this chapter, we describe *real-time human proxy*, a new concept for avatar-based interaction, which makes it easy to acquire and present participants' information, and which provides an efficient and natural communication environment via a virtual space.

15.2 Concept of Real-time Human Proxy

In avatar-based interaction, an avatar is expected to reflect a participant's activities into a virtual space as if he/she were there. To achieve this goal, it is necessary to reproduce not only his/her speech utterances but his/her body actions, which are quite important for communication. In this context, legacy input devices, such as keyboards and mice, are not sufficient to acquire participant's activities in aspects of quality and quantity. Therefore, we have developed and used a vision-based motion capture system (VMCS) (Date *et al.* 2004) as an input device, for acquiring proper information of participants without any annoying operations.

15.2.1 What is Real-time Human Proxy?

Real-time human proxy (RHP) is a new concept for avatar-based interaction (Arita *et al.* 2004), which makes the avatar act meaningfully referring to action information acquired by VMCS. As the first step, we have focused on acquisition and representation of human body actions, which take a major role in nonverbal information. In RHP, we symbolize the human action information, and represent the symbolized action information via an avatar. Figure 15.1 shows the basic scheme.

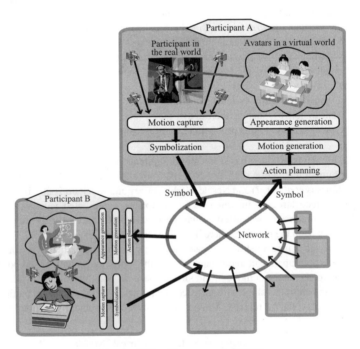

Figure 15.1   The concept of RHP

In general, it is not possible to reflect all the participants' actions into avatars even if we use a VMCS to acquire the human motion information. Such information as changes of hand shapes and movement of eyebrows and a lip cannot be fully acquired, and the lack of such detailed information may sometimes cause unnatural avatar representation. Therefore, we do not exactly measure how he/she moves his/her body parts but estimate the meaning of his/her actions in a given context or circumstance. Then, we appropriately represent the action meaning by generating the avatar's motion based on the context, and the generated motion includes secondary motion which is not directly measured by a VMCS. Our basic idea is that it is not necessary for an avatar to move its body exactly the same as its corresponding participant, since participants want to know the meaning of the others' actions, not how the others are moving their bodies. Based on this idea, RHP consists of three major parts: human motion sensing, human action symbolization, and human action presentation.

Human Motion Sensing

We have used a vision-based motion capture system to acquire human body actions. There are many motion capture systems, but most of them force us to attach special markers precisely on our body. It is neither natural nor comfortable for daily-life situations. Therefore, we have developed a vision-based motion capture system which does not require any special marker attachments. It gives us a natural motion sensing environment.

Action Symbolization

With RHP, we acquire human action meanings instead of human raw motions. We categorize motion sequences into predefined actions and express them as symbols. Each symbol consists of a label of an action and its parameters, such as "walking (p_x, p_y, v_x, v_y)", where p_x and p_y represent the position and v_x and v_y represent the velocity of a participant. After recognizing human actions from captured motion data, the system transfers the symbols to the presentation side of a virtual space. Currently action symbols are recognized by a simple rule-based algorithm, which classifies input body posture parameters into predefined action categories.

Avatar Presentation in Virtual Space

We define that an avatar is an object which is a participant's substitute in a virtual space. An avatar has predefined knowledge to generate its motion and appearance from the symbols. However, in general, it is time-consuming job to construct or modify the knowledge. Therefore, the predefined knowledge is to be described in a reusable and extensible form. Our predefined knowledge is described in a layered structure, which consists of a Behavior model, a Motion model, and a Figure model. An avatar plans the next action based on the Behavior model, generates a motion corresponding to the next action based on the Motion model, and generates the avatar's appearance performing the motion[1] based on the Figure model.

Through these phases, the avatar's appearance is generated and presented in a virtual space. Every participant is able to see the virtual space, in which all the participants are represented as avatars. The viewpoint of each participant is anchored where his/her avatar is presented, and each participant can see any views through a head-mounted display (HMD) or a 2D display with a view-angle control mechanism.

---

[1] Here, motion means posture sequences of the avatar's body parts.

15.2.2 Benefits of RHP

In human communication, it is well-known that nonverbal information takes quite an important role. Body action is an important aspect of nonverbal information, and, therefore, conveying its information enriches the communication. Especially when the communication environment is rather complex, such as when more than two people are talking, body actions play many roles such as triggers of turn-taking, topic management, target pointing, etc. Therefore, in avatar-based communication, adequate presentation of nonverbal information, or body actions, is essential, and the use of a vision-based motion capture system liberates participants from controlling an avatar: an avatar acts as the user does without any explicit control. On the symbolization process, the system recognizes the action that a participant makes. This operation is intuitive and natural since participants make quite similar actions as they make in the real world.

The concept (i.e., symbolization of actions and presentation via an avatar) makes the representation process seen simple. Usually acquired motion information about a participant does not have detailed information of several body parts, such as angles of fingers, and this must be compensated to provide natural nonverbal representation. Using RHP, actions are recognized in a given context, and it is possible to generate full body motion sequences in an arbitrary manner referring to predefined knowledge of the given context such as a lecture scene. In addition, symbol-based action representation does not suffer from network congestion. If raw motion parameters of body parts acquired by a VMCS are transfered via a congested network (especially the Internet), the timing of raw motion data representation is uneven and, as a result, represented avatar motion sometimes seems unnatural.

When an avatar reflects a participant's motions, the avatar should be designed to physically fit the participant so that 3-dimensional positions of body parts acquired by a VMCS are correctly represented in computer graphics. If the sizes of the avatar body parts are different from those of the participant, unnatural motions are often generated and the avatar's actions can be misunderstood. In the case of RHP, the symbolized action is transmitted, not the 3-dimensional body parameters directly, and consequently RHP largely relaxes the constraints. We can use any kind of avatar, including a taller body, shorter arms, bigger head, etc, suitable for a given context. This improves not only usability of the system, but also the variety of avatars in interaction. For example, when RHP is used for communication among kids, avatars can be designed to have cute shapes.

In this chapter, we describe major technical issues of our RHP, which are vision-based human motion sensing, avatar representation from action symbols, and development of prototypical systems.

15.3 Acquisition of Human Motion

In RHP, acquisition of human motion is the first step, and vision-based motion sensing is adopted for this purpose, because it does not impose any physical restrictions on a participant, and provides a smart and natural way of motion sensing. Although there are several important issues in vision-based human motion sensing, from the viewpoint of interactive applications, a real-time feature is quite important. In human communication via a virtual space, or in avatar-based human communication, sending messages, which is one way of communication, consists of three steps – acquisition, transmission, and presentation of the messages (verbal and nonverbal) – and the delay of messages tends to become longer. This means the real-time feature becomes more important in system design.

Real-time proceesing here means that human motion should be smoothly captured, which indicates that images should be processed at the speed of a TV camera signal, 20–30 frames/second. Therefore, computation-intensive approaches (Deutscher et al. 2001; Nunomaki et al. 2000) are not realistic even though they provide a general framework of motion sensing. To realize such real-time systems, the key issues are: robust image features, which are easy to extract; and fast human posture estimation from the image features.

Usually, as image features, blobs (coherent region) (Etoh and Shirai 1993; Wren et al. 1997) or silhouette contours (Takahashi et al. 2000) are employed. However, image features which can be robustly detected are limited, and, therefore, the estimation of 3D human postures from the limited cues are quite essential. To solve this problem, we have introduced vision-based inverse kinematics. In addition, to deal with the view dependency and the self-occlusion problem when a human makes various poses, we have employed an approach of multi-view image analysis. We have implemented our vision-based human motion sensing on a PC-cluster to realize its online and real-time processing.

15.3.1 System Overview

The basic algorithm flow of our real-time motion capturing is as follows:

1. *Detection of visual cues*:
   - Silhouette detection, skin color blob detection, face direction estimation.
   - Calculation of 3D positions of features using multi-view fusion.
2. *Human motion synthesis*:
   - Generation of human figure full-body motion and rendering in the virtual space including calculation of the interaction.

The key point of our system is *human motion synthesis* referring to a limited number of visual cues, which are acquired by the perception process. Several real-time vision-based human motion sensing methods have been developed. However, they are based on a rather simple human model and its generated human motion sensing is not natural. For example, the direction of the face is not detected, and the number of articulations is limited. Here, we make the human figure model more complex and develop a vision-based algorithm for human motion sensing based on the model (details are discussed in section 15.3.3).

When a real-time and on-line system is designed, an error recovery mechanism is quite important. If the system is based on feature tracking, once features fail to be tracked, the posture estimation process may reproduce unrealistic human postures, and it cannot escape from the erroneous situation. Therefore, in order to avoid having to reset the system in such a case, we have introduced a simple error recovery process, which executes concurrently with the main human posture estimation process, and which always checks whether the human silhouette makes predefined shapes which are easy to recognize precisely. When the process finds the silhouette makes such shapes, it notifies the human posture estimation process, and the estimation process adopts the recognition result regardless of the estimation result.

According to these considerations, we have designed a vision-based real-time human motion sensing as shown in Figure 15.2. To deal with the view dependency and self-occlusion, we have adopted a multi-view approach, and images acquired by each camera are, at first, processed individually by multiple PCs (dotted rectangles in Figure 15.2). In this early stage, silhouette detection, 2D blob detection, pose matching for error recovery mentioned above, and face direction estimation are applied to the input image. Then, in the following feature integration stage, the acquired information in the previous stage is integrated; that is, 3D positions of the blobs are calculated and human posture is estimated by vision-based inverse kinematics referring to the blob positions and the face directions. These processes are executed in another PC. Details of the processings will be described in the following sections.

15.3.2 Acquisition of Perceptual Cues

Feature Detection

We detect color blobs for a face, hands, and feet, whose colors are acquired in advance, and calculate 2D positions of the color blobs as the centroids of the blobs. Color blob detection is based on color similarity evaluation, which is done in the HSV (hue–saturation–value) color space. The thresholds to detect skin

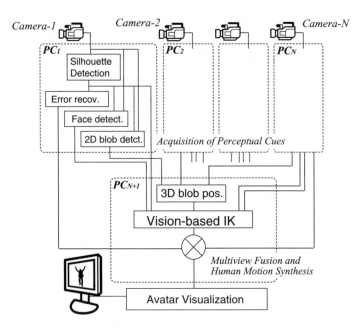

Figure 15.2   Software structure of real-time vision-based human motion sensing

color are decided based on pre-acquired skin color samples,[2] and the color conversion and thresholding are simultaneously executed using a color look-up table to speed up the calculation. We also extract a human silhouette region by background subtraction, which is used by an error recovery process and a human posture estimation process.

In principle, the 3D positions of the blobs can be calculated from two views. However, due to self-occlusion, we cannot always calculate the 3D positions with only two views, and we have used a certain number of cameras. Therefore, we have to solve a stereo-pair correspondence problem in a multi-view situation. We first extract possible stereo-pair candidates of the 2D blobs,[3] and then classify their 3D positions into five clusters of feature points: head, right hand, left hand, right foot, left foot. Classification is done based on the distances from the feature points detected in the previous frame. In each cluster, we estimate the feature point position as the average position of the 3D position candidates after a dense part of the cluster is selected.

Estimation of Face Direction

Face direction is indispensable for interactive applications. The problem here is the low resolution of the face region, because cameras are arranged to capture a full-body and a face becomes small in the field of view. Therefore, features such as eyes and mouths cannot be clearly detected, and then feature-based or structure-based techniques of face direction estimation (e.g., Yao *et al.* 2001) cannot be applied. We have employed a template-matching method preparing face templates with multiple aspects.

---

[2] For extracting feet, we can use shoe/sock color instead.
[3] When the distance of the two lines of sights passing through the centroids of two 2D blobs is small, the two blobs are a stereo-pair candidate. In this case, a point, which is the nearest to the both lines in a sense of the least square error, is judged to be their 3D point.

Currently, we prepare 300 or more templates for each person in advance. Making templates is quite easy: a 3D rotation sensor is attached to the top of the user's head and the user makes a variety of face directions in front of a camera before using the motion sensing system. Recording the output of the rotation sensor and the face image frames synchronously, we can get face templates with a variety of face directions. It takes only a couple of minutes. This approach is not sophisticated but quite practical and robust.

To reduce the computation time, we have employed the eigen-space method (Murase and Nayar 1995) with several speed-up tactics (Nene and Nayar 1997). The size of templates is 45×50 pixels, and the dimension of the eigen-space is set to be 60. After detecting a face region, or skin-color region which corresponds to a face, the face region is normalized to have the same size as that of the template. Then, the normalized face region is mapped on the eigen-space, and the most similar template is searched for in the eigen-space. Pre-acquired face direction of the selected template becomes the face direction of the input face. The estimation accuracy is not very high because of the low-resolution images, but in most interactive applications we can get certain feedback from the system and we can modify the face direction based on this. We think the accuracy acquired by this approach is high enough to use.

15.3.3 Vision-based Inverse Kinematics

Our problem is to estimate human postures from a limited number of perceptual cues, which are blobs corresponding to hands, feet, and head. This problem can be explained in a framework of inverse kinematics (IK) in the field of robotics. IK is to determine joint angles θ_i of a manipulator so that the position of an end effector, or a final point, $\mathbf{P}_n$, coincides with a given goal point $\mathbf{G}$: $\mathbf{P}_n(\theta_1, ..., \theta_n) = \mathbf{G}$: where the manipulator has n segments. The difficulty here is that even if the goal is attainable,[4] there may be multiple solutions and, thus, the inverse problem is generally ill-posed.

In our problem, end effectors are hands, feet, and a head, and the goals are the blob positions acquired by the perceptual process. The posture estimation, which is to decide the positions of joints of the human model, is achieved by calculating the joint angles in the framework of IK. In human posture sensing, each joint position acquired by IK should be coincide with a joint position of a given human posture, so we have to find the unique and correct solution.

Our method to solve this problem is divided into two phases: *acquisition of initial solution* and *refinement of initial solution*. For simplicity, here, we explain human posture estimation of the upper body.

Acquisition of Initial Solution

Inverse kinematics is solved by an ordinary numerical method (Wang and Chen 1991) and initial candidates of 3D positions of shoulders and elbows are calculated. Here, we assume that the lengths of the bones in Figure 15.3 are given in advance. At time t, a hand position $(x(t), y(t), z(t))$ is represented as

$$(x(t), y(t), z(t)) = \mathbf{P}(T_x(t), T_y(t), T_z(t), \theta_1(t), \theta_2(t), \cdots, \theta_N(t)) \tag{15.1}$$

where

- $T_x(t), T_y(t), T_z(t)$ indicate the head positions in the world coordinate
- $\theta_1(t), \theta_2(t), \theta_3(t)$ indicate rotation angles between the world coordinate and the local coordinate of the head, which are calculated from the face direction
- $\theta_j(t)(4 \leq j \leq N)$ indicate rotation angles among connected part.

---

[4] If the distance of the goal to the initial point of the manipulator is larger than the sum of the lengths of the segments, the goal is not attainable.

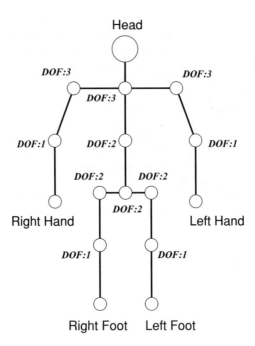

Figure 15.3   Human model

We suppose that, at time $t + 1$, the hand position moves to $(x(t + 1), y(t + 1), z(t + 1))$, the head position moves to $(T_x(t + 1), T_y(t + 1), T_z(t + 1))$, and the head direction changes to $(\theta_1(t + 1), \theta_2(t + 1), \theta_3(t + 1))$. Here, we slightly modify $\theta_j(t + 1)$, $(4 \leq j \leq N)$ so that the hand position – i.e., the position of the end effector, $\mathbf{P}(T_x(t + 1), T_y(t + 1), T_z(t + 1), \theta_1(t + 1), \cdots, \theta_N(t + 1))$ – approaches the goal position $(x(t + 1), y(t + 1), z(t + 1))$. Repeating this process until the end effector position coincides with the goal position, we acquire the positions of a shoulder and an elbow. In order to exclude impossible postures, we have imposed a possible range on each angle θ_j.

Refinement of Initial Solution

The posture estimated in the previous step is just a solution of inverse kinematics, and it is not guaranteed that it coincides with the actual posture. This is due to ill-poseness of the inverse kinematics. To estimate the posture more accurately, we refine the acquired solution by referring to input image data. The basic idea is simple: if the shoulder and elbow positions acquired by the previous phase are correct, they should be inside the human region in the 3D space. Otherwise, the acquired solutions are not correct and they should be modified so as to be included in the human region. Here, we empirically assume that the shoulder position is acquired by solving the basic inverse kinematics, and we mainly refine the elbow position. Its basic algorithm is as follows:

1. We have the shoulder position by solving the inverse kinematics and the hand position by color-blob analysis.
2. When the lengths of its upper arm and forearm are given, the position of its elbow is restricted on a circle in the 3D space. The circle is indicated by C.
3. When the elbow is searched on the circle, we exclude impossible values, with which the arm gets stuck in the torso.

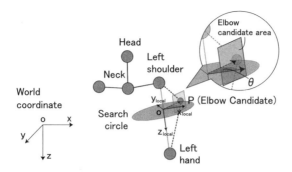

Figure 15.4 Elbow position estimation. (Reproduced by permission of Information Processing Society of Japan)

4. As shown in Figure 15.4, an elbow detection rectangle is established in a plane which is constructed by the shoulder, an hypothesized elbow, and the hand.
5. Then, in each view, the rectangle is reverse projected on the image plane and correlation between the projected rectangle and the human silhouette region is calculated.
6. Then, by varying the position of the hypothesized elbow, the correlation R can be parameterized by θ, which is the angle around the center of the circle C. We search for θ giving the maximum R, which indicates the elbow position.

If the refinement process fails to find the correct solutions, the system restarts the modification process by changing the initial values. Since this system is required to work in real time, this iteration is stopped when the deadline comes, and an intermediate result is returned.

15.3.4 Implementation and Result

In this experiment, we have used nine sets of IEEE1394-based color cameras (Sony DFW-V500) with f:4mm lenses, which are geometrically calibrated in advance. The images are captured with the size of 640×480 pixels, and the frame rate is 15 fps.[5] We have implemented our vision-based human motion analysis on a PC-cluster with 3.0-GHz Pentium IVs, where detection of perceptual cues in each view is executed in parallel, and where solving IK and human figure generation are executed in a succeeding PC.

Figure 15.5 shows a typical example of human action sensing, and it shows that the system has enough performance for human motion acquisition in RHP. The biggest problem for this system is latency. Currently, the latency is about 200 ms, which should be reduced to achieve more efficient interaction.

15.4 Presentation of the Avatar

15.4.1 Basic Framework of Avatar Generation

Bruce and co-workers built a virtual pet which is an autonomous and directable CG character (Bruce and Tinsley 1995). They achieved these features by describing predefined knowledge of the virtual

---

[5] Due to camera specification.

Figure 15.5   Result of human posture estimation. (Reproduced by permission of Information Processing Society of Japan)

pet in a layered structure. Their aim is interaction between an autonomous CG character in a virtual world and a human in the real world. However, in RHP, a CG character, or an avatar, is used as a surrogate of a participant to interact with other participants, and the action specified by a given symbol must be represented immediately. Therefore, it is not easy to apply their technique directly to our system.

Our predefined knowledge is described in a three-layered model structure, which provides flexibility and reusability of the knowledge description. Our idea is that an avatar's appearance should be changed according to the applications, because affinity to the participants is quite important in smooth communication. For example, in a teaching environment for kids, an avatar of a teacher is set to be a cute one, such as, a cartoon-like character. Therefore, the system has been designed so that the appearance of the avatar can be changed easily.

Basically, an avatar plans the next action based on the Behavior model, generates a motion corresponding to the next action based on the Motion model, and generates an appearance performing the motion based on the Figure model. The outline of our avatar generation is illustrated in Figure 15.6.

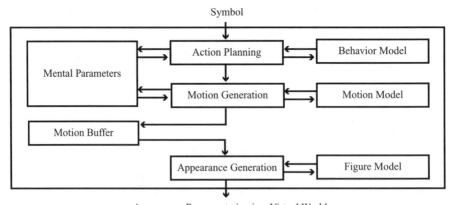

Figure 15.6   Process flows in the avatar

15.4.2 Behavior Model and Action Planning

There is an action planner in an avatar, which outputs the avatar's next action, such as "walking" or "raising hand", based on received symbols, the Behavior model, and mental parameters.[6] Such outputs are mostly independent of the avatar's physical structure. This allows model constructors to modify or replace the Behavior model.

Action Planning

A human can perform multiple actions at the same time if these actions do not require the same body part. For example, "walking" (an action using right and left legs) and "raising hand" (an action using right or left arm) can be done simultaneously. Therefore, the action planner should handle such simultaneous actions. Moreover, in RHP, the vocabulary of the action symbols depends on applications, or interaction environments, and the Behavior model should be modified easily. Consequently, we have designed the Behavior model so that it consists of a set of state transition graphs. Each graph corresponds to each of the actions defined in the RHP, such as "walking" and "raising hand", and has only two states, "ON" and "OFF". "ON" means that the action is activated and "OFF" means that the action is deactivated. When the avatar's posture is a neutral posture, or the base posture of starting actions, all states in the Behavior model are set to be "OFF". For instance, for a human avatar, the neutral posture is a standing posture with his/her arm down.

Moving from state "OFF" to "ON" means that the avatar makes an action transiting from the neutral posture to a posture of the action activated. Such action is called *outward action*. Moving from state "ON" to "OFF" is called *homeward action*.

An outward action can be planned when the posture of the avatar's body parts used for the action is the same as the neutral posture. When the posture of the avatar's body parts is not the same as the neutral posture (i.e., it conflicts with the current posture), the outward action cannot be planned. However, if the conflicting posture is in a homeward action, the outward action can be planned with a little delay, because the avatar's posture is to become the neutral posture quite soon. A homeward action, on the other hand, is planned to follow its corresponding outward action.

An action is mainly planned according to a received symbol. However, an avatar sometimes freezes if the avatar acts only when symbols are transmitted. This is because no symbols are transmitted when a participant does not make any predefined actions, or when the predefined actions cannot be recognized by the system. Needless to say, such avatar's behavior does not seem natural. To solve this problem, the action planner plans some actions spontaneously such as "folding arms" or "sticking hand into a pocket", which have little influence on interaction. These actions are planned probabilistically.

To realize integratively the symbol-based and the probabilistic action generation, each of the outward and homeward actions has a transition probability, and each probability is controlled according to the communication situation. Transition probabilities corresponding to symbol-based actions are usually 0, but when an action symbol is received, the probability corresponding to that action becomes 1. This means that the corresponding action is activated immediately. On the other hand, probabilities corresponding to spontaneous actions are not 0, and are not changed according to the received symbols. Thus, the actions are planned probabilistically.

Importance of Actions

Each action has a degree of importance. An action with a higher degree of importance is planned preferentially. For example, consider the situation that an outward action with a higher degree of importance

---

[6] In RHP, an avatar can have a set of mental parameters which represent an avatar's mental state such as "happy" and "disgusting", and motion and appearance generated varies depending on the mental state.

is selected when an outward action with a lower degree of importance is presented. In this case, at first, the homeward action with a lower degree of importance is planned. Then, the outward action with a higher degree of importance is planned immediately. In the opposite case, an action with a lower degree of importance is just ignored. Fundamentally, the highest importance is attached to actions according to action symbols. On the other hand, lower importance is given to actions which are unrelated to the interaction. Currently, the degrees of importance are assigned heuristically by hand, but calculating them automatically from accumulated motion data is an important and interesting problem.

15.4.3 Motion Model and Motion Generation

Basic Scheme

The motion generator generates an avatar's motion based on a planned action, the Motion model, and mental parameters. The Motion model, stores detailed motion information corresponding to each action to be planned, and it is a list of correspondences between actions defined and their motion information. Each motion is generated from its keyframe sequence, and each motion information consists of the following:

- set of keyframes (key postures): $Q_1, Q_2, \ldots, Q_{N_k}$
- the number of frames of the motion to be generated: N_f
- the positions of the keyframes: $p_1, p_2, \ldots, p_{N_k} (0 = p_1 < p_2 < \ldots < P_{N_k} = N_f)$
- interpolation function.

The generated motion (posture sequences) is inserted into the motion buffer, a kind of queue. In general, newly generated motions are added to the tail of the motion buffer.

However, when a generated motion does not conflict with an already stored motion, the generated motion is not added to the tail but a unified motion is stored where the old motion was stored. This makes these two motions be represented simultaneously. For example, when a motion "walking (using right and left leg)" is already stored and a motion "pointing with right finger (using right arm)" is generated, these motions do not conflict with each other. Therefore, these two motions can be represented simultaneously, and they are unified and turned into one motion "walking and pointing with right finger".

Motion Fusion

As described above, actions planned are outward and homeward actions. This can reduce the animator's job, and make action planning simpler. However, according to this scheme, in principle, every action is generated after a homeward action of its previous action is finished; that is, the avatar returns to the neutral posture. This possibly causes delay of action presentation. Moreover, it is unnatural when all the actions start from the neutral posture. To solve this unnaturalness, the motion generator *fuses* a homeward action and an outward action which collide with each other. Figure 15.7 shows an example of motion fusion.

15.4.4 Figure Model and Appearance Generation

The figure model stores the avatar's geometrical and texture data, which consist of the following:

- surface shapes constituting each body part and their texture
- size parameters of each body part
- a tree representing physical structure among body parts.

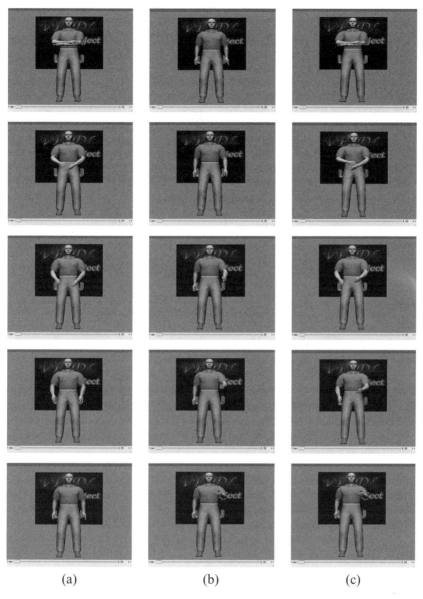

(a)   (b)   (c)

Figure 15.7 Fusing motion: (a) the motion of a homeward actions, (b) that of an outward action; (c) the fused motion. (Reproduced by permission of Technische Universität Graz)

In our system, every avatar is represented by tree-structured body parts, whose the root node corresponds to the waist of the avatar (Figure 15.8). The geometry of each body part is represented in a local coordinate system attached to its parent node, and the change of the posture of a body part is automatically propagated to its descendant body parts. Thus, the posture of an avatar is easily controlled.

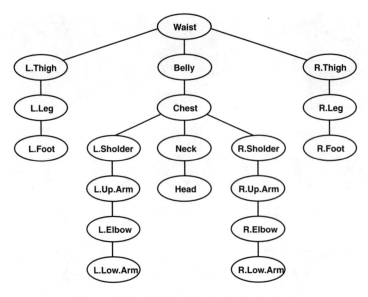

Figure 15.8   Outline of avatar body structure

15.5 Prototype of Real-time Human Proxy

15.5.1 Preliminary Experiment

To verify the effectiveness of our proposed techniques, we evaluate whether participants can recognize actions in the virtual space naturally. We apply RHP to an interactive game, a simplified version of a famous gesture game in Japan, whose rules are as follows:

1. One of participants becomes a leader.
2. The leader says "A" and points to another participant.
3. The participant who is pointed to at step 2 says "B" and points to another participant.
4. The participant who is pointed to at step 3 says "C" and puts his/her hands up.
5. The participant who puts his/her hands up becomes a leader.
6. Return to step 2 until someone fails.

Verbal and nonverbal information uttered by participants is acquired by a vision-based motion capture system and microphones and is transmitted via a network with one another. The labels of actions which are parts of predefined knowledge are as follows:

- *Finger pointing*: an action activated by a transmitted symbol
- *hands up*: an action activated by a transmitted symbol
- *head turn*: an action activated by a transmitted symbol
- *folding arms*: a probabilistic action
- *putting hand to waist*: a probabilistic action.

Some snapshots of the game are shown in Figure 15.9. Some impressions of participants about representation are as follows:

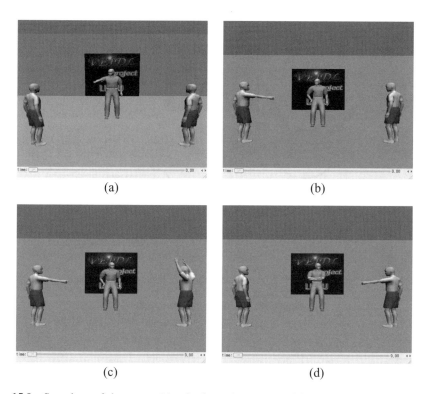

Figure 15.9 Snapshots of the game: (a) a leader points to a participant; (b) a participant points to another participant; (c) a participant puts his/her hands up; (d) a participant becomes a leader and the next turn starts. (Reproduced by permission of Technische Universität Graz)

- Avatars can properly represent predefined actions corresponding to participants' behavior. When participants do not make any predefined actions, it can present only actions that do not indicate participants' intentions (i.e., meaningless actions). Therefore the avatar's behavior seems natural.
- When a participant acts *finger pointing* while its avatar acts *arms folding*, the avatar immediately finishes folding arms and starts finger pointing.
- Delays of representation felt by participants seem to be similar to those of conventional techniques.
- Participants are able to easily understand where the avatars look and point.

15.5.2 Virtual Environment for Immersive Distributed Learning

We are also developing a system called Virtual Environment for Immersive Distributed Learning (VEIDL), which is a prototype system for evaluating the RHP. VEIDL is a virtual classroom environment where geographically dispersed people can attend through avatars of teachers and students, as shown in Figure 15.10.

With VEIDL, every participant has his/her own microphone and cameras as input devices, which acquire verbal and nonverbal information from him/her, and a display unit and speakers as output devices, which present a scene of the classroom generated by a computer. Acquired information from each site is transmitted via a network. Each participant can see a computer-generated scene of the classroom from

Figure 15.10   The concept of VEIDL

the viewpoint of his/her avatar, and, of course, can also see the other participants' behaviors, which are generated via avatars according to the transmitted information. This makes participants able to interact each other through the virtual environment. The advantage of dealing with a virtual classroom, as a prototype of the RHP, is that it is easy to decide which information (or symbol) should be transmitted, since the objective of interaction in a classroom is clear. Figure 15.11 shows snapshots of VEIDL, which is applied to an English conversation class. In this class, a simple conversation practice is done under a teacher's instruction.

Detailed evaluation of this system is under way, but we have had the following encouraging feedback from users:

- Intuitive and easy interaction has been realized, because commonly used actions can be used as the input to the system, and because no special control is required.
- Spatial relationships among the participants are clear and, then, it is easy to understand to whom the speaker is speaking, and at whom the speaker is pointing.
- Thanks to HMD or view-controllable display, a user can easily get a view which he/she wants to see in a virtual space.

On the other hand:

- Sometimes delay between presentation of a speech sound and corresponding body action becomes large, which causes an unnatural feeling.
- When a user wears the HMD for a long time, this can cause fatigue.

From this feedback, it can be said that we have achieved our prime goals, which were easy/intuitive interaction and consistency of the spatial relation among the participants. However, the system itself is not mature, and we have to improve it to achieve more natural interaction. Especially, synchronization of body action and speech utterance is an important issue. For example, when presentation of a pointing action

S1: Teacher: "Who will perform Mr.A's part?"

S2: Teacher: "Then, you perform Mr.A's part."

S3: Teacher: "You perform Mr.B's part."

S4: Teacher: "You perform Mr.C's part."

S5: Student (Blue): "This is Mr.Green."

S6: Student (Blue): "He is my friend. ···"

S7: Student (Blue): "This is Mr.White. ···"

S8: Students (Green, White): "Nice to meet you !"

Figure 15.11 Interaction example in VEIDL: *left:* virtual classroom; *right:* view from the teacher's avatar. (Reproduced by permission of Technische Universität Graz)

is delayed against its corresponding speech utterance, listeners sometimes have difficulty understanding what is being pointed out.

Current asynchrony between speech and body action is mainly caused by independent execution of the speech process (acquisition, transmission, and presentation) and the vision process (motion sensing, action recognition, transmission, and avatar generation). Particularly, the vision process takes longer than the speech process, and the computation time for the vision process varies depending on the action. This sometimes makes the delays of body actions apparent to a receiver. To technically solve the problem, we have to introduce an explicit mechanism to synchronize the two processes. We should recognize synchronization points in a speech sequence and a body action sequence, and when the avatar is presented in a virtual space the body action of the avatar should be generated according to synchronization points attached to the received speech. This synchronization is quite important especially when nuance in communication should be caught, which is left for future research. Also, the influence of the delay on communication should be thoroughly investigated.

15.6 Conclusions

In this chapter, we have proposed Real-time Human Proxy (RHP) for avatar-based communication systems, which virtualizes a human in the real world in real time, and which makes the virtualized human behave as if he/she were present. RHP has been designed to solve the problems of ordinary video-stream-based communication, which are the lack of spatial consistency, restriction of the number of participants, and privacy issues. Especially, we have used vision-based techniques to observe human actions in order to simplify observation of human actions, and have used a symbolization scheme to make the communication system flexible and robust. Using action symbolization, the system can represent human actions in various avatars quite naturally.

To verify the effectiveness of RHP, we have developed and evaluated an interactive game system. We have developed VEIDL, a virtual environment for immersive distributed learning, on RHP. Using a simple scenario of teaching, it is effective for interactive communication among participants due to the spatial consistency.

The RHP is not yet mature, and we are currently refining each component. We think the following are important issues:

- Improving synchrony between speech utterances and body actions should be investigated. We should introduce a mechanism to synchronize the modalities based on synchronization point recognition of sequences of speech and body actions.
- The Behavior model and the Motion model should be improved to make the avatar's behavior more natural. Especially, to virtually hide the system delay, a proactive or autonomous mechanism of action generation is quite important.
- Evaluation of the RHP using a larger environment is necessary, using larger numbers of participants and more complex communication scenarios.

The key issue is how to represent the meaning of motion referring to observed motion data, or how to symbolize the motion meaning. Particularly, the problem is what kind of motions should be symbolized. Of course, one method is to decide the symbols according to observation by ourselves, and when developing the prototype described in section 15.5.2 we have heuristically defined the action symbols based on observation of classroom situations. However, it requires much time to examine a large amount of accumulated communication data. Currently, instead, we are trying to automatically decide the symbols to be recognized referring to accumulated communication data (i.e., motion data) where we suppose frequently occurring motion patterns (motion motifs) convey meaningful information (Araki *et al.* 2006).

References

Araki Y., Arita D., Taniguchi R., Uchida S., Kurazume R. and Hasegawa T. (2006) Construction of symbolic representation from human motion information. In *Proceedings of International Conference on Knowledge-Based Intelligent Information and Engineering Systems*, pp. 212–219.

Arita D., Yoshimatsu H., Hayama D., Kunita M. and Taniguchi R. (2004) Real-time human proxy: an avatar-based interaction system. *CD-Rom Proceedings of International Conference on Multimedia and Expo.*

Bruce M.B. and Tinsley A.G. (1995) Multi-level direction of autonomous creatures for real-time virtual environments. In *Proceedings of 22nd Annual Conference on Computer Graphics and Interactive Techniques* (SIGGRAPH 95), pp. 47–54.

Date N., Yoshimoto H., Arita D. and Taniguchi R. (2004) Real-time human motion sensing based on vision-based inverse kinematics for interactive applicationson. In *Proceedings of International Conference on Pattern Recognition*, vol. 3, pp. 318–321.

Deutscher J., Davison A. and Reid I. (2001) Automatic partitioning of high dimensional search spaces associated with articulated body motion capture. In *Proceedings of Computer Vision and Pattern Recognition*, vol. 2, pp. 669–676.

Etoh M. and Shirai Y. (1993) Segmentation and 2d motion estimation by region fragments. In *Proceedings of International Conference on Computer Vision*, pp. 192–199.

Jeffrey P. and McGrath A. (2000) Sharing serendipity in the workplace. In *Proceedings of International Conference on Collaborative Virtual Environments*, pp. 173–179.

Murase H. and Nayar S. (1995) Visual learning and recognition of 3-d objects from appearance. *Internatonal Journal of Computer Vision* **14** (1), 5–24.

Nene S. and Nayar S. (1997) A simple algorithm for nearest-neighbor search in high dimensions. *IEEE Trans. on Pattern Analysis and Machine Intelligence* **19**, 989–1003.

Nunomaki T., Yonemoto S., Arita D., Taniguchi R. and Tsuruta N. (2000) Multi-part non-rigid object tracking based on time model-space gradients. In *Proceedings of International Workshop on Articulated Motion and Deformable Objects*, pp. 72–82.

Roussos M., Johnson A.E., Leigh J., Vasilakis C.A., Barnes C.R. and Moher T.G. (1997) Nice: combining constructionism, narrative and collaboration in a virtual learning environment. *ACM SIGGRAPH Computer Graphics* **31** (3), 62–63.

Takahashi K., Sakaguchi T. and Ohya J. (2000) Remarks on a real-time 3d human body posture estimation method using trinocular images. In *Proceedings of International Conference on Pattern Recognition*, vol. 4, pp. 693–697.

Wang L.C.T. and Chen C.C. (1991) A combined optimization method for solving the inverse kinematics problem of mechanical manipulators. *IEEE Trans. on Robotics and Applications* **7**, 489–499.

Wren C., Azarbayejani A., Darrell T. and Pentland A. (1997) Pfinder: real-time tracking of the human body. *IEEE Trans. on Pattern Analysis and Machine Intelligence* **19**, 780–785.

Yao P., Evans G. and Calway A. (2001) Using affine correspondence to estimate 3-d facial pose. In *Proceedings of International Conference on Image Processing*, vol. 3, pp. 919–922.

16

Lecture Archiving System

Satoshi Nishiguchi, Koh Kakusho, and Michihiko Minoh

16.1 Introduction

As opposed to a *speech*, in which one person's oral presentation is directed to an audience of one or more individuals, in a *conversation*, two or more people take turns in speaking about a particular subject. The conversation itself, as well as the memory of its content, is a very powerful source of learning, not only about the topic of interest but also about other persons as well as about ourselves. A conversation is driven by the interest in, and knowledge of, the topic that makes its object, of those participating to the conversation.

The classroom situation can be thought of as a very special type of verbal communication, moving between a speech and conversation. As in a speech, often this communication is directed by one person (the teacher) to a group of individuals (the students). As in a conversation, those participating in a classroom share (hopefully) a strong interest in the topic of the communication, although they have different degrees of knowledge of it. It can be thought that a successful class is that in which successful conversations are conducted, in which the participants taking turns in speaking establish mutually interesting, previously unknown connections.

The motivation for the research presented here comes from many directions. Going beyond mere recording of lecture contents, the ability to record and display the dynamics of the classroom situation, including the communication (conversations) between teacher and students, requires an inconspicuous multimodal recording system, capable of capturing information exchanged between participants via verbal and nonverbal expressions. The importance of recording conversation goes beyond producing learning material: recorded conversations can also be useful as research material for investigating human conversational behavior.

Moreover, taking into account the stated research purpose of conversational informatics as "to investigate human conversational behavior and to design conversational artifacts that can interact with people in a human conversational style", recording human conversation should also contribute to the field conversational informatics.

A conversation recording system must be unnoticeable to the participants, lest it should otherwise affect their behavior. This is also the reason why *disappearing or invisible computing* is becoming dominant in the research of human behavior.

To emphasize that multimodal recording of conversation proceeds such that participants are unaware of the recording process, we introduce the notion of *environmental media.*

A small and well-known example illustrates our ideas as follows. In the university set-up a classroom situation contains lectures as well as conversation between a teacher and students, or between students. Although our system records the class situation, including lectures and conversations, in the interest of simplicity, we call it a *lecture archiving system* (rather than classroom situation archiving system).

Our system records various kinds of expressions used for conversation (verbal and nonverbal expressions of the classroom participants) together with teaching materials including description of a whiteboard and images of slides, by recognizing dynamic situations in the classroom. The dynamic situations are defined based on the status of people who participate in conversation in the classroom and are used for remote-control capturing cameras installed in the classroom.

16.2 Concept of Environmental Media

Figure 16.1(a) illustrates basic face-to-face communication between humans (Minoh and Nishiguchi 2003). This is the origin of human communication.

In recent years, in addition to computing and transmitting data, computers play also the role of *information media* supporting communication, mediating information transfer between humans. A lot of research aimed at improving and supporting human communication via computers has been proposed. Figure 16.1(b) illustrates a typical *computer-mediated communication*. In this, humans directly interact with computers in order to communicate indirectly with other humans. Each of these humans exists in their own environment that consists of the contexts of communication. These environments are not shared between humans.

Even when the communication is mediated by computers, human behavior should not change. In order to realize natural communication similar to face-to-face communication, we need to realize more natural interaction between humans by making computers transparent for humans. This approach is

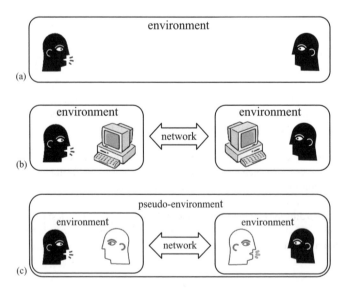

Figure 16.1   Three types of communication between humans: (a) between humans in the real world; (b) between humans, and machines; (c) between humans and the environment

Talk   Speech   Discussion

Figure 16.2   Three types of human conversation

similar to several projects like invisible or disappearing computers. In addition, we need to make possible sharing of individual environments in computer-mediated communication. These two requirements are also key features to be realized by environmental media. Figure 16.1(c) illustrates the computer-mediated communication that fulfills these requirements.

Generally, human conversation is classified into three types as shown in Figure 16.2 (Minoh and Nishiguchi 2003). These types affect the design of environmental media. In order to mediate information exchange in face-to-face conversation, the environmental media need to acquire complete and accurate verbal and nonverbal expressions of the participants and present them to the viewer. In order to realize these functions, the system needs at first to be aware of the dynamic communication situations; that is, to identify the kinds of expressions present during conversation. Therefore the environmental media must satisfy the following requirements:

1. Define dynamic situations (for controlling cameras) based on the natural status of participants.
2. Recognize the dynamic situations.
3. Relate recognized dynamic situations to the multiple cameras in order to capture suitable videos.
4. Create a database of videos indexed by the corresponding dynamic situations.

Hence, the important characteristics of the environmental media are:

• The ability to capture and record without interfering with the human activity.
• The ability to recognize the dynamic situations.
• Automatic decision of videotaping action based on the recognition of dynamic situations.
• Automatic generation of indexes for retrieval.

These characteristics are illustrated in Figure 16.3.

Our approach to the realization of environmental media is illustrated in a case study of university lectures. By contrast to mere videotaping of lectures, the *lecture archiving system* records the process of communication between a lecturer and students. The archived classroom communication as environmental media supports learners (who access this archive) to understand the classroom situation, enabling them to experience a state similar to that of being present in the classroom. We discuss definition of the dynamic situations of the classroom and how to recognize them. Evaluations of the system are done through experiments.

16.3 Related Works

In the earliest approaches of archiving and publishing lectures, actual lecture materials are provided on the Web to students in order to encourage self-learning. For example, the Open Course Ware Project at

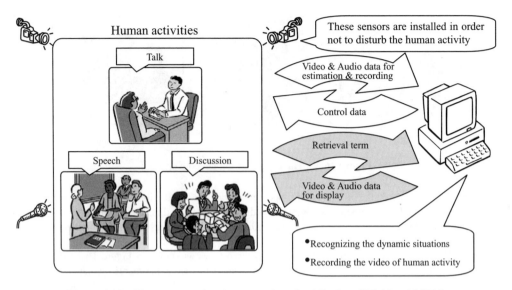

Figure 16.3   The concept of environmental media (Minoh and Nishiguchi 2003)

MIT publishes all of the lecture materials onto the Web.[1] Students accessing this material do not have access to the classroom situation, because the classroom interaction activity is not recorded in the course ware.

In other projects, lecture materials and videos of the lecturer are recorded in order to better convey the contents of teaching. For example, in the Web Lecture Archive Project (Bousdira *et al.* 2001) the video of a lecture and the PowerPoint slides are published together onto the Web. In the Lecture Browser project,[2] the contents, constructed with the videos of lecturers, the PowerPoint slides, and the title of their slides, for hundreds of lectures are recorded. These media are displayed separately on the computer screen. Learners can search the scenes of their interest using the slides as index.

These projects have in common the approach of displaying for learners the video of a lecturer and PowerPoint slides. Therefore, learners cannot experience the process of the communication in the classroom from the video. Moreover, in these projects, video cameras and video switchers are controlled by hand. In these systems, a method of simply videotaping the lecture is not used. Instead, human operators manipulate various devices such as video cameras and PCs in the classroom. The presence of such operators interferes with the classroom situation (lecture and conversation between classroom participants), so a system using human operators violates the environmental media requirements stated in section 16.2. In addition, the actual classroom situation is further altered by heavy postproduction work.

By contrast, the lecture archiving system achieves automatic archiving and releasing of the lectures and thus is an instance of the environmental media concept proposed here. We start from the strong belief, that in order to provide the user of the archive with a full understanding of the classroom activity, communication between a lecturer and students must be recorded. This is our motivation for advocating the automatic lecture archiving system as described in this chapter, and is the source of the major differences between exiting systems and ours.

---

[1] See http://ocw.mit.edu/index.html
[2] See http://bmrc.berkeley.edu/frame/projects/lb/

Table 16.1 Definitions of dynamic situations

Symbol	Dynamic situation
L1	A lecturer remains still and is speaking.
L2	A lecturer is moving and speaking.
L3	A lecturer remains still and is using materials.
S	One of the students is speaking.

16.4 Definition of Dynamic Situations in the Classroom

We start from the observation that the classroom situation can be captured by sensors, in a spatiotemporal feature space, with the actual sensor values referred to as *features*.

Conventional pattern recognition assumes that commonly understood concepts are described in terms of a collection of features. Then a new description in terms of the same features is presented, and the pattern recognition system classifies it (via a pattern matching operation) into one of the concept categories. In contrast, in the lecture archiving system, there is no such concept known in advance. Rather, introducing the purpose of processing, the feature space has to be partitioned accordingly. Each partitioned feature space is called a *dynamic situation* of a classroom. The purpose of processing here is multimodal classroom activity recording and archiving.

For example, speakers in the classroom become important since lectures are considered to be communication between a lecturer and students and among students. Hence, speakers take the main role in defining the dynamic situations. Table 16.1 shows the types of dynamic situations defined in connection with classroom activity. There are two main categories: the dynamic situation where the speaker is a lecturer (L1, L2, and L3), and the dynamic situation where the speaker is one of the students (S). Since we can assume that there is only one speaker in the classroom in general, the dynamic situations listed in Table 16.1 are disjoint and correspond to the symbols in the conventional pattern recognition problem.

When situation L1 is detected, shooting cameras take a lecturer in middle shot or close shot. In situation L2, shooting cameras have to track the lecturer's face. In situation L3, shooting cameras take both a lecturer and the materials to convey the atmosphere of the lecture. When situation S is detected, the system has to find where the speaking student is, and capture him/her by one of the shooting cameras.

16.5 Recognition of Dynamic Situations

The classroom is relatively large compared with targets in the literature of pattern recognition research. Hence, a lot of sensors are necessary to observe the whole classroom. This brings up the additional issue of how to integrate the sensor information from the multiple sensors. This problem is related to the problem of distributed processing and has often been discussed (e.g., Matsuyama 2001).

Since the dynamic situations are defined based on the speaker situations, the sensors that observe the classroom have to detect where the speaker is in the classroom. Three kinds of sensors are used to detect the speaker location: image sensors, voice sensors, and position sensors.

The most suitable sensor to detect a speaker is a voice sensor though the accuracy of voice sensors is not that good. In the lecture set-up, the only person who can walk and speak most of the time is the lecturer. Therefore, image sensors which can detect a moving object are used to detect the lecturer. If, in addition, the lecturer has a position sensor (e.g., on his/her shoulder) this sensor can also be used to detect the lecturer.

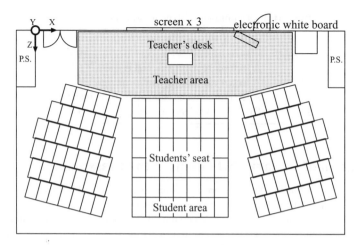

Figure 16.4   Partition of the classroom. (Reproduced by permission of © IEICE)

In order to further assist in accurate detection of the speaker (lecturer or student) the classroom is divided into two main areas: the lecturer area, and student area, as illustrated in Figure 16.4 (Nishiguchi *et al.* 2005b). If a speaker is detected in the lecturer area, the system assumes that the speaker is the lecturer. Otherwise, a speaker is a student. A position sensor and an image sensor always measure the location of the lecturer based on this assumption.

Even under this assumption, the main problem here is how to measure the speaker location robustly in a relatively wide area, the classroom.

16.6 Speaker Detection by Multimodal Sensors

This section describes our method for detecting a speaker in the classroom using multimodal sensors (Nishiguchi *et al.* 2004b).

16.6.1 Image Sensor

In this project an observing camera is used as an image sensor. Given the large size of the classroom, and the requirement of the stereo vision method that areas captured by two cameras must overlap, several cameras are used, although not as stereo vision. The image sensor must detect the location of the lecturer and the location of students in the classroom.

To identify the location of the lecturer, several clues are used as follows. If the feet of the lecturer are seen from the camera, then we can assume the floor as the horizontal plane under the feet. If the feet are not seen from the camera, then the height of a lecturer is assumed, based on some standard height (e.g., 160 cm). Once the feet or head are assumed, based on the lens center, we decide the view line as shown in Figure 16.5 (Nishiguchi *et al.* 2004a). The point where the view line intersects the horizontal plane located either at 0 cm or at the 160 cm height from the estimated floor plane, is estimated as the detected position of the lecturer.

This method gives an approximate estimation of the lecturer position. However, it can be applied to a large area of the classroom in which the position is detected. To improve accuracy in estimating the location of the lecturer, the same approach is used in conjunction with five observation cameras which are

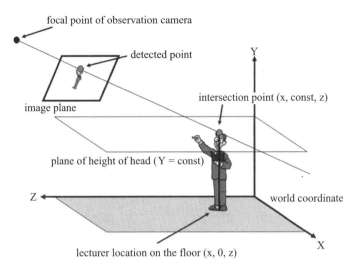

Figure 16.5   How to detect a lecturer location. (Reproduced by permission of © IEEJ)

installed in the classroom as shown later in Figure 16.10. The final location of the lecturer is calculated as the average of the locations estimated by these cameras.

In order to estimate the focus on students robustly and efficiently, a fisheye lens camera is installed in the center of the ceiling of the classroom. By observing the student area from above, the system is free from occlusion problems.

The seats are assumed to be arranged in a graduated auditorium style (see Figure 16.4) which enables the audience to clearly view the lecturer. Figure 16.6(a) shows the image captured by the fisheye lens camera, while Figure 16.6(b) shows inferred candidates, called *cells*, for individual student locations. In the following discussion we assume that the seats in the student area are identified in a rectangular coordinate system, in a plane parallel with the floor of the classroom. A seat is represented as occupied or not, by a binary variable that takes on the values *true* or *false*. The value of this variable is set according to a procedure based on the background differences and subtraction between continuous frames captured by the fisheye lens camera for a given cell. If the difference exceeds a threshold, the seat corresponding to the cell is assumed occupied; otherwise the seat is marked as empty.

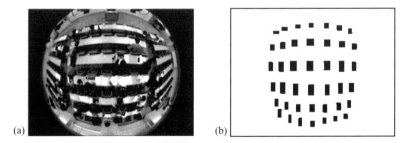

Figure 16.6   Student area captured by the fisheye lens camera: (a) example of image of student area; (b) cells in student area. (Reproduced by permission of © IEEJ)

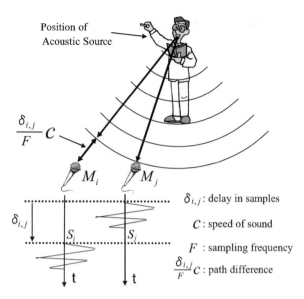

Figure 16.7   Principle of voice position detection

16.6.2 Voice Sensor

In order to estimate the position of a speaker, several microphones are mounted on the ceiling of the classroom. The position and direction of the sound can be estimated by several methods, including the MUSIC (Multiple Signal Classification) method (Pillai 1989) and the CSP (Cross-power Spectrum Phase analysis) method (Omologo and Svaizer 1997). For this project, we selected the CSP method as this requires only two microphones and is more computationally efficient.

The principle of the CSP method is illustrated in Figure 16.7. Two microphones, M_i and M_j, capture acoustic waves, S_i and S_j, from the same acoustic source by the sampling frequency F. The delay in samples calculated using the CSP method is $\delta_{i,j}$. The speed of sound is designated by c. The path difference of the same voice (DOA: delay of arrival) to the two microphones is calculated by $\frac{\delta_{i,j}}{F} c$. The hyperbolic curve with the two microphones as foci can be drawn and it is estimated that the source of sound (the mouth of the speaker) is situated on the curve. Since we have eight microphones in the classroom, there are' $(8!/6!2!) = 28$ pairings of the microphones.

Since a student's position is fixed at their seat, and the size of it is at least 50 cm, the accuracy of estimating the position of the student who is speaking is the seat size. To make the estimation robust to noise, the Hough transform is introduced, and hence this method can be used for real lectures. On the other hand, the accuracy of the position for a lecturer depends on the zoom factor of the shooting camera, which must be about 10 cm.

16.6.3 Position Sensor

A position sensor used here consists of several transceivers and markers. The measurement is obtained using ultrasonic and infrared signals. Under the ceiling, four X-shaped receivers are installed. A receiver transmits an infrared signal to order the markers to issue an ultrasonic signal. The markers are attached to the lecturer as shown in Figure 16.8 and transmit the ultrasonic signal. If more than three receivers detect the ultrasonic signal, the position of the markers is estimated.

Figure 16.8   Attachments of position sensor

The position sensor described above is suitable for this application since the markers are wireless and small, and therefore unobtrusive, thus satisfying one of the requirements of the environmental media.

16.6.4 Sensor Fusion

It is clear now that the lecturer's position can be estimated independently the three types of sensors described in the previous section. These sensors differ in accuracy: the position sensor is the best, followed by the voice sensor, and image sensor. Hence, when the position sensor detects the position of the lecturer, this is used as the real position. However, often, due to occlusion problems, the position sensor may fail to detect the lecturer's location. In this case, we have to use the position estimated by the image sensors. Usually, several cameras can detect the position of a lecturer. After removing outliers from the data, the mean of detected positions is estimated as the real position.

To detect a student who is speaking, only the image sensor and voice sensor are used (the student does not have the markers necessary for the position sensor). These two sensors collaborate with each other. Since the voice sensor detects students using the Hough-like method described above, the target seats are those in which students sit. If, based on the information estimated by the image sensor, the speaker seat estimated by the voice sensor is not occupied, the neighboring seat is considered as the next candidate for the student location. Thus the system solves contradicting estimates of the two types of sensors.

16.7 Experimental Results

This section illustrates the approach described above with various experimental results. The experiments were carried out in a classroom at the Academic Center for Computing and Media Studies (ACCMS), of Kyoto University (Figure 16.9).

16.7.1 Equipment of Our Lecture Archiving System

The coordinate system of the classroom is shown in Figure 16.10 (Nishiguchi *et al.* 2005b). The measurements of the classroom are about 15.5 m × 9.6 m × 3.2 m. Five cameras are installed near the ceiling of the classroom for measuring the lecturer position. One fisheye lens camera (with vertical axis of the lens) is mounted on the ceiling of the classroom. The lecturer and students are videotaped by eight cameras in addition to the six cameras used as image sensors. Eight microphones are installed for detecting the speaker position among students, and an ultrasonic position sensor system is employed for detecting the lecturer's position.

The specifications of sensors used for capturing the information of human activity in the classroom and computers used for archiving lectures are shown in Tables 16.2 and 16.3 (Nishiguchi *et al.* 2005b). These

Figure 16.9   A classroom at ACCMS

sensors are installed on the wall or ceiling of the classroom in order to minimize any interference with the usual classroom activities. As shown in Table 16.3, we do not use special or expensive computers. Therefore, our archiving system can be installed in the other lecture rooms at low cost. The Attachment diagram, describing the relationship between sensors and computers, is shown in Figure 16.11. All of the computers in Figure 16.11 are connected to a 100BaseT network. All of the sensors are connected to one of the computers in order to be controlled based on the estimation of the dynamic situation in the classroom.

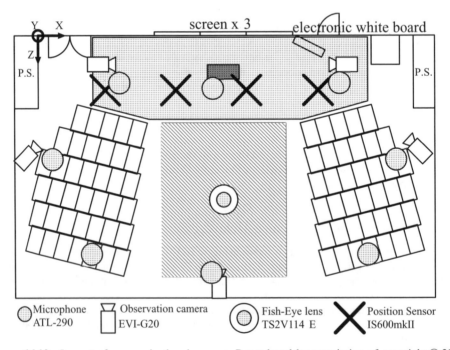

Figure 16.10   Layout of sensors in the classroom. Reproduced by permission of copyright © 2006 IEICE

Table 16.2 Use of sensors. (Reproduced by permission of © IEICE)

Model	Specification
IS-600 markII	Position sensor, serial control (9600bit/s)
EVI-G20	Serial control (VISCA)
TS2V114E	Fisheye lens
VK-C555R	C-mount camera
SoftBoard	Serial control (9600bit/s)
ATL-290	Microphone
PowerDAQ	Synchronous multiple audio data acquisition

16.7.2 Representation of Archived Lectures

Learners can browse the archived lectures using a Web browser. An example of the screen that is presented to learners is shown in Figure 16.12 (Nishiguchi *et al.* 2005a). The layout of the screen is described in HTML. The whole screen consists of three panes.

The top pane of the screen shows the date, the subject, and the name of the lecturer. The left pane is the main screen of the archived lectures. The lecture video, slides, and writing on the whiteboard are displayed on the left pane in SMIL. The upper left part of the left pane shows the lecture video that is automatically shot by our archiving system. The video format is Realvideo. The lower left part of the pane shows a slide of PowerPoint presented by the lecturer. The right part of the pane shows the writing on the electronic whiteboard by the lecturer.

The right pane is used for searching the scene of the lecture. The thumbnail slides in the pane are listed in the order displayed on the projector screen by the lecturer. Learners can select a scene of the lecture by clicking one of the thumbnail slides or by sliding the slider bar of the Realvideo player.

16.7.3 Evaluation of Recognizing Dynamic Situations

First, we compare the accuracy of the lecturer position estimated by two kinds of sensors, a position sensor and an image sensor. The result is shown in Table 16.4. From this table, it can be seen that the position sensor is about ten times more accurate compared with the image sensors everywhere in the classroom.

Second, the percentage of detecting the speaking person in the seat in the auditorium is calculated by repeating the detection experiments. The precision and recall ratios are listed in Table 16.5. In our

Table 16.3 Use of computers. (Reproduced by permission of © IEICE)

Use	Number	CPU	Clock
Controlling camera and capturing video	6	P-III	1.0 GHz
Recording selected video	1	P-III	1.0 GHz
Controlling whiteboard	1	P-4	2.53 GHz
Controlling PowerDAQ	1	P-4	2.53 GHz
PowerPoint presentation	1	P-III	800 MHz
Position estimation	1	P-III	1.0 GHz
Total	11		

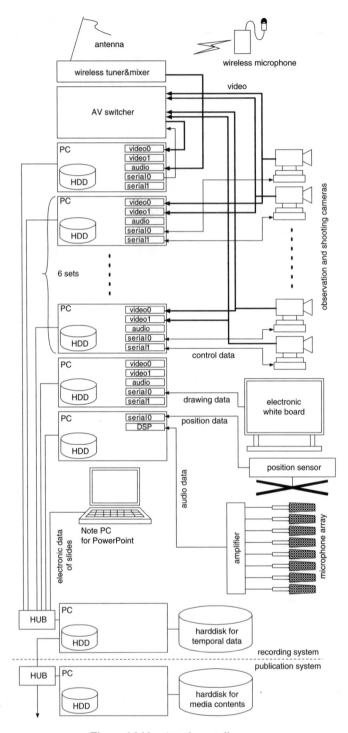

Figure 16.11   Attachment diagram

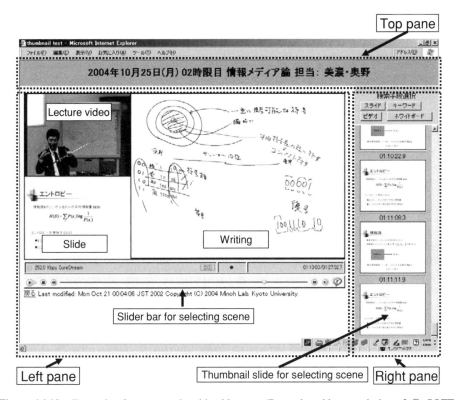

Figure 16.12 Example of represented archived lecture. (Reproduced by permission of © SOFT)

experiment, as the number of microphone pairs increases, the ratios become better. Hence, we use five microphone pairs.

We applied this system to two lectures. The dynamic situations were correctly estimated for 88.8% of the time of one lecture and 77.1% of the time of the other. Based on the recognition of the dynamic situations, the automatic shooting camera system can take the lecture video for lecture archiving (Kameda *et al.* 1999, 2000).

Table 16.4 Comparison of sensor errors

Position (m)	Position sensor errors (m)			Image sensor errors (m)		
	Mean	Var.	Max.	Mean	Var.	Max.
$X = 2.10$	0.078	0.0096	0.65	0.19	0.023	0.55
$X = 5.10$	0.052	0.0035	0.21	0.15	0.016	0.62
$X = 8.10$	0.050	0.0010	0.12	0.21	0.066	1.69
$Z = 0.90$	0.036	0.0015	0.23	0.26	0.042	0.84
$Z = 1.90$	0.043	0.0021	0.23	0.18	0.027	0.78

Table 16.5 Result of voice sensor

Number of row from the front	Number of experiments	Recall ratio (%)	Precision (%)
1	25	71.4	78.1
3	35	94.3	94.3
6	35	97.1	94.4

16.7.4 *Evaluation of the System as a Realization of Environmental Media*

The lecture archiving system consists not only of the automatic video shooting system but also of the teaching material recording system (i.e., electric whiteboard and presentation materials). Using this system, six lectures were archived in the period October 2002 to January 2003.

To evaluate this system, a survey of the students attending all the six classes was conducted. At the same time, we interviewed lecturers who taught these classes. The initial results suggest that class participants, lecturers and students, are aware of the equipment in the classroom and distracted by it to some extent, affecting students' ability to concentrate on the lecture. However, this phenomenon disappeared as the lecture was going on. Lecturers tried to use the equipment more effectively. Some equipment, such as the electric whiteboard, is not used in regular lectures so lecturers needed adjustment to their use. Some lecturers even said that they would try to change the way of teaching to adapt to the equipment in the classroom.

Eleven students using the archived lectures were asked to rank the media from the viewpoint of understanding lectures. Table 16.6 shows that the voice of the lecturer is the most useful for students followed by PowerPoint slides. Overall, each media component is highly ranked by some students. Therefore, all media archived by our system can be useful for various purposes of students.

We showed four lecturers their archived lectures and asked them if the archived lecture was useful or not. Most of them found some scenes that they did not remember during lectures. This result suggests that lectures archived by our system are useful for faculty development.

16.8 Conclusions

The concept of environmental media has been discussed, focusing on a lecture archiving system. The important characteristics of the environmental media were listed in section 16.2. The lecture archiving system we developed satisfies these characteristics.

Table 16.6 Result of student evaluation of media used in the archiving system

Media type	Vote count				
	1st	2nd	3rd	4th	5th
Motion picture	1	2	0	4	4
Audio	6	2	1	1	1
Whiteboard	0	2	4	1	2
PowerPoint	4	4	2	0	0
Index	0	1	3	3	2

We have discussed our method of recognizing dynamic situations of the classroom in order to take the video of the lecture automatically. The dynamic situations, identified as symbols for pattern recognition, can be defined by introducing the purpose of the processing. Here, the purpose is automatic lecture archiving. Hence, the recognition itself becomes relatively easy because the symbols are defined by considering the recognition process.

The initial survey of class participants showed that our system disturbed the lecture to some extent. In general, "observation" or "watching" is not acceptable to people since it implies that a person's activities are watched and evaluated (e.g., as bad or good). Resistance to such observation (similar to spying) is widely spread in our society. Thus, the use of video cameras in the street or inside the building give an uncomfortable feeling to people. Paraphrasing Lincoln's words, we should propose *"the cameras of the people, by the people, for the people"*. In other words, the use of such cameras should be restricted to the people that they observe. Such features should be included in the environmental media of the future.

Our system is one of the concrete implementations of environmental media for conversational informatics. Thus, the universal principles and techniques of environmental media must be explored in accordance with the purpose of conversational informatics.

Acknowledgments

The help of Dr Anca Ralescu of University of Cincinnati in the preparation of this chapter is gratefully acknowledged.

References

Bousdira N., Goldfarb S., Myers E., Neal H.A., Severance C., Storr K.M. and Vitaglione G. (2001) WLAP, the Web Lecture Archive Project: The development of a web-based archive of lectures, tutorials, meeting and events at CERN and at the University of Michigan. CERN-OPEN-2001-066; Geneva: CERN.

Kameda Y., Miyazaki H. and Minoh M. (1999) A live video imaging for multiple users. *Proc. ICMCS '99*, vol. 2, pp. 897–902.

Kameda Y., Ishizuka K. and Minoh M. (2000) A study for distance learning service: TIDE project. In *Proceedings of IEEE International Conference on Multimedia and Expo*, vol. 3, pp. 1237–1240.

Matsuyama T. (2001) Cooperative distributed vision: dynamic integration of visual perception, camera action, and network communication. *Proceedings of the Fourth International Workshop on Cooperative Distributed Vision*, pp. 1–25.

Minoh M. and Nishiguchi S. (2003) Environmental media: in the case of lecture archiving system. *Proceedings of 7th International Conference on Knowledge-Based Intelligent Information and Engineering Systems* (KES2003), LNAI 2774, Springer-Verlag, pp. 1070–1076.

Nishiguchi S., Higashi K., Kameda Y., Kakusho K. and Minoh M. (2004a) Audio and visual information integration for speaker localization in automatic video shooting system. *Transactions C of IEEJ* vol. 124, No. 3, pp. 729–739 (in Japanese).

Nishiguchi S., Kameda Y., Kakusho K. and Minoh M. (2004b) Automatic video recording of lecture's audience with activity analysis and equalization of scale for students observation. *Journal of Advanced Computational Intelligence and Intelligent Informatics* vol. 8, No. 2, pp. 181–189.

Nishiguchi S., Higashi K., Kameda Y., Kakusho K. and Minoh M. (2005a) Automatic lecture archiving system for recording bilateral communication by shooting students attending the lecture. *Transactions of SOFT* vol. 17, No. 5, pp. 587–598 (in Japanese).

Nishiguchi S., Kameda Y., Kakusho K. and Minoh M. (2005b) Automatic lecture archiving system for practical use at universities. *IEICE Transactions on Information and Systems* vol. J88-D-11, No. 3, pp. 530–540 (in Japanese).

Omologo M. and Svaizer P. (1997) Use of the crosspower-spectrum phase in acoustic event position. *IEEE Trans. on Speech and Audio Processing*, pp. 288–292.

Pillai S.U. (1989) *Array Signal Processing*. Springer-Verlag, New York.

Part IV

Conversational Measurement, Analysis, and Modeling

17

A Scientific Approach to Conversational Informatics: Description, Analysis, and Modeling of Human Conversation

Yasuharu Den and Mika Enomoto

17.1 Introduction

In this chapter, we describe a scientifically motivated approach to conversational informatics. The goal of this approach is to understand mechanisms underlying natural human conversation. We present a methodology for describing, analyzing, and modeling natural human conversation as well as the results achieved thus far in our project.

Dialogue research in the areas of informatics has developed in accordance with the development of dialogue data available to the community. In the first generation, researchers used text-based transcripts of dialogues and, often, artificial examples to investigate mainly linguistic phenomena such as plan recognition (Allen 1983), discourse structure (Grosz and Sidner 1986), and cohesion and anaphora (Grosz et al. 1995; Walker et al. 1994). In the second generation, they utilized recordings of spoken dialogues and focused on such issues as speech disfluency (Shriberg 1994) and the effects of speech prosody on turn-taking (Koiso et al. 1998) and dialogue acts (Shriberg et al. 1998). In the third generation, they utilized recordings of visual information conveyed in dialogues to examine communicative functions of gestures, eye gaze, and facial expressions (Chapter 8; Nakano et al. 2003; Whittaker 2003). And, in the fourth generation, they started making use of the sensed data of body and eye movements to precisely analyze and model human motions during conversation (Vertegaal et al. 2001).

Our project belongs to the third generation research, where nonverbal behavior in conversation is studied by means of recording and inspection of video data. What is unique to our project is the focus on multiparty, rather than dyadic, conversation. Dialogue studies in informatics have focused mainly on two-party dialogue, whether task-oriented or casual, due to technical restrictions on high-quality and precise recording as well as simplification of the issues to be dealt with by dialogue systems on computers.

Conversational Informatics: An Engineering Approach   Edited by Toyoaki Nishida
© 2007 John Wiley & Sons, Ltd

Our daily interactions, however, usually involve three or more participants. For instance, the family in which we grow up usually consists of more than two members, and in school or in the workplace we usually gather in a group of more than two members. Interactions with a single person are rather rare, with the exception of telephone conversation. This reason, among others, is why such research areas as conversation analysis, interactional sociolinguistics, and cultural anthropology – which are interested in ordinary human activities – have included multiparty conversation in the range of their research targets. This trend has started spreading to dialogue studies in informatics (e.g., Carletta *et al.* 2006; McNeill 2006; Shriberg *et al.* 2004).

Another feature of our project is the use of casual conversation rather than task-oriented dialogue. This is motivated by our interests in ordinary human activities, but it is also meaningful for possible engineering applications in multiparty dialogue studies. One emerging area of application in multiparty dialogue studies is the recognition and the summarization of natural meetings. Based on an evaluation of the interaction style in their meeting data, Shriberg *et al.* (2004) stated that the dialogue in meetings is less like human–human or human–machine task-oriented dialogue and more like human–human casual conversation. This fact implies the significance of the study of casual conversation not only for scientific purposes but also for engineering purposes.

Conversation is a complicated activity, involving comprehension and production of language and gesture, understanding of others' minds, management of interaction and the flow of topics, expansion and modification of personal and public knowledge, and so on. Since it is extremely difficult to fully understand these details, we start by investigating rather limited, but significant, behavior typical of human conversation. The types of behavior we study include eye gaze, head nods, hand movements, and laugh. These actions seem to be produced readily in response to the physical and mental context in which conversation participants are embedded, rather than built up deliberately. Nonetheless, they are influential in the management of conversation. Eye gaze is particularly important in the management of turn-taking as well as conversational exchange, as we show in this report.

In section 17.2, we describe our data collection methodology. We record three-party conversations in Japanese produced by 12 triads, and alongside speech transcriptions, we annotate them with nonverbal actions, creating a multimodal corpus of multiparty conversations. In section 17.3, based on a part of our data, we analyze the distributions of hearers' actions during a speaker's turn, comparing them between different statuses of hearers in terms of next speakership. We further analyze the relationship between eye gaze of a speaker and of hearers to precisely understand how eye gaze is used in the management of interaction in three-party conversation. In section 17.4, we discuss how our findings can be linked to the theory of conversation. We propose a model of conversation participants upon which our line of research is based, and show its relation to two influential theories of communication; that is, relevance theory and conversation analysis.

17.2 Recording and Description of Multimodal Three-party Conversations

17.2.1 Why Study Three-party Conversation?

The first step toward a scientific study of human conversation is recording and description of naturally occurring human–human conversations. Because of our interests in ordinary human interactions, which often involve more than two participants facing each other, we recorded three-party conversations in a face-to-face setting. Our choice of *three-party* from among the various possible numbers of participants was motivated by the following reasons.

First, three-party conversation is the first occasion in which different kinds of listeners play a role in interaction. When more than one listener is concerned with an utterance of the current speaker, they can be distinguished by whether or not they are expected to respond to the current utterance (Clark 1996;

Goffman 1981). The listener who is addressed by the current speaker and, hence, expected to respond to the current turn is called the *addressee*. The other listener, however, is expected to comprehend the content of the current turn, but is not necessarily expected to respond. This listener is called a *side-participant*. The presence of side-participants is usual in ordinary interactions, and three-party conversation serves as the simplest instance for investigating the variety of listeners' roles in ordinary interactions.[1]

Second, three-party conversation is simple enough to continue using a single turn-taking system. Conversations with more than three participants can occasionally be divided into several subgroups, each of which employs its own management of topic and turn-taking (Sacks *et al.* 1974). Such division, or *schisming*, of a conversation never occurs in three-party conversation. In this sense, three-party conversation is suitable for focusing on several fundamental issues in multiparty interaction without being influenced by more complicated issues.

Third, three, and not as many as five or six, participants are tractable for precise and high-quality recording of multimodal multiparty conversation. For fine-grained description and analysis of conversations, we need both high-quality audio recordings, where each participant's speech is recorded on a separate track, and high-quality video recordings, where each participant's picture is recorded at a reasonable resolution. For a large number of participants, such as more than four or five, it is difficult to obtain such high-quality recordings without specialized equipment and environment.

For these reasons, we decided to start our study with three-party conversation. In the following subsections, we describe how we collected our set of multimodal three-party conversations.

17.2.2 Participants

Twelve triads, consisting of 36 different individuals, participated in the recording. They were all native speakers of Japanese, and they ranged in age from 18 to 33 years. The participants of each triad were friends on campus and of the same gender. A half of the triads were male groups and the other half female groups.

17.2.3 Topics of Conversation

Each triad produced three conversations, each lasting approximately 10 minutes. The initial topic of each conversation was determined by a throw of a die whose faces showed six different topics, such as an embarrassing, surprising, or exasperating experience. We instructed the participants not to feel restricted to the initial topic, and the actual topics brought into the conversations were diverse.

17.2.4 Conversational Settings and Recording Environment

The participants sat in a quiet room, facing each other, in an equilateral-triangular form (Figure 17.1). They stayed about 1.5 m distant from each other, and a table was placed between them. On the table, soft drinks were served. The room was a large lounge for social use by graduate students. This conversational setting allowed the participants to converse as naturally as in an ordinary situation and promoted their relaxation, which is in contrast to the more widely used settings for high-quality recording where participants are confined in a small soundproof chamber.

---

[1] In general, there can be other sorts of listeners, referred to as *unratified participants* (Goffman 1981) or *overhearers* (Clark 1996), who are outside the conversation and not expected to contribute to it. These kinds of listeners do not exist in our setting of three-party conversation.

Figure 17.1   Conversational settings

In order to improve the audio recording conditions, a carpet of 3 m × 3 m was laid on the floor and soundproof curtains were hung behind the participants (Figure 17.2). Three digital video camera recorders (Sony DCR-VX2000) were arranged to record the bust shots of the participants (Cams. A, B, and C for participant A, B, and C, respectively). The curtains behind the participants also served as a monotone background for the pictures. Another digital video camera recorder (Cam. D) was used to record the whole view of the conversation as in Figure 17.1. The participants wore headset microphones (Sennheiser MKE104), and their voices were recorded on separate tracks of a digital hard disk recorder (TASCAM MX-2424). Due to the face-to-face setting, crosstalk sometimes occurred on the tracks, but

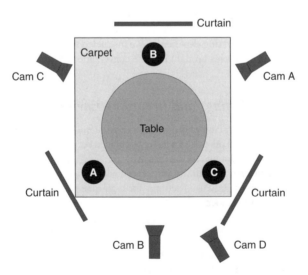

Figure 17.2   Recording environment

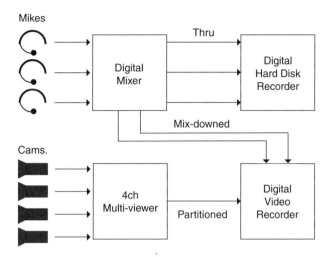

Figure 17.3   Recording equipment

did not severely reduce the recording quality. A four-partitioned video through a multi-viewer (For.A MV-40E), composed of the pictures from the four cameras, together with mix-downed audio through a digital mixer (TASCAM DM-24), were also recorded on a digital video recorder (Sony DSR-1500A) as a reference for the entire data. See Figure 17.3 for the equipment.

17.2.5 Procedures for Synchronization

We uploaded, in AVI format at 29.97 fps (frames per second), the three bust shots and the whole view recorded on the four digital video camera recorders and the four-partitioned video with the mix-downed audio recorded on the digital video recorder. We also uploaded, in WAV format at 48 kHz, the three-track audio sound recorded on the digital hard disk recorder. Since no real-time synchronization was made at the time of recording, we employed the following procedures to synchronize the videos and the audios.

First, we identified the frame in the four-partitioned video in which an electronic flash cuing the start of a conversation was recorded.[2] This video frame, and the corresponding audio frame, were used as the reference for the entire data. Second, we identified the corresponding frames in the four videos (i.e., the three bust shots and the whole view), using the starting flash as a clue. Third, we synchronized the three-track audio with the reference audio by maximizing the correlation between the waveforms extracted from an initial portion of the conversation. The extracted portion was a short utterance by one participant announcing the initial topic of the conversation (e.g., "The topic is a love affair") which we requested her/him to utter immediately after the starting flash. Using these procedures, the videos and the audio could be synchronized with a lag of maximally one frame, or 1/29.97 second.

All the video and audio data were cut off 9.5 minutes after the reference frame regardless of progress of the conversation. Thus, the amount of the entire data set, consisting of 36 conversations, was about 5 hours and 40 minutes.

---

[2] The flash for the starting signal was placed on the table so that it could be captured by all the cameras.

17.2.6 Transcription and Annotation

We transcribed all the audio data in standard Japanese orthography. Utterances were segmented by a perceivable pause, and the starting and the ending times of each segmented speech were identified. Each line in a transcript consists of the starting and the ending times, the speaker name (A, B, or C), and the content of the utterance, following the LDC CallHome format, as in Figure 17.4.

In the transcription of utterances, several special tags were employed. " (W *Pron; Word*) " was used for a wrong or unusual pronunciation; *Word* is an intended word written in standard orthography and *Pron* is the literal transcription of the actual pronunciation. " (D *Pron*) " was used for a phonetic sequence that cannot be identified as a word. When phonation itself was unintelligible, "<?>" was used instead. " : " indicates an audible lengthening of a word final syllable, and "<laugh>" a laugh occurring independently of a linguistic content. Laughs while speaking were not indicated. These tags were borrowed from the tag set of the *Corpus of Spontaneous Japanese* (Maekawa *et al.* 2000) and are useful for precise analysis of linguistic and interactional phenomena in conversation.

For six conversations produced by two triads, we performed more detailed annotations concerning both verbal and nonverbal behavior. This subset will be used in the analyses presented in the next section. As an annotation of verbal behavior, we re-segmented the utterances in terms of turn-constructional units (TCUs) (Sacks *et al.* 1974), which are widely used in the literature of conversation analysis. A TCU is a minimal unit by which a turn in conversation can be constructed. At the end of a TCU, a speaker change may, but need not, occur. A TCU may be a sentence, a clause, a phrase, or even a single word depending on the context. TCUs are said to be intuitively recognizable by ordinary speakers of the language, but for the current purpose we needed a more strict definition. Since TCUs are most influenced by syntax (Ford and Thompson 1996; Sacks *et al.* 1974), we identify a TCU according to the following criteria, with reference to the clause unit manual of the *Corpus of Spontaneous Japanese* (Takanashi *et al.* 2004). Place a TCU boundary when:

- a clause ends with the conclusive or imperative form of a verb, an adjective, or an auxiliary verb, or with an interjection or a final particle
- a clause ends with a coordinate-conjunctive particle *ga, kedo(mo), keredo(mo)*, or *si*
- a clause ends with a rising intonation or is followed by a pause longer than 300 ms
- a postposed constituent appears after a TCU having been identified by either of the above criteria
- an utterance is disrupted spontaneously or by an interruption from another participant.[3]

When a fragmental clause lacking a predicate served an equivalent discourse function as a full clause, it was subject to the above criteria. We also identified backchanneling utterances, or *aizuchi*, such as *hai* and *un*, which are considered not to constitute a turn.

As an annotation of nonverbal behavior, we marked every occurrence of several types of nonverbal behavior: eye gaze at participants, head nods, and hand movements. The nonverbal actions were identified by evaluating the bust shot of each participant on a frame-by-frame basis. In order to precisely locate these actions, we used the video annotation tool Anvil (Kipp 2004) with a configuration specifically designed for this purpose (Figure 17.5). Eye gaze was marked with the name of the participant whom the participant being labeled is gazing at. When a gaze starts shifting away from the current participant to the other participant or some other place, the current gaze ends. A new gaze starts when the participant being labeled starts shifting her/his gaze toward one participant. A look at a place or an object other than the conversation participants was not marked. Note that due to the equilateral-triangular setting, participants could gaze at only one participant at a time. Head nods were marked regardless of whether they were accompanied by verbal backchannels. Hand movements were roughly classified into

---

[3] Indeed, the final case is a premature end of a TCU, and will be excluded from the analysis described in section 17.3.

116.8060 118.8510 B: Soo bibitta hanasi-to ieba-ne sakki-no are
So, speaking of a happening that made me nervous,
it happened a short while ago.
117.6080 118.7820 A: Bikkuri-sita bibitta (W hanahi;hanasi)
A story about a surprising, a nervous experience.
118.8820 119.7980 A: Hai hai sakki-no
Yeah, yeah, a short while ago.
119.3890 120.7764 B: Soo itigootoo-no-sa:
Yes, at the building 1.
120.8800 121.1980 A: Un
uh-huh
121.0040 122.6750 B: Nanka yuugata-ka-na yozigoro-ka-na:
Well, it was in the evening, around 4 p.m.
123.0500 123.3730 A: Hai
uh-huh
123.0970 123.4910 B: ni:
At that time,
124.0100 126.5180 B: aano: sigoto kaettekita-no kenkyuusitu-ni
uh, I returned from the work to the lab.
126.8000 127.4130 B: So-sitara-sa:
Then,
127.5160 128.7400 B: ikkai-de matte-tara-sa:
when I was waiting on the first floor,
129.1970 129.7600 B: ikinari-sa
suddenly
129.7600 130.4775 B: <laugh>
130.5280 131.2530 B: Ki-tara-sa:
they came.
130.6780 130.9490 A: Nha
Ouch.
131.7610 133.6550 B: X-sensei-to Y-sensei-ga-sa:
Prof. X and prof. Y,
134.1960 135.0780 B: hutari-site oritekita-no
they were coming down in a pair.
135.4360 137.8970 B: Sikamo metyametya kowai kao siten-no, nanka
sin-nai-kedo-mo
And they looked very fierce, I didn't know why.
136.7000 137.4375 C: <laugh>
137.2350 137.9025 A: <laugh>
138.0175 138.8000 C: <laugh>
138.3625 138.6225 A: <laugh>
139.2130 140.1660 A: Kowa-ku nai-yo
They don't look fierce.

Figure 17.4   Excerpt from a transcript: *X* and *Y* substitute person names to maintain privacy

Figure 17.5 Annotation tool for nonverbal behavior

Ekman and Friesen (1969)'s categories: emblems, illustrators, adaptors, and beats. All these nonverbal actions were represented by the starting and the ending times of the action, the participant name, and additional information such as a gaze direction and the type of a hand movement.

17.3 Analysis of Multimodal Three-Party Conversations

In this section, we present two studies in which we analyzed nonverbal behavior of participants in three-party conversations. The analyses are both quantitative and qualitative. Such dualism is indispensable in approaching novel and challenging issues.

17.3.1 Background

In this subsection, we briefly review previous works on the analysis of nonverbal behavior of participants in conversation, particularly focusing on its relation to the management of interaction. One obvious phenomenon in which nonverbal actions of conversation participants play a significant role is *turn-taking*. When two or more people speak in turn in a conversation, there should be a certain system that regulates the size and the order of turns. Sacks *et al.* (1974) proposed a simple system for turn-taking for conversation, which describes a minimal possible unit of turn – i.e., the TCU (see section 17.2.6) – and turn-transfer rules operable at the end of such a unit. They mentioned certain techniques by which a current speaker can select a particular hearer as a next-speaker; such techniques include the affiliation of an address term or a gaze at one party to a class of utterances such as question, request, invitation, etc.

Other researchers looked for a wider range of actions that may signal a participant's relevance to the next speakership. Duncan and colleagues (Duncan 1972; Duncan and Fiske 1977; Duncan and Niederehe 1974) showed that some nonverbal actions, as well as verbal actions, are relevant to turn-taking: termination of any hand gesture or the relaxation of a tensed hand position by the speaker may signal his intent to yield a turn to the hearer; the turning of the speaker's head toward the hearer within a turn may request backchannels from the hearer; the turning of the speaker's head away from the hearer may signal the continuation of his turn; the speaker's hand gesture or his tensed hand position may sometimes suppress the hearer's attempts to take a turn; and the hearer's shift away in head direction or initiation of a gesture around the end of an utterance may signal her attempt to take a turn.

Kendon (1967) examined, in particular, eye-gaze behavior in conversation, and showed that the speaker with an intent to yield a turn gazes at the hearer around the end of his utterance and that the hearer shifts her gaze away from the speaker to indicate her acceptance of turn. Thus, initiation, followed by termination, of mutual gaze between the speaker and the hearer around the end of an utterance could be indicative of turn-taking.

These findings were based on analyses of dyadic conversations, and very few studies have investigated nonverbal behavior in conversations with more than two participants. Obviously, the range of hearers' actions in multiparty conversation would be much more diverse due to the presence of different kinds of hearers. This is what we explore in the following.

17.3.2 Hearers' Actions and Next Speakership

In this subsection, we examine how hearers' behavior during a speaker's turn is related to turn-taking. As typically shown by Kendon (1967)'s work, hearer's eye-gaze behavior in dyadic conversation often signals her relevance to next speakership. In the following, we show an important finding concerning the relationship between hearer's eye-gaze behavior in three-party conversation and next speakership; that is, mid-turn gaze-shift of a non-next-speaking hearer toward a next speaker.

Method

We used the data subset consisting of six conversations produced by two triads that had been labeled with rich annotations concerning both verbal and nonverbal behavior. One triad was a female group and the other a male group. The total amount of analyzed data was about one hour. Since detailed annotation of nonverbal behavior in multiparty conversations is very time-consuming, we used only a small portion of the entire data. Still, we believe that the total amount of one hour is sufficient to obtain a tentative conclusion.

We focused on the following five types of actions that hearers perform while they are listening to the current speaker's turn: (i) eye gaze at the speaker, (ii) eye gaze at the other hearer, (iii) hand movement, (iv) head nod, and (v) laugh. The data was segmented into TCUs, and labeled with these types of actions. We excluded from the analysis premature ends of TCUs, in which utterances were disrupted spontaneously or by an interruption from another participant. We also excluded TCUs that were started simultaneously with the other party's TCUs or after a long lapse. For temporally overlapping TCUs by two different speakers, we retained only cases where the latter TCU was regarded as an early start on the basis of the prediction of the completion of the former.

We obtained a total of 448 TCUs followed by smooth speaker changes. For these TCUs, we defined the corresponding *hearer units* as an interval between the starts of the current and the succeeding TCUs. Thus, a pause between two TCUs was included in the former hearer unit, and an overlapped part of two TCUs was included in the latter hearer unit but excluded from the former. Since our interest was to examine the distribution of hearers' actions within a hearer unit, we required that the units to be analyzed had similar durations. This requirement, however, was not satisfied with our 448 hearer units. Therefore, we re-sampled the following data from among these hearer units:

- *shorter units* with durations between 0.8 s and 1.5 s ($N = 150$, mean $= 1.1$ s)
- *longer units* with durations between 1.5 s and 2.8 s ($N = 117$, mean $= 2.0$ s).

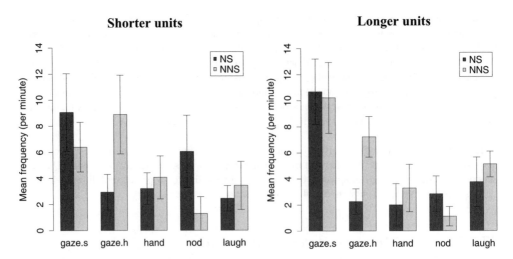

Figure 17.6   Mean frequency (per minute) of hearers' actions ($N = 6$). "gaze.s" and "gaze.h" indicate gaze at the speaker and at the other hearer, respectively. "hand" and "nod" indicate hand movements and head nods, respectively. Error bars show standard errors

For these hearer units, there were always two hearers. Since we concentrated on cases where speaker changes occurred, we could distinguish the two hearers by whether or not they were the speaker of the following TCU. This distinction gave us two different hearers' statuses, *next-speaking hearer* (NS) and *non-next-speaking hearer* (NNS). We compared the distributions of hearers' actions with respect to these statuses.

Results

For each hearer unit, we counted the occurrence of each type of hearers' action that was started in that unit,[4] and divided the counts by the total duration of the hearer units. This calculation was conducted for each participant and for each hearer status. Figure 17.6 shows the mean, over the six participants, of the frequencies (per minute) of the five types of hearers' actions according to the hearers' statuses, "NS" and "NNS". We can see from the graphs the following tendencies.

- NSs frequently gaze at the speaker.
- NNSs' gaze at the speaker is less than that of NSs in the shorter units, but they are almost equally frequent in the longer units.
- It is remarkable that NNSs frequently gaze at the other hearer, i.e., NS. NSs gaze at the other hearer, on the other hand, quite rarely.
- NSs use more head nods than NNSs do. This tendency is clearer for the shorter units.
- NNSs use slightly more laughs than NSs do.

In order to look more closely at these tendencies, we then examined the distribution of each type of action within a hearer unit. For each occurrence of hearers' action, the relative location, within a hearer unit, of the starting point of the action was calculated. That is, if the occurrence was close to the beginning

---

[4] This means that the actions that were started in a previous unit and that continue in the current unit were not counted. Compare this with the analysis in the next subsection.

of a unit, the value was close to 0; on the other hand, if the occurrence was close to the end of a unit, the value was close to 1. Figure 17.7 shows the histograms, accumulated over the six participants, of each type of hearers' actions for each speaker status. For both the shorter and the longer units, the upper five graphs show the distributions for NSs, and the bottom five graphs the distributions for NNSs. We can see from the graphs the following tendencies.

- NSs frequently gaze at the speaker in the early and the middle parts of a hearer unit. This tendency is clearer for the longer units.
- NNSs frequently gaze at the speaker in the middle of a hearer unit. The peak comes earlier in the longer units than in the shorter units.
- NNSs frequently gaze at the other hearer, i.e., NS, around the end of a hearer unit.
- NSs' head nods are mostly located around the end of a hearer unit.
- NNSs sometimes laugh in the middle or the final part of a hearer unit.

Discussion

The findings above suggest that hearers not only perform relevant actions during a hearer unit but also do so at a relevant location according to their different statuses in terms of next speakership. Particularly, eye-gaze behavior is characteristic. It was remarkable that NSs and NNSs frequently gaze at the speaker and that NNSs also frequently gaze at NS. NSs' and NNSs' gaze at the speaker occur mainly in the beginning of a hearer unit, while NNSs' gaze at NS occurs mainly around the end of a hearer unit. Note that according to our definition of hearer units, this gaze-shifting behavior of NNSs occurs before NS actually starts her turn. Thus, it can be seen as NNS displaying her orientation toward non-taking of a next turn.

These hearers' gaze behavior is nicely illustrated by an example shown in Figure 17.8. Participants B and C are talking about the grammar of Japanese.[5] When participant C utters a question 01 to B, "*E, arieru-tte arie-nai-no* (Huh, can't I say 'being possible'?)", B starts gazing at C in the beginning of C's turn and continues looking at C all through the turn. The speaker C keeps gazing at B, which signals that the question is addressed to B, and B's gaze seems to confirm this fact. Soon after B's initiation of gaze at C, the other participant A starts gazing at the speaker as well. At this moment, A would know that the speaker's gaze is toward the other hearer B and that he is not being directly addressed. The two hearers both continue to look at the speaker to the end of C's turn, at which A, who is the non-next-speaking hearer at this turn, shifts his gaze from C to B, who is the next-speaking hearer. After a short delay of about 400 ms, B answers "*Un* (you can't)," as a next speaker.[6] This scene clearly shows how hearers, A and B, perform relevant actions at relevant locations during a hearer unit according to their different statuses; that is, B as the next-speaking hearer and A as the non-next-speaking hearer.

17.3.3 Relationship between Eye Gaze of a Speaker and of Hearers

As shown in the previous analysis, the actions of two hearers may interact with each other. They may also interact with the speaker's actions. In order to understand more precisely the relationship among actions of all conversation participants, we next investigate the co-occurrence of eye-gaze behavior of a speaker and two hearers.

---

[5] *Arieru* is a theoretically expected form of a Japanese verb meaning "being possible", but for historical reasons, an alternative form *aruru* is usually used.

[6] Simultaneously with B's answer, A gives a laugh. This can also be seen as his contribution, without taking a turn, to C's question appropriately as a non-turn-taker.

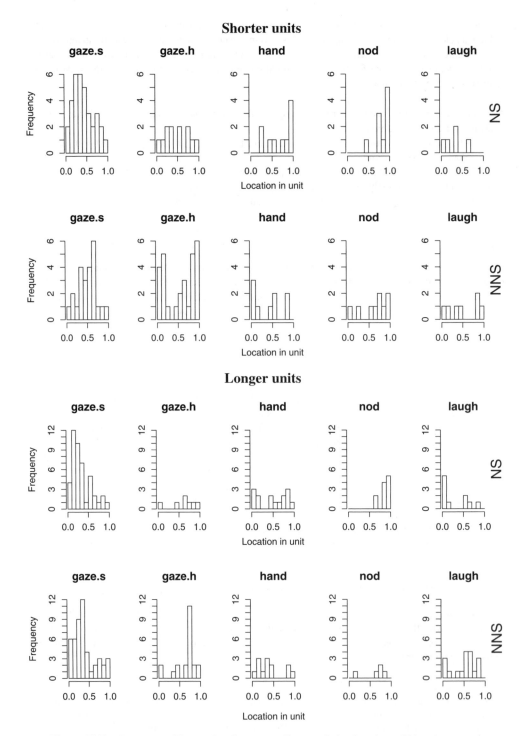

Figure 17.7 Frequency of hearers' actions according to relative location within a hearer unit

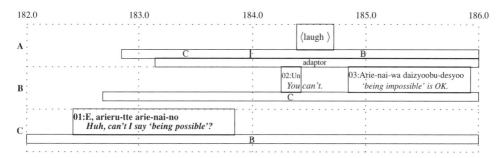

Figure 17.8 Example of hearers' behavior. The boxes with labels B and C indicate gaze directions, and the boxes with the label "adaptor" indicate hand movements. The numbers on the top axis represent the time (in seconds) measured from the beginning of the conversation

Method

The same six conversations as in the previous analysis were used. Among the 448 hearer units followed by smooth speaker changes, we used those units that were longer than 1 s and shorter than 10 s. This selection criterion left us 168 instances.

Unlike the previous analysis, in which only actions starting in the current hearer unit were counted, we considered, in the present analysis, actions that started before the current hearer unit and that continued into that unit. The following patterns of eye gaze during a hearer unit were relatively frequent.

- *Stay at S/NS/NNS*. A participant has started gazing at the (current) speaker (S), the (current) next-speaking hearer (NS), or the (current) non-next-speaking hearer (NNS) before the hearer unit, and continues gazing at that person all through the unit.
- *Shift from S to NS*. NNS has started gazing at S before the hearer unit, and shifts her gaze from that person toward NS somewhere in the unit.
- *Away from S*. NNS has started gazing at S before the hearer unit, and shifts her gaze away from that person (toward other places than the conversation participants) somewhere in the unit.
- *No one*. A participant gazes at no one during the hearer unit.

Results

The hearer units in which gaze patterns of S, NS, and NNS were all classified into either of the above four categories were counted. Table 17.1 shows a three-way contingency table according to the interactions among gaze patterns of the three participants. The results indicate the following:

- When S continues gazing at NS, NNS frequently continues gazing also at NS, or shifts her gaze away from S.
- It is also frequent that NNS shifts her gaze from S toward NS, when S continues looking at NS.
- When S continues gazing at NNS, NNS frequently continues gazing at S.

To see more concretely what happens, we then performed case studies for these frequent patterns.

Table 17.1 Co-occurrence of gaze patterns of the three participants ($N = 126$)

| | S's gaze | | | | | |
| | Stay at NS NS's gaze | | Stay at NNS NS's gaze | | No one NS's gaze | |
NNS's gaze	Stay at S	No one	Stay at S	No one	Stay at S	No one
Stay at S	33	3	19	6	8	1
Away from S	8	2	1	1	0	0
Shift from S to NS	7	5	2	0	0	0
Stay at NS	11	1	0	0	1	0
No one	8	1	1	0	6	1

S gazes at NS, and NNS shifts from S to NS

First, consider the case where the speaker continues gazing at the next-speaking hearer, and the non-next-speaking hearer shifts her gaze from the speaker to the next-speaking hearer. This pattern was observed in 12 instances in total; 7 instances when NS gazes at S, and 5 instances when NS gazes at no one.

In the excerpt shown in Figure 17.9, participants A and C are talking to B about a class they attend at *toodai* (the University of Tokyo). A and B exchange utterances at 11–13 and 15–16, but the other participant, C, also tries, at 12, to answer B's question 11, "*Zya, toodai-no gakusei-mo takusan ita wake-desu-ka?* (So, did it happen that there were many students at the University of Tokyo, too?)", which is addressed to A according to the speaker B's gaze direction. During the exchange 15–16 between A and B, in which A tells B that the class gives a credit for students at the University of Tokyo, C succeeds in taking a turn at 17, by providing to B additional information that in addition to the students, Prof. X at the University of Tokyo, as well as A and C from Chiba University, also attend the class. In this way, the exchange 17–18 is performed by B and C. Near the end of this exchange, participant B, who has been an accepter of the information from C, suddenly shifts her gaze from C to A, and raises a new question 19 to A, "*E, hoka-ni-mo sasottara iki-tai hito i-soo?* (Well, is it likely that other people want to go if invited?)". At this moment, A is the next-speaking hearer and C is the non-next-speaking hearer. Interestingly, the non-next-speaking hearer C shifts her gaze from her previous addressee B toward the expected next speaker A while the speaker B keeps looking at the next-speaking hearer A, and A actually speaks next.

What is remarkable in this example is that the participant who uses a mid-turn gaze-shift, in 19, was involved in the previous exchange but is excluded from the current exchange. (Note that exchange 17–18 is performed by B and C, while exchange 19–20 is performed by B and A.) This characteristic concerning a change of exchange members was repeatedly observed in the 12 instances in which NNS shifts her gaze from S to NS while S is gazing at NS.

Both S and NNS gaze at NS

Next, consider the case where both the speaker and the non-next-speaking hearer continues gazing at the next-speaking hearer. This pattern was observed in 12 instances; 11 instances when NS gazes at S, and 1 instance when NS gazes at no one.

In the excerpt shown in Figure 17.10, participant B is telling A and C about feminism. Before this segment, participant A, responding to B's question about when A knew the term feminism, had been talking for one minute, and just stopped. In 31, C raises a new question, "*B-san kyoogaku-desi-ta?* (Ms. B, did you go to a coed high school?)", explicitly addressing B by means of an address term. After B's affirmative answer, C continues asking a question 34 to B, "*De, zyugyoo-no, tyuugaku-kookoo-to-ka-no*

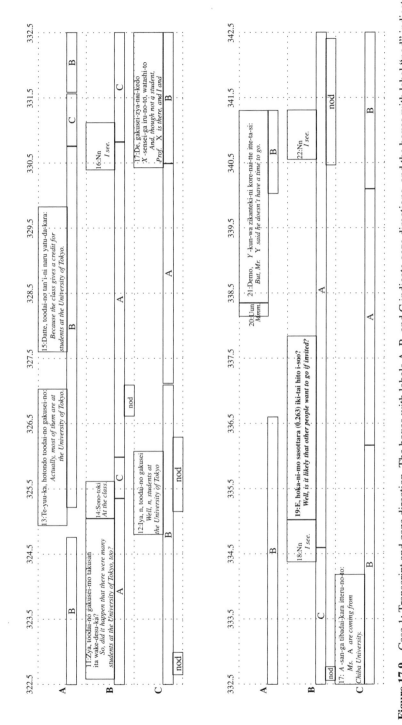

Figure 17.9 Case 1: Transcript and gaze direction. The boxes with labels A, B, and C indicate gaze directions, and the boxes with label "nod" indicate head nods. A, X, and Y substitute person names for reasons of privacy, where A corresponds to the name of participant A

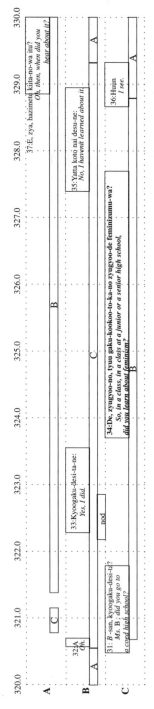

Figure 17.10 Case 2: Transcript and gaze direction. *B* in the transcript corresponds to the name of participants B

zyugyoo-de feminizumu-wa? (So, in a class, in a class at a junior or a senior high school, did you learn about feminism?)", which is answered by B negatively. That is, exchanges 31–33 and 34–35–36 are both in the format that C asks B a question and B answers it, both participants looking at each other. Since the last part of C's first question 31, A, a non-member of these exchanges, has been looking at B. At the moment of 34 uttered by C, she is the non-next-speaking hearer, and continues gazing at the next-speaking hearer, B.[7]

In this way, when two of the three participants continue exchanging turns, the other participant (i.e., the non-next-speaking hearer) continues looking at one of them. This characteristic was also repeatedly observed in our data.

S gazes at NNS, and NNS gazes at S

Finally, consider the case where the speaker continues gazing at the non-next-speaking hearer and the non-next-speaking hearer continues gazing at the speaker. This pattern was observed in 25 instances in total, which was relatively frequent considering an overall low frequency of S's continued gaze at NNS.

In the excerpt shown in Figure 17.11, participant B asks A and C about an entrance examination for a firm. B asks C, in 41, if he will take an exam for Sharp, which is one of the most famous electrical equipment manufacturers in Japan, and C answers, in 43–44, that he is not going to take an exam for Sharp because it is specialized in home electronics. B, in response, issues a confirmation question, "*E, datta, syaapu betu-ni: kaden-zya-nai-no-mo a-n-zya-nai?* (But, Sharp isn't specialized in home electronics, is it?)". This question is obviously addressed to C, judging from the context as well as B's gaze direction, but A responds to it in 47, "*Ekisyoo-to-ka* (Like LCDs)". As a result, C happens to have been the non-next-speaking hearer in 46, in which he and the speaker B continued looking at each other.

This is a deviant case in which the turn is not smoothly transferred to an expected next speaker but is seized by a participant who has not been expected to take a turn. In this sense, it is different from the first case, in which the turn-transfer occurred smoothly, although a change of exchange members is involved in both cases. The difference in the expectation of next speakership explains the difference of gaze behavior of a non-next-speaking hearer between the two cases.

Discussion

It has been shown that one of the most frequent patterns, where S and NNS both continue gazing at NS, occurs when the members of consecutive exchanges do not change. In those cases, two of the three participants speak by turns, and the other participant, as the non-next-speaking hearer, remains looking at one of them. On the other hand, another relatively frequent pattern, in which S continues gazing at NS and NNS shifts her gaze from S to NS during a hearer unit, occurs when a non-member of the previous exchange becomes a member of a new exchange. In those cases, the participant who has dropped out of the new exchange, as the non-next-speaking hearer, shifts her gaze from the speaker to the next-speaking hearer, who is now a new member of the exchange. In this way, mid-turn gaze-shifting behavior of a non-next-speaking hearer is relevant to the change of participation structure.

Gaze-shift is never observed in the pattern where S and NNS continue gazing at each other, although a change of exchange members is also involved in this pattern. A major difference between the previous case and this case is that in the former the change of participation structure is expected, and acknowledged, by all the conversation participants, while in the latter it is not. The change of exchange members in the

---

[7] If we pay attention to 33 instead, A is the non-next-speaking hearer, and continues gazing at speaker B. This pattern is also frequent in Table 17.1, but this would be because of overall high frequency of NNS gazing at S. More remarkable is NNS's continued gaze at S not while S continues gazing at NS but while S continues gazing at NNS, which will be discussed in the next case.

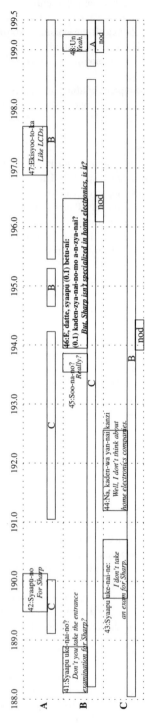

Figure 17.11 Case 3: Transcript and gaze direction

latter pattern is due to a "snatch" of an expected next-speakership by a non-addressed participant, and is not acknowledged by other participants.

In summary, eye-gaze behavior of conversation participants is relevant not only to the local management of next speakership but also to wider structural resources such as the change of participation structure.

17.4 Modeling Human-to-Human Conversation

In this section, we discuss how our findings can be linked to the theory of conversation. We propose a model of conversation participants upon which our line of research is based, and show its relation to two influential theories of communication, relevance theory and conversation analysis.

One obvious fact about conversation is that it is composed of actions produced by a group of conversation participants. Thus, an understanding of the mechanisms underlying human conversation involves, to some extent, an understanding of the mental processes of participants during a conversation. For the moment, we focus on the processes related to the management of interaction rather than more fundamental processes such as comprehension and production of language and gesture and understanding of others' minds. Such processes related to the management of interaction are, in some way, implemented in individual participants' minds, and, working in coordination, yield interactional phenomena such as turn-taking.

Consider one characteristic observation in three-party conversation, which we found in the analyses presented in the previous section:

(1) A non-next-speaking hearer shifts her gaze from the speaker toward the next-speaking hearer around the end of a hearer unit, when she has dropped out of the current exchange members.

How can we model the mental process of the non-next-speaking hearer that brings about this interactional behavior? A simple way is to represent, in the mental model of the participant, a causational rule that relates her awareness of her non-inclusion among the exchange members with her gaze-shifting behavior as in the following rule:

(2) Upon detection of having dropped out of the exchange members, shift eye gaze, around the end of the current turn, from the current speaker to the next-speaking hearer, provided that you are a non-next-speaking hearer.

Application of this rule, however, requires application of some other rules for identifying the exchange members, the next-speaking hearer, and the non-next-speaking hearer. Note that being a non-next-speaking hearer is not equivalent to having no intent to speak next. Even when having intent to speak next, you may have to give up that intent if the current speaker explicitly addresses another participant, in which case you are the non-next-speaking hearer. In this way, even recognition of one's own status would need application of rules that involve perception and understanding of other participants' behavior.

So, is it plausible to model the mental process of a conversation participant to represent a set of rules in a form like (2)? Is our daily interaction really regulated by a chain of applications of such rules? Maybe not. Our actions concerning the management of interaction seem to be produced readily in response to the physical and mental context in which we are embedded, rather than built up deliberately. It would be more plausible if participants' actions were directly guided by the change of context, such as occurrence of a salient event or other party's action.

Consider, again, the observation in (1). A typical context of this phenomenon, as shown in Figure 17.9 in the previous section, involves gaze-shift not only of the non-next-speaking hearer but also of the speaker. (The speaker of turn 18–19, B, shifts his gaze from C to A.) This is because gaze-shift of the speaker is one way to invite a new participant into the exchange and to expel an old participant from it. Furthermore, according to several works including Kendon (1967)'s, when the speaker addresses a particular hearer, by shifting his gaze toward her, the addressed participant is likely to take a new turn; that is, she is the next-speaking hearer.

Combining these, we could implement the following rule, which realizes an interactional behavior in (1):

> (3) Upon detection of the speaker shifting his gaze away from you toward other participant, shift eye gaze, around the end of the current turn, from the speaker to the other participant.

This rule, unlike (2), does not refer to concepts like exchange members, next-speaking hearer, and non-next-speaking hearer, which should further be derived by means of applications of other rules. Instead, it directly maps observation of another's action to one's own action. This is somewhat similar to the implementation employed in behavior-based AI (e.g., Brooks 1991), but it is also different from behavior-based AI in that it refers to categories such as gaze-shift, speaker, and other participant while behavior-based AI uses only direct mapping between perception of the physical world and motor execution.

An implication of this modeling is that mental processes related to the management of interaction are, to some extent, automatic. In this aspect, our model has some parallels with *relevance theory* (Sperber and Wilson 1986; Wilson and Sperber 2004). In explaining a fundamental problem of communication (i.e., how a message, or meaning, in one's mind is transferred into another's mind), Sperber and Wilson (1986) developed a communication model based on a simple basic assumption that the expectations of relevance raised by an utterance, or other ostensive stimulus, whether verbal or nonverbal, are precise enough, and predictable enough, to guide the hearer towards the speaker's meaning. They stated that human cognition tends to be geared to the maximization of relevance (Sperber and Wilson 1986), meaning that the human cognitive system has developed in such a way that our perceptual mechanisms tend automatically to pick out potentially relevant stimuli, our memory retrieval mechanisms tend automatically to activate potentially relevant assumptions, and our inferential mechanisms tend spontaneously to process them in the most productive way (Wilson and Sperber 2004, p. 610). With this universal cognitive tendency, communication is possible; by knowing the hearer's tendency to pick out the most relevant stimuli in the environment and process them so as to maximize their relevance, the speaker may be able to produce a stimulus which is likely to attract the hearer's attention, activate an appropriate set of contextual assumptions, and point the hearer toward an intended conclusion (Wilson and Sperber 2004, p. 611).

Our model shares with relevance theory the idea of the universal cognitive tendency for relevance maximization. It is, however, different from relevance theory in that we use this universal cognitive tendency to model not only ostensive-inferential communication but also interaction that may seem much more responsive or that may not involve communicative intention. For instance, speaker's gaze-shift from the non-next-speaking hearer to the next-speaking hearer may not involve a message, to the non-next-speaking hearer, that he intends to expel her from the current exchange, although it may involve a message, to the next-speaking hearer, that he intends to invite her into the current exchange. Or, the non-next-speaking hearer may or may not infer the speaker's meaning, if such a meaning is present, when she performs her gaze-shifting behavior. For the speaker and the non-next-speaking hearer to coordinate their actions in context, it is sufficient that they both follow the universal cognitive tendency, which maximizes the relevance of their actions in that context, with no care about messages that may be involved in those actions. This motivates our implementation of rules in a form like (3).

At this point, we want to mention another influential discipline in communication studies, *conversation analysis* (for introductory text, see e.g., Psathas 1995). Conversation analysis studies the regularity, or orderliness, of social actions, particularly in everyday interaction. Researchers in conversation analysis try to discover order that is repeatedly observed in their data. It is assumed that order is produced by the participants on a particular occasion and that they orient to that order by themselves. Thus, when conversation analysts formulate the rules that produce some orderliness being observed, they do so not aiming to generalize the observations, which might be useful for the analysts to account for the phenomena, but aiming instead to describe the organized practices that conversation participants themselves follow. The most famous instance of such rules is those for turn-taking (Sacks *et al.* 1974), which restrict possible opportunity for speaker-change to a possible completion point of a turn-constructional unit and which provide three options for turn-allocation, in an ordered fashion: (a) selection, by the current speaker, of the next speaker, (b) self-selection by the next speaker, and (c) continuation of the current speaker. Note that they do not claim that these rules are, in some way, implemented in our minds nor that we explicitly apply them at any possible completion of a turn-constructional unit. These are the practices that participants follow in regulating turn-taking in conversation, but not the principles, or constraints, operating in their minds.

How is this view of conversation analysis related to our model of conversation participants? Obviously, our approach is different from conversation analysis, since we aim at modeling mental processes of participants, which conversation analysts avoid. We would, however, claim that we are, to some extent, in line with their idea of *rules as practices*, in the sense that we implement rules in a form of action-to-action mapping, which might be acquired through experiences in real situations where such rules are demanded for optimal behavior. Here, we cannot go into the issue of the learning mechanism of such practices, but we expect it to be important in our line of modeling, as it is a central issue in behavior-based AI and its successors.

Still, this is not the full story. When conversation analysts talk about rules, they do not consider that participants automatically and unconsciously obey the rules, but that participants are aware of the rules. When we produce smooth turn-transfer in conversation, it seems that we are not aware of the turn-taking rules in operation at every possible completion of a turn-constructional unit. However, once the rules are violated, we will soon become aware of the rules. For example, when someone starts a new utterance before the current turn has reached a possible completion, we immediately notice that a violation of the rules has occurred and seek for a reason for it; we may surmise that the new speaker disagrees with the content of the current turn. In doing so, we refer to the turn-taking rules in order to know what is expected behavior at that moment, and, based on that expectation, we detect a deviance, which triggers our inference of the reason for the deviance.[8] Thus, these rules are not just the description of phenomena or the summary of tendencies, but are practical knowledge that conversation participants are aware of.

For our observation (1) as well, this mode of the use of rules may be possible. For instance, the non-next-speaking hearer may continue her gaze at the speaker even when she has dropped out of the current exchange members, in order to signal her disagreement with being expelled from the exchange. Or, a participant may produce the gaze-shifting action peculiar to the non-next-speaking hearer when she is addressed by an utterance from the current speaker, in which situation other types of gaze actions would be adequate for her; by this, she may convey to the speaker a message that she is not willing to be the next speaker and that she, instead, wants the party designated with her gaze to respond. In these examples, participants utilize knowledge about their own practices, which may be stated as in (1), to find a deviance and to effect a reason for it. In this sense, the description of participants' practices, by itself, is used as a rule. (Note that this mode of rule application cannot be achieved for causational rules like (3).)

We are now in a position to present our tentative model to realize two modes of rule application. In one mode, participants automatically and unconsciously obey the rules. These rules would be implemented in a form of action-to-action mapping, and, through interaction among the rules applicable in the physical and mental context in which they are embedded, optimal behavior emerges. With the assumption that participants all have a similar set of rules and operations (i.e., the universal cognitive tendency), we would expect that participants can show highly coordinated behavior relevant to the context they are in. The mental processes working in this mode would be responsive, rather than deliberate. We call this mode of rule application *responsive mode*. In the other mode, participants utilize the rules in order to know an expectation, to find a deviance, and to infer a reason for the deviance, by which other party's

---

[8] In this respect, these rules are rather similar to Grician maxims (Grice 1975).

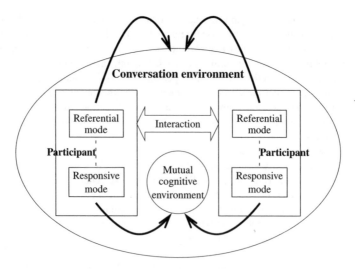

Figure 17.12   The dual model of conversation

belief, intention, message, etc. may be conveyed. The rules in this mode are regarded as knowledge about practices of the participants themselves, and would be meta-representations of their practices. The mental processes working in this mode would refer to those meta-representations as well as to an analytic view of the environment they are in, possibly including mental states of self and others. We call this mode of rule application *referential mode*.

Figure 17.12 depicts how conversation between two participants can be modeled by our *dual* model of interaction. When both participants are in responsive mode, they access the mutual cognitive environment, which is part of the memory and the physical environment easily accessible by the two participants. The presence of the mutual cognitive environment, along with the universal cognitive tendency, would improve chances for the two participants to spontaneously coordinate with each other. On the other hand, when both participants are in referential mode, they have access to the whole conversation environment, both physical and mental. They obtain an analytic view of the environment in real time, depending on the situation. They may refer to the meaning of an action and mental states of self and the other, such as belief, desire, intention, etc. Each mode is activated in an appropriate situation, and participants may switch their mode from one to the other depending on the situation.[9]

Unfortunately, we do not yet have a clear idea of the computational mechanism of our dual model. We might have to wait for advances in such areas as embodied cognitive science, cognitive developmental robotics, constructive brain science, etc.

17.5 Conclusions

In this chapter, we have described a scientifically motivated approach to conversational informatics. We have presented a methodology for describing, analyzing, and modeling natural human conversation, as well as the results achieved thus far in our project.

We have described our data collection methodology, and presented our multimodal corpus of three-party conversations in Japanese. We then analyzed the distribution of hearers' actions during a speaker's turn, comparing them between different statuses of hearers in terms of next speakership. We showed

---

[9] At this moment, we have no good model to account for when and how this mode switch occurs.

how hearers' eye-gaze behavior is used in the management of interaction in three-party conversation. We finally discussed how our findings could be linked to the theory of conversation. We proposed a new model of conversation participant, and showed its relation to two influential theories of communication, relevance theory and conversation analysis.

There are many problems to be tackled. As for data description, we need more data with annotation of nonverbal behavior. Besides this need, we also require annotation of phenomena at a more abstract level, such as dialogue acts, discourse structure, and addressee. Such annotations would introduce new perspectives into our data analysis. Our current analysis, based on a very small portion of the entire data, should be enhanced by using the rest of the data. Our model should be further elaborated, particularly in terms of the computational aspect, and verified by controlled experiment or corpus analysis. We leave these problems for future studies.

References

Allen J.F. (1983) Recognizing intentions from natural language utterances. In M. Brady and R.C. Berwick (eds), *Computational Models of Discourse*. MIT Press, Cambridge, MA, pp. 107–166.

Brooks R.A. (1991) Intelligence without representation. *Artificial Intelligence* **47**, 139–159.

Carletta J., Ashby S., Bourban S. *et al.* (2006) The AMI meeting corpus: a pre-announcement. In *Machine Learning for Multimodal Interaction: Second International Workshop*, MLMI 2005, Edinburgh, UK, July 11–13, 2005. *Revised Selected Papers* (ed. S. Renals and S. Bengio), vol. 3869 of *Lecture Notes in Computer Science*. Springer, Berlin, pp. 28–39.

Clark H.H. (1996) *Using Language*. Cambridge University Press, Cambridge.

Duncan S. (1972) Some signals and rules for taking speaking turns in conversations. *Journal of Personality and Social Psychology* **23**, 283–292.

Duncan S. and Fiske D.W. (1977) *Face-to-face interaction: research, methods, and theory*. Lawrence Erlbaum, Hillsdale, NJ.

Duncan S. and Niederehe G. (1974) On signalling that it's your turn to speak. *Journal of Experimental Social Psychology* **10**, 234–247.

Ekman P. and Friesen W.V. (1969) The repertoire of non-verbal behaviour: categories, origins, usage and coding. *Semiotica* **1**, 49–98.

Ford C.E. and Thompson S.A. (1996) Interactional units in conversation: syntactic, intonational, and pragmatic resources for the management of turns. In E. Ock, E.A Schegloff and S.A Thompson (eds), *Interaction and Grammar*. Cambridge University Press, Cambridge, pp. 134–184.

Goffman E. (1981) *Forms of Talk*. University of Pennsylvania Press, Philadelphia, PA.

Grice H.P. (1975) Logic and conversation. In P. Cole and J.L. Morgan (eds), *Syntax and Semantics 3: Speech Acts*. Academic Press, New York, pp. 41–58.

Grosz B.J. and Sidner C.L. (1986) Attention, intentions, and the structure of discourse. *Computational Linguistics* **12**, 175–204.

Grosz B.J., Joshi A.K. and Weinstein S. (1995) Centering: a framework for modeling the local coherence of discourse. *Computational Linguistics* **21**, 203–225.

Kendon A. (1967) Some functions of gaze direction in social interaction. *Acta Psychologica* **26**, 22–63.

Kipp M. (2004) *Gesture Generation by Imitation: From Human Behavior to Computer Character Animation*. Dissertation.com, Boca Raton, FL. Also see http://www.dfki.de/~kipp/anvil/.

Koiso H., Horiuchi Y., Tutiya S., Ichikawa A. and Den Y. (1998) An analysis of turn-taking and backchannels based on prosodic and syntactic features in Japanese Map Task dialogues. *Language and Speech* **41**, 295–321.

Maekawa K., Koiso H., Furui S. and Isahara H. (2000) Spontaneous speech corpus of Japanese. *Proceedings of the 2nd International Conference of Language Resources and Evaluation*, Athens, pp. 947–952.

McNeill D. (2006) Gesture, gaze, and ground. In *Machine Learning for Multimodal Interaction: Second International Workshop*, MLMI 2005, Edinburgh, UK, July 11–13, 2005. *Revised Selected Papers* (ed. S. Renals and S. Bengio), vol. 3869 of *Lecture Notes in Computer Science*. Springer, Berlin, pp. 1–14.

Nakano Y.I., Reinstein G., Stocky T. and Cassell J. (2003) Towards a model of face-to-face grounding. *Proceedings of the Annual Meeting of the Association for Computational Linguistics*, Sapporo, Japan, pp. 7–12.

Psathas G. (1995) *Conversation Analysis: the study of talk-in-interaction.* Sage, Thousand Oaks, CA.

Sacks H., Schegloff E.A. and Jefferson G. (1974) A simplest systematics for the organization of turn-taking for conversation. *Language* **50**, 696–735.

Shriberg E.E (1994) *Preliminaries to a Theory of Speech Disfluencies.* PhD thesis, University of California at Berkeley.

Shriberg E., Bates R., Stolcke A., Taylor P. *et al.* (1998) Can prosody aid the automatic classification of dialogue acts in conversational speech? *Language and Speech* **41**, 443–492.

Shriberg E., Dhillon R., Bhagat S., Ang J. and Carvey H. (2004) The ICSI meeting recorder dialogue act (MRDA) corpus. In *Proceedings of the 5th SIGdial Workshop on Discourse and Dialogue* (ed. M. Strube and C. Sidner), Cambridge, MA, pp. 97–100.

Sperber D. and Wilson D. (1986) *Relevance: Communication and Cognition.* Harvard University Press, Cambridge, MA.

Takanashi K., Uchimoto K. and Maruyama T. (2004) *Manual of Clause Unit Annotation in Corpus of Spontaneous Japanese* Version 1.0 (in Japanese). National Institute for Japanese Language. http://www.kokken.go.jp/katsudo/seika/corpus/public/manuals/clause.pdf

Vertegaal R., Slagter R., van der Veer G.C. and Nijholt A. (2001) Eye gaze patterns in conversations: there is more the conversational agents than meets the eyes. *Proceedings of ACM Conference on Human Factors in Computing Systems*, Seattle, WA. pp. 301–308.

Walker M.A., Iida M. and Cote S. (1994) Japanese discourse and the process of centering. *Computational Linguistics* **20**, 193–232.

Whittaker S. (2003) Theories and methods in mediated communication. In A.C Graesser, M.A Gernsbacher and S.R Goldman (eds), *The Handbook of Discourse Processes.* Lawrence Erlbaum Associates, Mahwah, NJ, pp. 243–286.

Wilson D. and Sperber D. (2004) Relevance theory. In L.R Horn and G. Ward (eds), *The Handbook of Pragmatics.* Blackwell, Oxford, pp. 607–632.

18

Embodied Synchrony in Conversation

Chika Nagaoka, Masashi Komori, and Sakiko Yoshikawa

18.1 Introduction

18.1.1 Basic Definition of Embodied Synchrony

In interactions among humans, rhythmic synchronization of a listener's body movements with those of the speaker (Condon and Ogston 1966; Kendon 1970) is often observed, as is synchronization of breathing (Watanabe and Okubo 1998). In addition, it is often observed that participants' nonverbal behaviors, including posture (e.g., Scheflen 1964), mannerisms (e.g., Chartrand and Bargh 1999), facial actions (e.g., Gump and Kulik 1997), and speaking style (e.g., Nagaoka *et al.* 2005a; Natale 1975b) can be mutually similar. In this chapter, we refer to the phenomenon of synchronization or similarity of nonverbal behaviors among participants as "embodied synchrony". It is considered that the biological basis of human beings that synchronizes with the external world, including others, would emerge as the synchronization and similarity of nonverbal behavior. Across disciplines, a variety of terms are employed, including convergence (e.g., Giles and Smith 1979), congruence (e.g., Feldstein and Welkowitz 1978; Scheflen 1964), matching (Cappella and Planalp 1981), coordinated interpersonal timing (e.g., Jasnow *et al.* 1988), mimicry (e.g., Chartrand and Bargh 1999; Gump and Kulik 1997), mirroring (e.g., LaFrance and Ickes 1981), imitation (Meltzoff and Moore 1977), and interpersonal synchrony (Condon and Ogston 1966; Kendon 1970). In this chapter, when referring to individual cases, the above terms are used, but when making general references, the term "embodied synchrony" is used.

Embodied synchrony can be observed from the early stages of a child's development (Figure 18.1). In interactions between a neonate and an adult, for example, the neonate's body movements synchronize with the flow of the adult's speech (Condon and Sander 1974), and the neonate imitates the adult's facial gestures (Meltzoff and Moore 1977). In interactions among children, imitation of other children's body movements and gestures (Eckerman *et al.* 1989) is observed, along with congruence of vocal behavior (Garvey and BenDebba 1974; Welkowitz *et al.* 1976). The various forms of embodied synchrony mentioned above are also observed in interactions among adults. In an interaction between conversation partners from different linguistic backgrounds, speaking style such as accent, dialect, language, and speech rate are often accommodated and may converge. Thus, the communication channels

Conversational Informatics: An Engineering Approach   Edited by Toyoaki Nishida
© 2007 John Wiley & Sons, Ltd

Neonate

In an adult–neonate interaction
-Entrainment
-Neonate imitation

Infant/Toddler

In an adult–infant interaction
-Coordinated interpersonal timing
In an infant–infant interaction
-Imitation of nonverbal action
-Congruence of paralanguage

Adult

In an adult–adult interaction
-Congruence (or convergence, matching) of paralanguage
-Postural mirroring (or postural sharing, congruent
 posture)
-Mimicry of facial expressions

In intercultural context
-Convergence of language, accent, speech rate

Figure 18.1 Development and embodied synchrony. The communication channels which display embodied synchrony become more diversified with development. Partially revised from Nagaoka *et al.* (2005b). (Reproduced by permission of © 2005 IEEE)

which display embodied synchrony become more diversified with development. Furthermore, the degree of embodied synchrony also strengthens with development (Garvey and BenDebba 1974; Welkowitz *et al.* 1976).

Embodied synchrony is thus thought to provide a series of cues for the investigation of interactions. First, embodied synchrony is influenced by various social factors, and as such, can provide an index of the quality of interactions as described in section 18.3. Second, embodied synchrony plays an important role in establishing and maintaining interpersonal relations as will be described in section 18.4.

Embodied synchrony can be observed in various forms as mentioned below, and research on this topic is ongoing in a variety of different disciplines. There are only a handful of accounts, however, which systematically describe the phenomenon (e.g., Cappella 1981; Daibo 1985; Feldstein and Welkowitz 1978; Hess *et al.* 1999), and the scope of these investigations has been quite limited. This chapter provides an overview of previous studies of embodied synchrony from a broader perspective, integrating viewpoints from a variety of disciplines.

18.1.2 Embodied Synchrony in Various Communication Channels

Embodied synchrony is observed in various communication channels. In this section, we will describe embodied synchrony in the following four categories of nonverbal behavior: (1) body movement and gestures, (2) facial behavior, (3) vocal behavior, and (4) physiological reactions. We focus here on dyadic interactions, which are considered the most basic form of interactions.

Body Movement and Gesture

It has been observed that, in interactions through vocal sounds, interactant's body movements (including movements of body parts such as the head, eyes, torso, elbows, and fingers) are simultaneous with the body movements of the interaction partner (Condon and Ogston 1966; Kendon 1970). This phenomenon has been called "interpersonal synchrony" and argued to occur because the speaker's body movements occur simultaneously along with flow of his speech sounds (called "intrapersonal synchrony") while the addressee's body movement also occurs simultaneously along with the flow of the speaker's speech sounds. It has also been observed that infants move their limbs along with the speech sounds of their caretakers (Bernieri *et al.* 1988; Condon and Sander 1974). In addition, Shockley *et al.* (2003) have reported that the temporal pattern of postural sway between two people who are standing and talking to one another tends to synchronize, even if they are not facing each other, if the two are participating in the same interaction.

Furthermore, when two interactants are facing each other and one of them takes on a particular posture or a certain mannerism such as rubbing the face or shaking the legs, the partner may take on a congruent posture (e.g., Bernieri *et al.* 1996; Charny 1966; LaFrance 1979; LaFrance and Ickes 1981; Maurer and Tindall 1983; Scheflen 1964; Schmais and Schmais 1983) or a congruent mannerism (e.g., Chartrand and Bargh 1999). For instance, if one is crossing his legs with the left leg on top of the right, the other may also cross his legs with the right leg on top of the left leg (called "mirroring") or with the left leg on top of the right leg (called "postural sharing").

Facial Behavior

When one of two interactants facing each other takes on a certain facial action, the partner may take on a congruent action (e.g., Gump and Kulik 1997; Lanzetta and Englis 1989). For instance, if one is crying, the other may also cry. It's known that when images of a facial expression displaying a particular emotion like happiness and sadness are presented, we display similar expression, even if those images are still images (e.g., Dimberg and Lundqvist 1988; Lanzetta and Englis 1989; Lundqvist 1995; Vaughan and Lanzetta 1980, 1981).

Vocal Behavior

Often, when two people are speaking together, they engage in similar styles of vocal behavior. This is observed when they are exchanging speech, even when they are not facing each other. For example, when two people who are close to one another are speaking, if one speaker makes use of short response latencies (the time from the conversational partner stopping talking to the speaker starting to talk), the conversational partner may do so as well. On the other hand, another pair of very close conversation partners may use long response latencies. This seems to hold even when the level of interest and difficulty for the conversational topic is controlled between the two pairs. In short, mean durations of response latency are different across pairs but quite similar within pair (e.g., Nagaoka *et al.* 2002: see Figure 18.2; Street 1984). Similar observations have also been made for other aspects of vocal behavior, including utterance duration (the mean duration of a turn by one speaker: see Feldstein 1968; Staples and Sloane 1976), the duration of switching pauses (silences between the utterances of two speakers, the period of

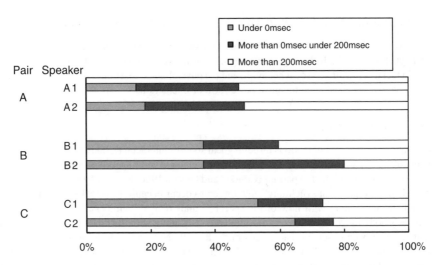

Figure 18.2   Inter-speaker similarity in duration of response latencies. Duration of response latencies in dialogues by three dyads of very close relationship was analyzed. This figure represents distributions of the duration between speakers. Latencies below 0 ms indicate simultaneous speech. Reproduced by permission of Japan Ergonomics Society

time between "the speaker's stopping talking and the partner's starting"[1]: e.g., Natale 1975b; Welkowitz and Kuc 1973; Welkowitz et al. 1976, 1990), the duration of pauses (silences within the utterances of a single speaker), total number of utterances (fragments of speech by an identical speaker that are divided by pauses: Garvey and BenDebba 1974), vocal intensity (Meltzer et al. 1971; Welkowitz et al. 1972), fundamental frequencies (Gregory 1990; Gregory et al. 1997, 2001), and speech rates (Street 1984). Furthermore, in intercultural context, convergence of speaking style like dialects, accents and speech rate is also observed (e.g., Ball et al. 1985; Gallois et al. 1995; Giles and Smith 1979; Willemyns et al. 1997).

In addition, it is observed that one speaker changes her/his vocal behavior in response to changes in the vocal behavior of the conversation partner. For instance, Jasnow and Feldstein (1986) on switching pause duration in mother–infant interactions reported that, when infant's switching pause durations lengthen/shorten, subsequent pauses on the part of the mother also lengthen/shorten, and that when mothers' switching pause durations lengthen, infant's subsequent pauses shorten only slightly (also see Newman and Smit 1989). Such interspeaker influences have been reported in terms of duration of utterance (Matarazzo et al. 1967), response latencies (Matarazzo et al. 1965), and vocal intensity (Natale 1975a).

Physiological Reactions

In dyadic interactions through vocal sounds, it is reported that a change in the listener's breathing pattern (i.e., from expiration to inhalation, or vice versa) occurs nearly simultaneously with the speaker's breathing pattern (Watanabe and Okubo 1998). Such entrainment of breathing has often been observed in musical settings, both among groups of performers and between performers and the audience (Nakamura 2002).

---

[1] With one important exception, "response latency" is similar to the "switching pause" introduced by Jaffe and Feldstein (1970) who credited the silence between turns to the first speaker who relinquishes the floor. "Response latency" credited to the second speaker assuming the floor.

18.1.3 Phenomena that should be Excluded from Embodied Synchrony

It should be noted that there are cases which should not be considered embodied synchrony, even though they involve congruent or simultaneous communication behaviors between two people. For instance, embodied synchrony does not include cases in which people show congruent behaviors due to some external physical factors. People may talk more loudly in a noisy environment, for example, or take on a congruent posture in an environment limiting their postures, such as in a small chair, but these cases are not considered as embodied synchrony.

18.2 Measurement and Quantification of Embodied Synchrony

Because embodied synchrony appears in various forms, it is important for researchers to choose appropriate measurement and quantification techniques. In this section we will outline each of the measurement and quantification techniques that have been employed in previous studies. We must be aware of each method's strengths and weaknesses in terms of reliability and validity.

18.2.1 Body Movement and Gesture

Coding and psychological evaluation have been used as methods to evaluate the embodied synchrony of body movements and gestures. Using a motion capture system is also effective for measuring the embodied synchrony of body movements.

Condon and Ogston (1966) recorded the interactions on 48 frame per second film (one frame equals approximately 0.02 seconds) and coded the movements of each body part (including head, eyes, mouth, torso, elbow, finger, etc.) in each frame. It was observed that participant A's hand started to move at the same time that participant B's head started to move, or that participant A's mouth stopped moving when participant B's mouth stopped moving. In Chartrand and Bargh (1999) and Van Baaren *et al.* (2004a), two judges coded participants' mannerisms such as rubbing their face and shaking their legs, and recorded the number and duration of occurrences. Coding is also useful to evaluate whether posture is matched between interactants at the same time. For example, Charny (1966) defined postural congruence as sharing or mirroring of postures between two interactants for more than 0.4 seconds (equivalent to ten frames in 24 frames/second film). Congruent postures were identified by the judges, and the total time spent in postural congruence during a given period was calculated.

Bernieri and Rosenthal (1991) proposed a way to comprehensively quantify these kinds of embodied synchrony using psychological evaluation. This method is known as the pseudosynchrony experimental paradigm and it is used to compare the level of embodied synchrony between an actual conversational interaction and a comparison group, for example through composite footage of speaker A at time1 and speaker B at time2 during their conversation, or composite footage of speaker A and speaker C from an unrelated interaction.

These composite films were presented as stimuli to participants, who then rated the level of embodied synchrony on a nine-point scale according to the following three items: simultaneous movement, tempo similarity, and coordination and smoothness. The average value of the rating values for the three items was calculated and compared between the actual interaction group and the comparison group.

There have also been studies on the coordination of postural sway and hand waving motions between interactants using a motion capture system (Richardson *et al.* 2005; Shockley *et al.* 2003). Shockley *et al.* (2003) used a magnetic tracking system to measure postural sway and analyzed this movement using cross-recurrence quantification analysis, a kind of nonlinear data analysis, to evaluate the shared activity between two postural times series.

In addition, Komori *et al.* (2007) used a novel computer-based method to analyze body movement synchrony during psychological counseling. The intensity of the participants' body movement was respectively measured using a video-based system. The cross-correlation function between the two time series of the intensity was used as the body synchrony index. Two educational counseling sessions and four psychotherapeutic counseling sessions were analyzed using this system. Results of both types of counseling sessions indicated that body movement synchrony was observed when experts rated the counseling session positively, but not when they rated it negatively. The advantages of this method are (1) high objectivity because it is based on an objective indicator, (2) that the privacy of participants is protected, and (3) it is not expensive and needs only a video camera and a computer.

18.2.2 Facial Behavior

Mimicry of facial expressions, between adults, can be studied by measuring electric potential in facial muscle (e.g., Dimberg and Lundqvist 1988; Lanzetta and Englis 1989; Lundqvist 1995; Vaughan and Lanzetta 1980, 1981). However, natural interaction between participants can be hindered because of the attached measurement apparatuses. Researchers who emphasize natural interactions may prefer to avoid this method.

An alternative method is to code the facial behavior of participants (Gump and Kulik 1997; Meltzoff and Moore 1977). In Meltzoff and Moore (1977), neonates were shown adult facial gestures and then filmed for their facial reactions. The movements of the babies' tongues and mouths were coded by judges. It was found that when adults stuck out their lips, there was a high frequency of babies doing the same thing in response. Likewise, when adults closed their mouths, there was a high frequency of babies also closing their mouths.

For the protection of the participants' privacy and improved objectivity, it is desired to develop an alternative method that does not depend on human perception. A computer-automated system for the analysis of facial expressions, body movements and gesture would be ideal for this purpose.

18.2.3 Vocal Behavior

It is quite common for embodied synchrony in vocal behaviors such as duration of pauses and fundamental frequency (F_0) to be quantified through physical measurement. The analyzed physical quantity is statistically processed to indicate the level of congruence in the vocal behavior between two speakers or to show how the vocal behavior of a speaker influences that of the conversational partner.

The intraclass correlation coefficient (one-way model) or the absolute value of the difference between two speakers (unlike the intraclass correlation coefficient, this allows the characteristics of each pair to be represented) can be calculated to quantify the level of congruence between two speakers. For example, Nagaoka et al. (2005a) analyzed A-weighted sound pressure levels [dB(A)] and the median response latencies for each speaker were calculated to arrive at an intraclass correlation coefficient.

How the vocal behavior of a speaker may change in response to that of the conversation partner can be examined by calculating Pearson's product-moment correlation coefficient between the analytic value of one speaker's speech sound and that of the partner's subsequent speech sound (Matarazzo et al. 1967), or intercorrelation coefficient of two time series data (Jasnow and Feldstein 1986; Street and Cappella 1989).

Cording and psychological evaluation have also been used to examine the embodied synchrony of vocal behavior. For example, accent and dialects may be coded by trained judges (e.g., Al-Khatib 1995; Tong et al. 1999). Nagaoka et al. (2006) used psychological evaluation like Bernieri and Rosenthal (1991), using the following three items: voice strength similarity, speech tempo similarity, and coordination and smoothness of response timing. This evaluation is thought to be useful for choosing the best parameters for physical measurement.

18.3 Influences on Embodied Synchrony

Since Condon and Ogston (1966) and Matarazzo (1965) pointed out that embodied synchrony does not always occur in an interactions, a number of studies have examined the factors that influence embodied synchrony. In general, empathetic and cooperative relationships among interactants seem to be linked to embodied synchrony. Here, we will outline three factors that determine interpersonal relationships: (1) individual factors, (2) situational factors, and (3) cultural and ethnic factors. We now turn to studies examining the influences of these factors on embodied synchrony, as well as correlations between these factors and embodied synchrony.

18.3.1 Individual Factors

Empathy and Social Skills

Many studies have indicated the influence of interactant's empathy on embodied synchrony. Empathy has two major components, namely, emotional empathy (an emotional reaction that occurs when an interactant perceives the emotional state that the other interactant is experiencing or trying to experience) and cognitive empathy (related to perceiving entities from the other person's point of view). These two components seem to influence on different communication channels.

Emotional empathy may mainly influence embodied synchrony in emotional expressions. For instance, Sonnby-Borgstrom *et al.* (2003) have reported that the high-empathy participants showed a significant mimicking reaction when individuals were exposed to pictures of angry or happy faces for very short exposure times, while the low-empathy participants did not display mimicking. Schmais and Schmais (1983) have reported a positive relationship between dance therapy graduate students' ability to image their emotions in someone else's position and their ability to correctly mirror the emotional body expressions of another.

On the other hand, embodied synchrony of vocal behaviors, body movements and gestures may be influenced by cognitive empathy. For instance, positive correlations have been reported between therapist's empathy and the degree of embodied synchrony in vocal behaviors and body movements by the therapist and the client (Nagaoka *et al.* 2006; Staples and Sloane 1976). A positive correlation was found between an interactant's perspective taking ability which is a part of cognitive empathy and is related to anticipate the behavior and reaction of others, and their tendency to mimic the mannerisms of the partner (Chartrand and Bargh 1999).

In addition to empathy, social skills, and self-monitoring ability, which are essential skills to smoothly carry out social tasks with others, and social desirability which is the degree to which the behaviors and characteristics of an individual are considered to be desirable in a given society, may also influence embodied synchrony, mainly in vocal behavior and body movement and gesture (e.g., Cheng and Chartrand 2003; Nagaoka *et al.* 2003; Natale 1975a,b). This is also supported by the report that elder infants use similar vocal behaviors to their partners more often than younger infants (Garvey and BenDebba 1974; Welkowitz *et al.* 1976). Furthermore, speakers with high social skills show embodied synchrony in a cooperative interaction, while they do not do so in a competitive interaction (Nagaoka *et al.* 2003), suggesting that embodied synchrony may be used as a social strategy.

Gender

Gender is another factor that influences the degree of embodied synchrony (e.g., Feldstein and Field 2002; LaFrance and Ickes 1981). LaFrance and Ickes (1981) studied dyadic posture mirroring between strangers. Results showed that sex-typed female pairs tended to show more posture mirroring than sex-typed male pairs did. Considering that females have a higher level of emotional empathy than males

(Mehrabian and Epstein 1972), the results can be interpreted to indicate the influence of emotional empathy on embodied synchrony.

Disorder

While normal populations have been observed to show embodied synchrony, some early studies have observed that patient populations do not do so. For example, in contrast to normal subjects, schizophrenic patients were not observed to increase or decrease the duration of their utterances according to the duration of their partner's utterances (Matarazzo 1965). Interpersonal synchrony was also not observed between a normal participant and an aphasic patient, as aphasics do not display intrapersonal synchrony (i.e., synchrony between their speech flow and their body movement)(Condon and Ogston 1966). More research on patient populations is necessary to establish the embodied synchrony behaviors that are common in these populations and to devise methods possibly to reduce the incidence of the disorders. (See Yaruss and Conture (1995) for work on individuals with stuttering problems.)

18.3.2 Situational Factors

Cooperativeness

Embodied synchrony is observed more frequently in cooperative interactions than in competitive interactions (e.g., Bernieri et al. 1996; Gump and Kulik 1997; Lanzetta and Englis 1989; Nagaoka et al. 2005a). Cooperative interactions include cooperative games and dialogues to discuss matters such as both speakers' common ground or travel plans, while competitive interactions include competitive games and debates. For instance, in cooperative dialogues, in which dyads of speakers holding opposing opinions reached a conclusion through discussion and consideration of their partner's opinion, the response latencies of the conversational partners became similar over time, while in debate dialogues, in which dyads of speakers holding opposing opinions imposed their opinions on their conversational partners, the response latencies of the partners did not become similar (Figure 18.3; Nagaoka et al. 2005a). This tendency has also been shown in terms of mimicry of facial expressions (e.g., Gump and Kulik 1997; Lanzetta and Englis 1989). Furthermore, in cooperative dialogues, embodied synchrony is observed from early on in the dialogue (Nagaoka et al. 2005a).

Rapport

A number of studies have been conducted to examine the relationship between rapport, meaning a mutual trust relationship between interactants, and embodied synchrony, especially in postures and body movements. For instance, the postural sharing between a teacher and their students is positively correlated with the degree of rapport with the teacher reported by the students (LaFrance 1979), and speakers who establish strong rapport in a cooperative dialogue tend to show congruent postures and simultaneous body movements (Bernieri et al. 1996). However, it should be noted that no relationship has been observed between degree of rapport and simultaneous body movement or congruent postures in debate situations (Bernieri et al. 1996).

Additional findings show that topics of conversation, which can change according to the degree of rapport, can influence postural congruence. For instance, in psychotherapy sessions, congruent postures tend to be observed when the counselor and the client are talking interpersonal, positive, and concrete topics, while incongruent postures tend to be observed when they are talking self-oriented, negative, or nonspecific topics (Charny 1966).

Thus, rapport seems to have a mutual relationship with embodied synchrony, and as such, well-considered artifice in an experiment might be needed in order to examine its causal relationship in isolation.

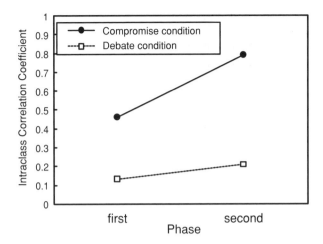

Figure 18.3 Congruence of response latencies in cooperative and competitive dialogues. Six dyads of participants with opposing views about a given topic reached a conclusion through discussion in the Compromise condition (cooperative dialogue) and another six dyads imposed their opinion on their partner in the Debate condition (competitive dialogue). The 15-minute dialogues were divided into two phases. Congruence in the duration of response latency between the two speakers in a dyad was calculated for the first and second phases in both conditions. Revised from Nagaoka *et al.* (2005a). (Reproduced by permission of Ammons Scientific, Ltd.)

Time

An increase in the number of interactions and the passage of time in an interaction are factors that strengthen embodied synchrony (e.g., Feldstein 1968; Nagaoka *et al.* 2005a; Natale 1975b; Welkowitz *et al.* 1976). For instance, in psychotherapy interviews, the total duration of periods in which postural congruence between a therapist and a client was observed increases as the interview progresses (Charny 1966).

However, the degree of embodied synchrony does not always increase along with the time. Nagaoka *et al.* (2006) have reported that counseling interviews that were highly evaluated by experts tended to involve an increasing degree of embodied synchrony in body movements and vocal behaviors (except for speech rates) as the interview progressed, whereas in another interview that was evaluated poorly by experts tended to involve a decreasing degree of embodied synchrony toward the last half of the session, especially at the closing (Figure 18.4). Based on these results, embodied synchrony of body movements and speech sounds is considered to be sensitive to degree of rapport, which changes across time (see above section for the relationship between rapport and embodied synchrony).

18.3.3 Cultural and Ethnic Factors

There have been very few studies on cross-cultural comparisons of embodied synchrony. However, it is supposed that there may be cultural differences between the West and the East with regard to the character and amount of embodied synchrony they display. Because there are differences in cognitive styles in information processing, judgment, and behavioral patterns between Westerners and Asians (e.g., Markus and Kitayama 1991; Nisbett 2003), and the cognitive styles influence mimicry of mannerisms (Van Baaren *et al.* 2004a). There have been only a small number of studies based on such interests (e.g., Kimura and Daibo 2006), and this certainly is an important issue that needs to be further investigated in the future.

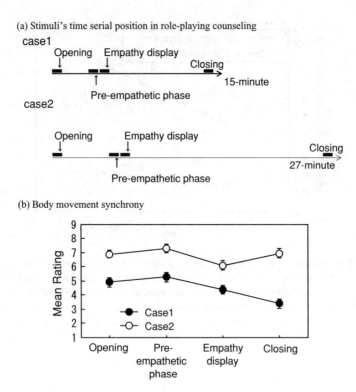

Figure 18.4 Embodied synchrony in body movement and its time serial changes. Experts evaluated case 1 as though "the enthusiasm of the counselor was conveyed to the client", "the client was not able to fully express his or her emotions". On the other hand, case 2 was evaluated as "having a warm atmosphere in general" and "an example of good counseling in which the client became able to imagine his or her future plan to a certain extent". (a) Four scenes from each case were used for stimuli: Opening (after sitting down), Pre-empathetic phase, Empathy display, and Closing (the last 1 minute was excluded). (b) 24 judges evaluated embodied synchrony of body movements and gestures, after having watched each stimulus. The judges gave ratings on 10-point scales, from 1 (very low level) to 10 (very high level). Partially revised from Nagaoka *et al.* (2006). (Reproduced by permission of Cognitive Science Society)

18.4 Embodied Synchrony and Human Behavior

This section examines the differences between interactions in which embodied synchrony is observed and those in which it is not. We focus specifically on the influence of embodied synchrony on (1) relationship maintenance and (2) performance, which are though to be the most important functions for successful group activity.

18.4.1 Relationship Maintenance

There are three major factors that are considered to be crucial for maintaining a good relationship, namely, (1) understanding other's emotions, (2) building empathy and rapport, and (3) giving positive impressions. Based on empirical studies, we discuss the influence embodied synchrony has on each of these areas. As indicated by the following discussion, the influence of embodied synchrony may vary depending on communication channels.

Understanding Other's Emotions

Mimicry of others' facial and vocal expressions of emotion is said to aid in the understanding of those emotions (Hess *et al.* 1998) because the innate reaction patterns of the observer's own facial muscles and glands feed back to their central nervous system to cause the corresponding emotions (Tomkins 1982). In addition, Hatfield *et al.* (1995) reported that mimicking recorded emotional vocal expressions could change one's emotional state, and suggested that vocal feedback influences not only one's facial expressions and postures but also one's emotional state.

Thus, mimicking others' emotional expression allows an observer to experience the same emotions as the other person (Hatfield *et al.* 1995; Hess *et al.* 1998). This process is a chain reaction that is called "emotional contagion" (e.g., Neumann and Strack 2000; Sonnby-Borgstrom *et al.* 2003). The fact that even newborns start crying when they hear other babies crying suggests that the basic method of understanding other's emotions through mimicry is gained at the very early stages of development.

Building Empathy and Rapport

Interactant's empathy toward the partner is thought to be essential for establishing rapport between them, and both of empathy and rapport are important in psychological counseling and is considered to be strengthened by embodied synchrony.

Maurer and Tindall (1983) have demonstrated that clients' perceptions of their counselor's level of empathy are positively affected by the amount of postural congruence the counselor displays. In the experiment, clients rated their counselor's level of empathy after a counseling session. Those counselors mirrored those client's arm and leg positions (congruent condition) were judged as having a higher level of empathy than those counselors who did not mirror the client's arm and leg positions (noncongruent condition). It is also reported that counselors who used more embodied synchrony in vocal behaviors and body movements were evaluated, by third-party observers, to be more empathetic (Staples and Sloane 1976) and trusted more by the client (Nagaoka *et al.* 2006). There has also been a study that investigated the relationship between congruent postures and empathy in a psychoanalytic dance therapy (Siegel 1995). Furthermore, in dance therapy, imitating clients' movement is considered to be the basic means of building a relationship with them (Schmais and Schmais 1983).

It is not always possible to establish a rapport with someone even if one shows the same postures as their partner, however. LaFrance and Ickes (1981) reported a negative correlation between rapport and posture mirroring in dyadic interactions between strangers in a waiting room. On the other hand, they reported a positive correlation between rapport and the total duration and frequency of utterances. Based on these results, LaFrance and Ickes (1981) concluded that the form of embodied synchrony that is appropriate in an interaction varies depending on the context of the interaction. In the waiting room interactions between strangers, they observed that it was speaking, rather than posture mirroring, which led to the building of rapport. There has also been a study that showed that the matching of linguistic style (how to use words) between chat participants is not related to the ratings of rapport reported by the participants (Niederhoffer and Pennebaker 2002).

Unfortunately, there are very few studies like Maurer and Tindall (1983) that empirically examine the causal relations between embodied synchrony and empathy or rapport. Much more work in this issue is needed.

Giving Positive Impressions

In general, those who display embodied synchrony are evaluated positively. Sociolinguistic studies have repeatedly reported that the convergence of accents or dialects in an intercultural communication positively influences the perception of interpersonal skills (e.g., Bourhis *et al.* 1975; Genesee and Bourhis

1982, 1988; Tong *et al.* 1999). Similar observations have been made on embodied synchrony in terms of speech rates, fundamental frequencies, and response latencies (e.g., Feldstein *et al.* 2001; Gregory *et al.* 1997, 2001; Street 1984). For example, Feldstein *et al.* (2001) reported that listeners tended to perceive those who spoke at about the same speech rate as themselves as more competent and attractive.

Among interpersonal impression, friendliness is considered to be a key to building strong relationships with others, and embodied synchrony in mannerisms and postures may be particularly related to the perception of friendliness (Bates 1975; LaFrance and Broadbent 1976; Navarre 1982). For instance, interaction partners who display identical mannerisms tend not only to be more positively evaluated (Chartrand and Bargh 1999) but also to receive larger donation (Van Baaren *et al.* 2004b) than those who did not display identical mannerisms.

It should be noted that those who display embodied synchrony are not always evaluated positively, however. If people find out that the interaction partner's postural imitation was actually the result of instruction from a third party, they do not evaluate the imitator highly (Bates 1975). In addition, it is not the case that imitating all characteristics of one's conversational partner always results in a good impression. Some researchers have pointed out that there is an optimal level of embodied synchrony (e.g., Berger and Bradac 1982; Giles and Smith 1979).

18.4.2 Performance

Unfortunately, little empirical research has been done on the influence of embodied synchrony on performance. However, one study has reported that people tend to think their interactive task was conducted more smoothly when their partners mimicked their nonverbal behaviors than when they did not (Chartrand and Bargh 1999). Therefore, it would be worthwhile to study the relationship between embodied synchrony and performance in the future, using objective indices such as performance efficiency or outcome as dependent variables.

Baaren *et al.* (2004a) implied that embodied synchrony may function effectively in certain tasks. They had experimenters mimic the posture and behavior of participants during an experimental task in one condition, while they did not do so in the other condition. They compared the cognitive styles of participants right after the task and reported that participants whose behaviors had been mimicked by an experimenter subsequently processed information in a more context-dependent manner (a tendency to depend more on external visual cues than one's own body when conducting a task) than did nonmimicked participants. Cognitive styles are related to individual types in information processing, judgment, and behavioral patterns, and are thus considered to be an important factor for performance efficiency and work volume. Therefore, this study suggests that mimicry is effective in tasks that require context dependence, such as memorizing a location.

18.5 Model and Theory of Embodied Synchrony

As discussed above, various factors affect embodied synchrony, as well as embodied synchrony affecting human behavior in various ways. As a result, a number of theories and models of embodied synchrony have been proposed. In this section, we will outline the main two models: (1) the model of mimicry in emotion communication (Hess *et al.* 1999), (2) the model of embodied synchrony on temporal pattern of speech (Nagaoka 2004), and a main theory: the communication accommodation theory (e.g., Giles and Smith 1979; Shepard *et al.* 2001). Those theories and models do not oppose each other; they are not mutual rivals. It can be supposed that they are schemes which respectively focus on one of various aspects that embodied synchrony possesses (a detailed explanation about this is given in section 18.6).

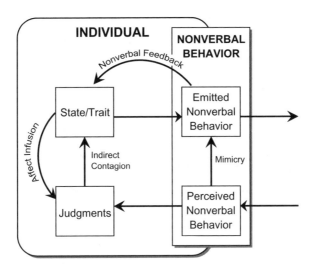

Figure 18.5 The model of emotion communication (Hess *et al.* 1999). The model explains mimicry in emotion communication, and argues that other's expressive display, directly and automatically, influence individual's emitted nonverbal behavior (mimicry). Furthermore, the model can explain the fact that embodied synchrony leads to understanding of other's emotions, since the model describes that emitted nonverbal behaviors influence individual's affective state through intrapersonal feed back (nonverbal feedback). (Reproduced by permission of Cambridge University Press, Hess *et al.*, 1999)

18.5.1 Model of Mimicry in Emotion Communication

The model of mimicry in emotion communication (Figure 18.5) was proposed by Hess *et al.* (1999), who have been investigating mimicry of facial expressions. In addition to mimicry of facial expressions, this model treats embodied synchrony in gestures, postures, and vocal behavior. It also includes the imitation of specific behavior such as wincing when observing other people's pain and increased forearm muscle tension when observing arm wrestling. As shown in Figure 18.5, this model argues that if we perceive an expressive display, it directly affects our nonverbal behavior, and considers the process to be automatic.

Similar ideas have been proposed by other researchers. Regarding mimicry of mannerisms, Chartrand and Bargh (1999) proposed the idea of the perception–behavior link, which considers mimicry as an automatic process and argues that watching a behavior by others will lead to the same behavior by the observer. In addition, Neumann and Strack (2000) showed that the mimicry of emotional vocal expressions does not change according to the attractiveness of the speaker or the cognitive load of the listener, and concluded that this process must be nonintentional and automatic.

18.5.2 Model of Embodied Synchrony on Temporal Pattern of Speech

Nagaoka (2004), who have investigated embodied synchrony of response latency, proposed a model of embodied synchrony on temporal patterns of speech (Figure 18.6). This model suggests that the degree of embodied synchrony is determined by two social factors: the speaker's receptiveness to the conversational partner and the speaker's social skills. Receptiveness to the conversational partner is the willingness to accept the conversational partner and it is the basis of empathy and rapport, which are factors that are built into counseling interviews and cooperative dialogues among others. Therefore, the model can explain why embodied synchrony is deeply related to empathy and rapport. The model emphasizes that if a

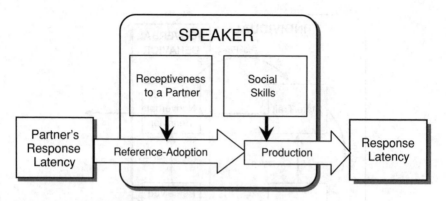

Figure 18.6 Model of embodied synchrony for temporal speech patterns (Nagaoka 2004). This model describes that speakers' receptiveness to their partner affects the process of reference-adoption of the partner's style, and that the speaker's social skills affect the process of speech production

speaker is receptive to the partner and has good social skills, the duration of the speaker's response latency would be similar to that of the partner. However, if a speaker has no receptiveness to the partner, or poor social skills, the duration of the speaker's response latency would be not similar to that of the partner. Thus, this model can explain why embodied synchrony is influenced by various social factors, while that is not really the case for Hess *et al.* (1999). And the model is predicted to be applicable to other aspects involving embodied synchrony, such as speech rates and posture, in addition to response latency.

18.5.3 Communication Accommodation Theory

The sociolinguist Giles and associates proposed the Communication accommodation theory (CAT; e.g., Giles and Smith 1979; Shepard *et al.* 2001), which explains convergence of language, accent, and dialect in intercultural communication (Figure 18.7). A large number of investigations on embodied synchrony in the field of sociolinguistics are based on the CAT theory. CAT argues that convergence is influenced by the socio-historical context, which is the factor that defines which group is the dominant one to whom one should converge, and from whom one should diverge.[2] In CAT, it is thought that convergence is driven by the need to gain approval from the partners, and divergence is conversely motivated to display distinctiveness from the partners. The absolute difference between CAT and the two models described above is in the frame of reference. In the above models, interactants conform to the nonverbal behavior of the partner. On the other hand, the frame of reference in CAT is considered to be the socio-historical context.

18.6 Conclusions

18.6.1 Embodied Synchrony Function as a Basis of Communication

As discussed above, embodied synchrony is observed in various communication channels, and it also represents various aspects of interactions and relationships among interactants. In Table 18.1, findings

---

[2] "Divergence" means to change one's own speech style to be different from the partner's. For example, in man–woman conversations in Jordan, male speakers use masculine pronunciation, while female speakers use feminine pronunciation (Al-Khatib 1995).

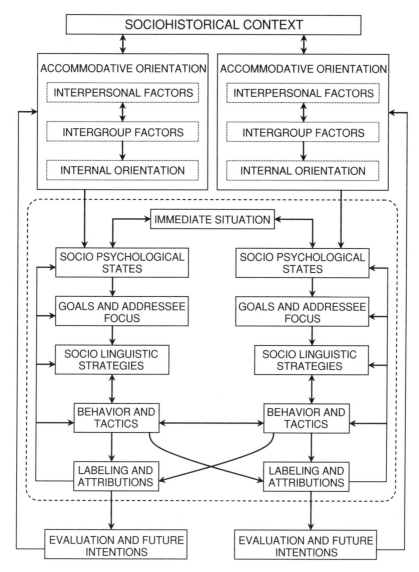

Figure 18.7 Communication accommodation theory in intercultural contexts. This figure represents two interactants in an intercultural context. It can be seen that the socio-historical context is the most critical factor, although various other variables dynamically affect the interactants' speech style. (Reproduced by permission of Sage Publications, Inc.)

from past studies are arranged according to the communication channel. This outline suggests that embodied synchrony includes two aspects: an inherent automatic aspect (left) and an acquired, more controlled aspect (right). This conceptualization helps to summarize our understanding of the relationship between different models and theories that have been proposed from various perspectives.

Examples that clearly illustrate the inherent automatic aspect are mimicry of emotional, facial, vocal and bodily expressions, as well as neonate imitation. Embodied synchrony of this aspect is influenced

Table 18.1 Outline of embodied synchrony. (Reproduced by permission of © 2005 IEEE)

	Embodied synchrony			
	Facial behavior (and emotional vocal or body expression)	Body movement and gesture	Vocal behavior	
			e.g., pause duration vocal intensity	e.g., dialect, accent
Development				
From early development	●	●		
Increasing degree with development			●	●
Factors				
Individual differences				
Emotional empathy	●	●		
Gender		●	●	
Disorder		●	●	
Cognitive empathy			●	
Social skills				
Situational factors				
Cooperativeness	●	●	●	
Rapport		●	●	
Time		●	●	
Effects				
Relationship maintenance				
Understanding other's emotions	●			
Building empathy and rapport		●	●	
Giving positive impressions		●	●	●
Model or theory				
Hess, Philippot & Blairy (1999) model	●			
Nagaoka (2004) model		○	●	
Communication Accommodation Theory			●	●

● represents established findings (adaption range of 'Models and Theories').
○ represents the predicted possible adaptation range.

by emotional empathy, and leads to emotional relationships with others through understanding their emotions and through building empathy and rapport. The model by Hess and co-workers provides a good explanation of this phenomenon. On the other hand, the acquired, more controlled aspect of embodied synchrony is illustrated by the convergence and congruence of the duration of response latencies, pauses, utterances, accents and so on, and accents. Embodied synchrony of this aspect seems to be influenced by relatively complicated socio-cognitive factors, such as cognitive empathy and social skills, and leads to socio-cognitive relation with others. Nagaoka's model and CAT provides an explanation of this aspect.

What is the function of embodied synchrony in communication? When verbal (i.e., symbolic) and nonverbal communication are compared, embodied synchrony seems to have traits that are peculiar to nonverbal communication, quite different from symbolic communication. Therefore, in order to consider the function of embodied synchrony, it might be necessary to turn focus on the pre-conscious factors.

Now, we would like to discuss the function of embodied synchrony. First, it is considered that embodied synchrony displays the participant's willingness to continue the interaction with the partner, though here the will is thought to be unconscious. Indeed, it seems that expression of the willingness is indicated unconsciously, and that partners cannot perceive it consciously. For example, when looking at a neonate imitating our facial gestures, we might recognize the neonate as having the mechanisms needed to interact with us. In psychological counseling, when the counselor's nonverbal behaviors display embodied synchrony, clients might feel that they are accepted by the counselor, even though they do not perceive embodied synchrony consciously.

Second, it is considered that embodied synchrony assures and emphasizes the sense that an interaction partner is in the same interaction as oneself, and at the same time as oneself. In other words, it assures and enhances the partner's social presence. Of course, the sense and social presence are usually not expected to reach consciousness. Social presence is defined as the "degree of salience of the other person in the interaction and the consequent salience of the interpersonal relationships" (Short *et al.* 1976, p. 65). This refers to the degree to which a participant can recognize the other as being present in the interaction in which one has participated at just the same time as oneself. The influence of embodied synchrony on social presence can be demonstrated by comparison between audiovisual communications, such as video telephone, in which the transmission is delayed and those in which it is not. When the media that allow seeing a conversation partner's face, such as a video telephone (very close to a face-to-face situation) is used, if the transmission is delayed, even if the delay is slight, the smoothness of an interaction will be severely reduced (e.g., Kurita *et al.* 1993). This might result from absence of embodied synchrony that should properly exist in natural situations. That is, under the situation where transmission is delayed, participants can observe neither the interpersonal synchrony of body movement, nor the quick mimicry of facial expression, but would observe the response that is delayed, rather than the expected response. For this reason, a participant might feel the partner as a presence who interacts with oneself in the past (i.e., a presence who is in a different time from oneself). When this happens, it is hard to interact. This remains a hypothesis, because the conventional embodied synchrony studies have been conducted in situations without including transmission delays. Investigations of embodied synchrony under transmission delay situations remain to be conducted.

It is possible that the feeling an interacting partner experiences the same time frame as oneself and is going to continue the interaction will engross a participant in an interaction, though the feeling does not usually reach consciousness, and then, the interaction will become firmer. Therefore, embodied synchrony is considered to function as a basis of communication, enabling the symbolic conveyance of information through verbal and nonverbal behaviors, and making the communication smoother, though embodied synchrony itself does not have symbolic function.

18.6.2 The Application of Embodied Synchrony and Future Tasks

It is considered that embodied synchrony will provide important clues to the future investigations of human interactions, mainly in two ways. The first is as an indicator of cooperativeness and empathy, among others. The second is its application as a means to enriching communication.

The presence/absence of embodied synchrony acts as an index to distinguish a group having cooperative (or empathetic) interaction from a crowd. The degree of embodied synchrony acts as an index indicating how cooperatively participants are going to continue the interaction.

Embodied synchrony as a means to enrich communication can be done by training skills to display embodied synchrony, or by making an embodied agent behave so that the nonverbal behaviors become similar, or synchronize with a user (e.g., Bailenson and Yee 2005; Ward and Nakagawa 2004). Of course, this requires making informed decisions on when to apply embodied synchrony, as well as detailed examinations regarding behavior after the informed decisions, because, as mentioned above, embodied synchrony may be negatively evaluated according to the situation.

There are multiple topics on embodied synchrony that need to be investigated further. As discussed above, particular topics include the influence of cultural and ethnic factors, as well as the influence of embodied synchrony on the level of empathy, rapport, or performance in an interaction. Furthermore, it is necessary to investigate embodied synchrony from the perspective of multiple communication channels. Nagaoka *et al.* (2006) is a pioneering study that quantified embodied synchrony in multi-communication channels. Results showed that embodied synchrony in body movement and speech sounds (except for speech rates) occur simultaneously. However, detailed research is needed on whether this would stand as a general principle. Finally, it will be necessary to investigate the relationship between linguistic content (e.g., conversation topics) and embodied synchrony in nonverbal behavior. This has been also pointed out by Charny (1966) and LaFrance and Ickes (1981).

References

Al-Khatib M. (1995) The impact of interlocutor sex on linguistic accommodation: a case study of Jordan radio phone-in programs. *Multilingua* **14** (2), 133–150.

Bailenson J.N. and Yee N. (2005) Digital chameleons: automatic assimilation of nonverbal gestures in immersive virtual environments. *Psychological Science* **16**, 814–819.

Ball P., Giles H. and Hewstone M. (1985) Interpersonal accommodation and situational constraints: an integrative formula. In H. Giles and R. St Clair (eds), *Recent Advances in Language, Communication & Social Psychology*. Lawrence Erlbaum, pp. 263–286.

Bates J.E. (1975) Effects of a child's imitation versus nonimitation on adults' verbal and nonverbal positivity. *Journal of Personality and Social Psychology* **31**, 840–851.

Berger C.R. and Bradac J.J. (1982) *Language and Social Knowledge: Uncertainty in interpersonal relations*. Edward Arnold.

Bernieri F.J., Reznick J.S. and Rosenthal R. (1988) Synchrony, pseudosynchrony and dissynchrony: measuring the entrainment process in mother–infant interactions *Journal of Personality and Social Psychology* **54**, 243–253.

Bernieri F.J. and Rosenthal R. (1991) Interpersonal coordination: Behavioral matching and interactional synchrony In R.S. Feldman and B. Rime (eds), *Fundamentals of Nonverbal Behavior*. Cambridge University Press, pp. 401–432.

Bernieri F.J., Gillis J.S., Davis J.M. and Grahe J.G. (1996) Dyad rapport and accuracy of its judgment across situations: a lens model analysis. *Journal of Personality and Social Psychology* **71**, 110–129.

Bourhis R.Y., Giles H. and Lambert W.E. (1975) Social consequences of accommodating one's style of speech: a cross-national investigation. *International Journal of the Sociology of Language* **6**, 55–71.

Cappella J.N. and Planalp S. (1981) Talk and silence sequences in informal conversations III: Interspeaker influence. *Human Communication Research* **7**, 117–132.

Cappella J.N. (1981) Mutual influence in expressive behavior: adult–adult and infant–adult dyadic interaction. *Psychological Bulletin* **89** (1), 101–132.

Charny M.D. (1966) Psychosomatic manifestations of rapport in psychotherapy. *Psychosomatic Medicine* **28** (4), 305–315.

Chartrand T.L. and Bargh J.A. (1999) The chameleon effect: the perception behavior link and social interaction. *Journal of Personality and Social Psychology* **76**, 893–910.

Cheng C.M. and Chartrand T.L. (2003) Self-monitoring without awareness: using mimicry as a nonconscious affiliation strategy. *Journal of Personality and Social Psychology* **85**, 1170–1179.

Condon W.S. and Ogston M.B. (1966) Sound film analysis of normal and pathological behavior patterns. *Journal of Nervous Disease* **143** (4), 338–347.

Condon W.S. and Sander L.S. (1974) Neonate movement is synchronized with adult speech: interactional participation and language acquisition. *Science* **183**, 99–101.

Daibo I. (1985) A brief review of synchrony tendency in interpersonal communication: a case of vocal behavior. *The Yamagata Psychological Reports* **4**, 1–15.

Dimberg U. and Lundqvist L.O. (1988) Facial reaction to facial expressions: sex differences. *Psychophysiology* **25**, 442–443.

Eckerman C.O., Davis C.C. and Didow S.M. (1989) Toddler's emerging ways of achieving social coordinations with a peer. *Child Development* **60**, 440–453.

Feldstein S. (1968) Interspeaker influence in conversational interaction. *Psychological Reports* **22**, 826–828.

Feldstein S. and Field T. (2002) Vocal behavior in the dyadic interactions of preadolescent and early adolescent friends and acquaintances. *Adolescence* **37**, 495–513.

Feldstein S. and Welkowitz J. (1978) Conversational congruence: correlates and concerns. In A. Siegman and S. Feldstein (eds), *Nonverbal Behavior and Communication*. Lawrence Erlbaum Associates, pp. 358–378.

Feldstein S., Dohm F.A. and Crown C.L. (2001) Gender and speech rate in the perception of competence and social attractiveness. *Journal of Social Psychology* **141**, 755–806.

Gallois C., Giles H., Jones E., Cargile A.C. and Ota H. (1995) Accommodating intercultural encounters: elaborations and extensions. In R.L. Wiseman (ed.), *Intercultural Communication Theory*. Sage Publications, pp. 115–147.

Garvey C. and BenDebba M. (1974) Effect of age, sex, and partner on children's dyadic speech. *Child Development* **45**, 1159–1161.

Genesee F. and Bourhis R. (1982) The social psychological significance of code switching in cross-cultural communication. *Journal of Language and Social Psychology* **1**, 1–27.

Genesee F. and Bourhis R.Y. (1988) Evaluative reactions to language choice strategies: the role of socio-structural factors. *Language and Communication* **8**, 229–250.

Giles H. and Smith P. (1979) Accommodation theory: optimal levels of convergence In H. Giles and R.N. St. Clair (eds), *Language and Social Psychology*. Basil Blackwell, pp. 45–65.

Gregory S.W. (1990) Analysis of fundamental frequency reveals covariation in interview partners' speech. *Journal of Nonverbal Behavior* **14**, 237–251.

Gregory S.W., Dagan K.A. and Webster S. (1997) Evaluating the relation of vocal accommodation in conversation partner's fundamental frequencies to perceptions of communication quality. *Journal of Nonverbal Behavior* **21** (1), 23–43.

Gregory S.W., Green B.E., Carrothers R.M., Dagan K.A. and Webster S. (2001) Verifying the primacy of voice fundamental frequency in social status accommodation. *Language and Communication* **21**, 37–60.

Gump B.B. and Kulik J.A. (1997) Stress, affiliation and emotional contagion. *Journal of Personality and Social Psychology* **72**, 305–319.

Hatfield E., Hsee C.K., Costello J., Weisman M.S. and Denney C. (1995) The impact of vocal feedback on emotional experience and expression. *Journal of Social Behavior and Personality* **10**, 293–313.

Hess U., Philippot P. and Blairy, S. (1998) Facial reactions to emotional facial expressions: affect or cognition? *Cognition and Emotion* **12**, 509–532.

Hess U., Philippot P. and Blairy S. (1999) Mimicry: fact and fiction. In P. Philippot and R.S. Feldman (eds), *The Social Context of Nonverbal Behavior*. Cambridge University Press, pp. 213–241.

Jaffe J. and Feldstein S. (1970) *Rhythms of Dialogue*. Academic Press.

Jasnow M.D. and Feldstein S. (1986) Adult-like temporal characteristics of mother-infant vocal interactions. *Child Development* **57**, 754–761.

Jasnow M.D., Crown C.L., Feldstein S., Taylor L., Beebe B. and Jaffe J. (1988) Coordinated interpersonal timing of Down-syndrome and nondelayed infants with their mothers: evidence for a buffered mechanism of social interaction. *Biological Bulletin* **175**, 355–360.

Kendon A. (1970) Movement coordination in social interaction: some examples described. *Acta Psychologica* **32**, 101–125.

Kimura M. and Daibo I. (2006) Interactional synchrony in conversations about emotional episodes: a measurement by the between-participants pseudosynchrony experimental paradigm. *Journal of Nonverbal Behavior* **30**, 115–126.

Komori M., Maeda K. and Nagaoka C. (2007) A video-based quantification method of body movement synchrony: an application for dialogue in counseling. *Japanese Journal of Interpersonal and Social Psychology* **7**, 41–48.

Kurita T., Iai S. and Kitawaki N.N. (1993) Effects of transmission delay in audiovisual communication. *Transactions of the Institute of Electronics, Information and Communication Engineers* **J76-B-I** (4), 331–339.

LaFrance M. and Broadbent M. (1976) Group rapport: posture sharing as a nonverbal indicator. *Group and Organization Studies* **1**, 328–333.

LaFrance M. (1979) Nonverbal synchrony and rapport: analysis by the cross-lag panel technique. *Social Psychology Quarterly* **42** (1), 66–70.

LaFrance M. and Ickes W. (1981) Posture mirroring and interactional involvement: sex and sex typing effects. *Journal of Nonverbal Behavior* **5**, 139–154.

Lanzetta J.T. and Englis B.G. (1989) Expectations of cooperation and competition and their effects on observers' vicarious emotional responses. *Journal of Personality and Social Psychology* **56**, 543–554.

Lundqvist L.O. (1995) Facial EMG reactions to facial expressions: a case of facial emotional contagion? *Scandinavian Journal of Psychology* **36**, 130–141.

Markus H. and Kitayama S. (1991) Culture and the self: implications for cognition, emotion and motivation. *Psychological Review* **98**, 224–253.

Matarazzo J.D. (1965) The interview. In B.J. Wolmaa (ed.), *Handbook of Clinical Psychology*. McGrawHill, pp. 403–450.

Matarazzo J.D., Wiens A.N., Saslow G., Dunham R.M. and Voas R. (1967) Speech durations of astronaut and ground communicator. *Science* **143**, 148–150.

Matarazzo J.D. and Wiens A.N. (1967) Interviewer influence on durations of interviewee silence. *Experimental Research in Personality* **2**, 56–69.

Maurer R.E. and Tindall J.F. (1983) Effect of postural congruence on client's perception of counselor empathy. *Journal of Counseling Psychology* **30** (2), 158–163.

Mehrabian A. and Epstein N. (1972) A measure of emotional empathy. *Journal of Personality* **40**, 525–543.

Meltzer L., Morris W.N. and Hayes D.P. (1971) Interruption outcomes and vocal amplitude: explorations in social psychophysics. *Journal of Personality and Social Psychology* **18** (3), 392–402.

Meltzoff A.N. and Moore M.K. (1977) Imitation of facial and manual gestures by human neonates. *Science* **198**, 75–78.

Nagaoka C. (2004) Mutual influence of nonverbal behavior in interpersonal communication. Unpublished doctoral dissertation, Osaka University.

Nagaoka C., Komori M. and Nakamura T. (2002) The interspeaker influence of the switching pauses in dialogues. *Japanese Journal of Ergonomics* **38** (6), 316–323.

Nagaoka C., Komori M. and Nakamura T. (2003) The inter-speaker influence in dialogues: study of temporal aspects. *Technical Report of IEICE* **103** (HCS2003-6), 19–24.

Nagaoka C., Komori M., Nakamura T. and Draguna M.R. (2005a) Effects of receptive listening on the congruence of speakers' response latencies in dialogues. *Psychological Reports* **97**, 265–274.

Nagaoka C., Komori M. and Yoshikawa S. (2005b) Synchrony tendency: interactional synchrony and congruence of nonverbal behavior in social interaction. In *Proceedings of the 2005 International Conference on Active Media Technology* (AMT2005), pp. 529–534.

Nagaoka C., Yoshikawa S. and Komori M. (2006) Embodied synchrony of nonverbal behaviour in counselling: a case study of role playing school counselling. In *Proceedings of the 28th Annual Conference of the Cognitive Science Society* (CogSci 2006), pp. 1862–1867.

Nakamura T. (2002) Sensuous information transmitted by Ma. *Journal of Japan Society for Fuzzy Theory and Systems* **14** (1), 15–21.

Natale M. (1975a) Convergence of mean vocal intensity in dyadic communication as a function of social desirability. *Journal of Personality and Social Psychology* **32** (5), 790–804.

Natale M. (1975b) Social desirability as related to convergence of temporal speech patterns. *Perceptual and Motor Skills* **40**, 827–830.

Navarre D. (1982) Posture sharing in dyadic interaction. *American Journal of Dance Therapy* **5**, 28–42.

Neumann R. and Strack F. (2000) Mood contagion: the automatic transfer of mood between persons. *Journal of Personality and Social Psychology* **79**, 211–223.

Newman L. and Smit A. (1989) Some effects of variations in response time latency on speech rate, interruptions, and fluency in children's speech. *Journal of Speech and Hearing Research* **2**, 635–644.

Niederhoffer K.G. and Pennebaker J.W. (2002) Linguistic style matching in social interaction. *Journal of Language and Social Psychology* **21**, 337–360.

Nisbett R. (2003) *The Geography of Thought: How Asians and Westerners think differently, and why*. Free Press.

Richardson M.J., Marsh K.L. and Schmidt R.C. (2005) Effects of visual and verbal interaction on unintentional interpersonal coordination. *Journal of Experimental Psychology: Human Perception and Performance* **31** (1), 62–79.

Scheflen A.E. (1964) The significance of posture in communication systems. *Psychiatry* **27**, 316–331.

Schmais C. and Schmais A. (1983) Reflecting emotions: the movement-mirroring test. *Journal of Nonverbal Behavior* **8**, 42–54.

Shepard C.A., Giles H. and LePoire B.A. (2001) Communication accommodation theory. In W.P. Robinson and H. Giles (eds), *The New Handbook of Language and Social Psychology*. John Wiley, pp. 33–56.

Shockley K., Santana M.V. and Fowler C.A. (2003) Mutual interpersonal postural constraints are involved in cooperative conversation. *Journal of Experimental Psychology: Human Perception and Performance* **29**, 326–332.

Short J., Williams E. and Christie B. (1976) *The Social Psychology of Telecommunications*. John Wiley.

Siegel E.V. (1995) Psychoanalytic dance therapy: the bridge between psyche and soma. *American Journal of Dance Therapy* **17**, 115–128.

Sonnby-Borgstrom M., Jonsson P. and Svensson O. (2003) Emotional empathy as related to mimicry reactions at different levels of information processing. *Journal of Nonverbal Behavior* **27**, 3–23.

Staples F.R. and Sloane R.B. (1976) Truax factors, speech characteristics, and therapeutic outcome. *Journal of Nervous and Disease* **163** (2), 135–140.

Street R.L. (1984) Speech convergence and speech evaluation in fact-finding interviews. *Human Communication Research* **11** (2), 139–169.

Street R. and Cappella J. (1989) Social and linguistic factors influencing adaptation in children's speech. *Journal of Psycholinguistic Research* **18**, 497–519.

Tomkins S.S. (1982) Affect theory. In P. Ekman (ed.), *Emotion in the Human Face*. Cambridge University Press, pp. 353–395.

Tong Y.Y., Hong Y.Y., Lee S.L. and Chiu C.Y. (1999) Language use as carrier of social identity. *International Journal of Intercultural Relations* **23**, 281–296.

Van Baaren R.B., Horgan T.G., Chartr T.L. and Dijkmans M. (2004a) The forest, the trees, and the chameleon: context dependency and nonconscious mimicry. *Journal of Personality and Social Psychology* **86**, 453–459.

Van Baaren R.B., Holl R.W., Kawakami K. and Van Knippenberg A. (2004b) Mimicry and pro-social behavior. *Psychological Science* **15**, 71–74.

Vaughan K.B. and Lanzetta J.T. (1980) Vicarious instigation and conditionings of facial expressive and automatic responses to a model's expressive display of pain. *Journal of Personality and Social Psychology* **38**, 909–923.

Vaughan K.B. and Lanzetta J.T. (1981) The effect of modification of expressive displays on vicarious emotional arousal. *Journal of Experimental Social Psychology* **17**, 16–30.

Ward N. and Nakagawa S. (2004) Automatic user-adaptive speaking rate selection. *International Journal of Speech Technology* **7** (4), 235–238.

Watanabe T. and Okubo M. (1998) Physiological analysis of entrainment in communication. *Trans Info Proc Soc of Japan* **39**, 1225–1231.

Welkowitz J., Feldstein S., Finklestein M. and Aylesworth L. (1972) Changes in vocal intensity as a function of interspeaker influence. *Perceptual and Motor Skills* **35**, 715–718.

Welkowitz J. and Kuc M. (1973) Interrelationships among warmth, genuineness, empathy, and temporal speech patterns in interpersonal interaction. *Journal of Consulting and Clinical Psychology* **41** (3), 472–473.

Welkowitz J., Cariffe G. and Feldstein S. (1976) Conversational congruence as a criterion of socialization in children. *Child Development* **47**, 369–272.

Welkowitz J., Bond R.N., Feldman L. and Tota M.E. (1990) Conversational time patterns and mutual influence in parent–child interactions: a time series approach. *Journal of Psycholinguistic Research* **19**, 221–243.

Willemyns M., Gallois C., Callan V.J. and Pittam J. (1997) Accent accommodation in the job interview: impact of sex and interviewer accent. *Journal of Language and Social Psychology* **16**, 3–22.

Yaruss J.S. and Conture E.G. (1995) Mother and child speaking rates and utterance lengths in adjacent fluent utterances: preliminary observations. *Journal of Fluency Disorders* **20**, 257–278.

19

Modeling Communication Atmosphere

Tomasz M. Rutkowski and Danilo P. Mandic

19.1 Introduction

Whereas "communication atmosphere" is a common term, it is not straightforward to define this notion as a research objective. This is largely due to the fact that humans have a well understood and developed ability to assess, create and recognize the psycho-social aspects of this problem; the mapping of these onto a set of machine learning features is rather demanding and still in its infancy. To that end, we set out to formalize this process in order to investigate the three observable characteristics of the communication process: dimensions of a climate of a meeting, which are determined, based on *social features*; *mental states* of the communicators; and *physical features* of the environment where the event takes place. In our opinion these are essential to create and re-create the "climate" of a meeting (broadly understood as communication atmosphere). The results in this chapter build upon the concept of a *communicative interactivity* discussed in Rutkowski and Mandic (2005, 2006) and Rutkowski *et al.* (2004b,c, 2005a), and represent a unifying generalization of our previous results.

Two main generic application areas for the introduced approach are in intelligent analysis of the communication process, and also for evaluation of the communication atmosphere. One direct side-application would be the indexing of multimedia archives, based on the query-specific level of the interaction of the participants. A second (and more challenging) application would be in the creation, maintenance, and predictive modification of the communication atmosphere and telepresence, such as in virtual chat rooms. The machine learning aspects of this research include the tracking of the features that describe human communication, which are then used to evaluate the behavioral (interaction) aspects of this paradigm. This virtue of our approach may prove very useful for distance communication or learning applications, where monitoring for the quality and transparency of the system is of crucial importance. A communication analysis system is presented in Figure 19.1, where two video and audio streams are captured separately but further analyzed together in order to extract necessary features for communication atmosphere estimation.

Conversational Informatics: An Engineering Approach   Edited by Toyoaki Nishida
© 2007 John Wiley & Sons, Ltd

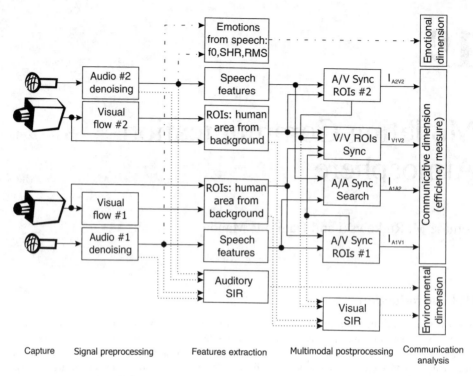

Figure 19.1   The communication atmosphere analysis system chart

19.2 Communication Atmosphere

The need for monitoring (and subsequently modeling) the communication atmosphere arises when a machine learning application is required to evaluate:

- the extent to which the communicators' behavior (interactions) influences the communications climate
- the extent to which the external environment where the communication takes place (e.g., a room) affects the communication efficiency
- the extent to which emotional states of the communicators shape the final assessment of the climate of the event.

To address these factors, which ultimately build up the communication atmosphere, we perform evaluation and fusion of information of data streams related to this phenomenon. To achieve this we propose a threefold approach in which the dimensions of the communication atmosphere are defined as follows:

1. *Environmental*, describing the physical attributes of the meeting's environment (noise, street, office, cafe). This dimension is important in order to comply with the fact that communication "episodes" (i.e., meetings) are not conducted in a vacuum but in real places; these can largely bias or at least influence the communication climate and can thus enhance or disturb the automatic features detection, estimation, and selection.
2. *Communicative*, characterizing the behavior of the participants. This dimension is introduced in order to reflect the communicators' ability to interact and is therefore related to the communication

efficiency, a qualitative measure of the communication process directly related to the attention level and to the dynamic involvement of the communicators (Kryssanov and Kakusho 2005).

3. *Emotional*, estimating emotional states of the participants. This dimension is introduced in order to account for the fact that the behavior related to emotional states of the communicators to a large extent determines the communication atmosphere. It is expected that emotions expressed by the communicating parties are correlated (or anti-correlated), or else emotions shown by the sender are somewhat reflected (after a delay) by the receivers (Dimberg *et al.* 2000; Pease and Pease 2004) creating a corresponding emotional state of the situation.

From the technological point of view, in the case of multimedia content, such as video captured during a meeting or discussion, analysis of the communication atmosphere along these dimensions appears to be easier to perform. In this case, the above three dimensions can be dealt with independently, and for consistency any arising interdependencies are beyond the scope of this chapter.

In his study of kinetics, anthropologist Birdwhistell (1974) discussed nonverbal communication. According to his results, humans can recognize about 250 000 facial expressions, which is a number too large for an artificial intelligence-based classification to be feasible (Pantie and Rothkrantz 2000; Ralescu and Hartani 1997). In addition, Mehrabian (1971) found that the verbal component of face-to-face communication comprises less than 35% of the whole communication volume, and that over 65% of communication may be conducted nonverbally. Experiments reported in Pease and Pease (2004) also demonstrated that, in business-related encounters, body language accounts for between 60% and 80% of the impact made around the negotiating table.

19.2.1 Environmental Dimension in Communication Atmosphere

It is important to notice that highly visually or auditory intensive environments affect the overall impression of the observed/perceived communication situation. On the other hand, since the communicators have usually a limited ability to change the environmental features (i.e., the level of external audio or video), this study recognizes the environmental characteristics as a distinct feature set taken from the overall evaluation of communication. Physical features of the environment can be extracted after the separation of the recorded information streams into two categories: items related to the communication process and items unrelated to it (not useful) (Jelfs *et al.* 2006; Rutkowski *et al.* 2004a). The general idea therein is to split the audio and video streams into the background noise and useful signals produced by the communicators.

To assess the influence of the environmental dimension on communication atmosphere, we shall first detect the presence of auditory and visual events that are not related to the communicators' actions (i.e., background audio and video). Therefore, the analysis of the environmental dimension is performed in two stages:

- noise and non-speech power level difference extraction
- non-communication-related visual activity (background flow) estimation.

The amount of environmental audio energy (classified as noise) is estimated as a segmental signal-to-interference ratio (SIR), since calculation of an integral signal-to-noise ratio would not reflect temporal fluctuations of the situation dynamics (e.g., when communicators speak louder or quieter). The segmental auditory SIR, denoted by A_{SIR}, is evaluated as:

$$A_{SIR}(m,t) = 10 \log_{10} \frac{\sum_{n=m}^{m+N} s_e^2(n,t)}{\sum_{n=m}^{m+N} s_n^2(n,t)} \tag{19.1}$$

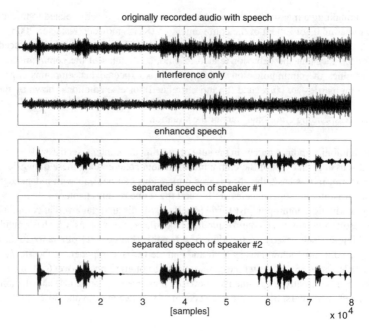

Figure 19.2 Environmental dimension: the wide-band noise might completely cover the speech usable frequencies. Observe: recorded signal in top box, interference only in second box, enhanced speech in middle box, and finally separated activities of two speakers. Notice the middle segment, when both speakers were active

where $s_n(n) = s_o(n) - s_e(n)$ is the noise estimate after removing the denoised version from the original signal; $s_o(n)$ is the recorded audio signal (with noise); N stands for the number of audio samples. For visual information, we compare the activity features detected in the communicators' areas with the remaining background. We calculate the amount of visual flow in the signal as an interference-like coefficient, V_{SIR}, given by:

$$V_{SIR}(m, t) = 10 \log_{10} \frac{\sum_{n=m}^{m+N} v_h(n, t)^2}{\sum_{n=m}^{m+N} v_b(n, t)^2} \tag{19.2}$$

where $v_b(n)$ represents the background visual flow features and $v_h(n)$ is related to the extracted motion features of the active communicators. Both A_{SIR} and V_{SIR} are then summed up to form a single audiovisual SIR measure characterizing the environmental conditions.

Figure 19.2 illustrates the evaluation of the environmental dimension, where the audio components of the environment are filtered from the information related to the communication process.

19.2.2 Role of the Communicative Dimension in Communication Atmosphere

The third and, perhaps, most important component of the communication atmosphere analysis is the communicative dimension, which is referred to the communicators' audiovisual behavior and to their ability to "properly" interact during the conversation we here apply. The synchronization and interaction measures (efficiency-like) developed in the authors' previous research are applied here (Rutkowski *et al.*

2003a, 2004a). The communication model used – the hybrid linear/transactional – is linear in short time windows (Adler and Rodman 2003). The active (in short time windows) communicator – the sender – is supposed to generate more audiovisual flow with breaks, when the receiver responds. On the other hand, the passive (in short time windows) communicator – the receiver – is expected to react properly, not disturbing (overlapping with) the sender's communication activity. Turn-taking (role changing) between the senders and receivers is a critical assumption in the hybrid communication model. For convenience only, the case of intentional communication, which occurs when all the communicators are willing to interact, is considered here. All the situations when so-called metacommunication (Adler and Rodman 2003) occurs are beyond the scope of this chapter.

Rutkowski *et al.* (2003a) defined the communication efficiency as *a measure that characterizes the behavioral coordination of communicators*. This measure describes the communication process from the point of view of the interactive sending and receiving of messages by the communicators, as observed through the audiovisual channel. Since, in terms of machine learning, there are no means to evaluate the understanding of the messages by the communicators, we restrict ourselves to the feedback or the receiver's reaction during communication.

19.2.3 *Emotional Dimension in Communication Atmosphere*

Experiments reported in Dimberg *et al.* (2000) and Pease and Pease (2004) showed that the unconscious mind exerts direct control of facial muscles, yet contemporary approaches to emotional state estimation usually deal with a static analysis of facial expressions and seldom consider a dynamic or multimodal analysis. It is intuitively clear, and was also demonstrated by experiments, that facial emotions presented by a sender (e.g., a smile) were reciprocated by a smile by the receiver. The experiments of Dimberg *et al.* (2000) were conducted using electromyography, in order to capture actual muscle activity of communicators. Results of those experiments revealed that the communicators often do not have total control over their expressed facial emotions. These findings clearly suggest that important aspects of "emotional face-to-face" communication could occur at the unconscious level.

In our approach we estimate emotional states from available nonverbal features only, which actually reflect (partially) the mental state of the communicator. Emotions of the communicators can be estimated from their speech only (Dellaert *et al.* 1996), since the video features are harder to analyze, owing to insufficient resolution and problems related to a simple definition of emotions captured from moving faces and bodily expressions. In our approach, the emotions from speech are estimated based on three features: *the voice fundamental frequency* with durational aspects of the stable fundamental frequency periods; *the speech harmonic content*; and the *signal energy expressed* as an averaged root mean square (RMS) value in voiced frames (Rutkowski *et al.* 2004a). Primary emotions, such as neutral, sad, angry, happy, and joyful, which the communicator might experience during the communication event, are determined using a machine learning algorithm (Vapnik 1995).

19.3 Automatic Assessment of Communication Atmosphere: Machine Learning Methods

This section presents the analysis of low-level sensory features followed by higher level intelligent associations among sensory features. The underlying aim of this analysis is to identify those audiovisual features of the (human) communication process that can be tracked and which, from an information processing point of view, are sufficient to create and recreate the climate of a meeting ("communicative interactivity"). This communicative interactivity analysis provides a theoretical, computational and implementation-related framework in order to characterize the human-like communicative behavior.

The analysis of spoken communication is already a mature field, and following the above arguments, our approach will focus on the dynamic analysis of nonverbal components of the communication. In the proposed model of communicative interactivity, based on interactive (social) features of captured situations, two sensory modalities (visual and auditory) of communicative situations are utilized (Rutkowski *et al.* 2005b).

The working hypothesis underlying our approach is therefore that observations of the nonverbal communication dynamics contain sufficient information to allow us to estimate the climate of a situation; that is, the communicative interactivity. To that end, the multimodal information about the communication environment must be first separated into the *communication-related* and *environmental* components. In this way, the audio and video streams can be separated into the information of interest and background noise (Mandic 2004; Mandic and Chambers 2001; Rutkowski *et al.* 2003b, 2004a).

19.3.1 Audio Features

Since the audio signal carries much redundant information, techniques used in speech recognition (Becchetti and Ricotti 1999) can be used for feature extraction and compression, for instance the Mel-frequency cepstrum coefficients (MFCC) (Furui 2001). The basic frequency resolution is based on the Mel approximation to the bandwidths of tuned resonators of the ear of humans. A digital cosine transformation (DCT) used to obtain MFCC coefficients produces highly uncorrelated features, making the stochastic characterization of the feature extraction process simpler. A zero-order MFCC coefficient is approximately equivalent to the *log* energy of the signal frame: it is generally discarded, since the energy of the signal is usually computed directly from the signal. In our work we found that the first 24 MFCC coefficients are sufficient for fair comparison of the speech; see Figure 19.3 for simultaneous comparison of MFCC and video features extracted from face-to-face conversation.

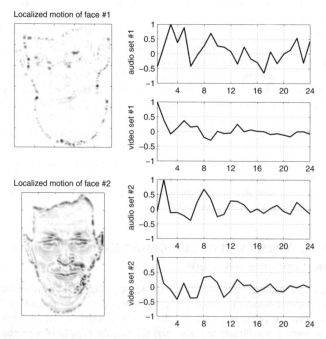

Figure 19.3 Both audio (MFCC) features and video motion (DCT) features extracted from two pairs of video and audio streams recorded during person-to-person communication. Face 1 presents limited activity and both audiovisual features are less synchronized in comparison to face 2, which has better synchronized multimodal features

19.3.2 Video Features

We desire to obtain video features that carry information about the communication-related motion, and which are also compatible with the audio features. Two modalities, the search for faces and moving contours, are combined to detect communicating humans in video. This is achieved as follows. For two consecutive video frames $\mathbf{f}_{[h \times w]}(t-1)$ and $\mathbf{f}_{[h \times w]}(t)$, the temporal gradient is expressed as a smoothed difference between the images convolved with a two-dimensional Gaussian filter $\mathbf{g}$ with the adjusted standard deviation σ. The pixel $\mathbf{G}_{[h \times w]}(t, n, m)$ of the gradient matrix at time t is calculated as:

$$\mathbf{G}_{[h \times w]}(t, n, m) = \left| \sum_{i=1}^{x} \sum_{j=1}^{y} \mathbf{d}_{[h \times w]}(t, n-i, m-j) \mathbf{g}_{[x \times y]}(\sigma, i, j) \right| \qquad (19.3)$$

where $\mathbf{d}_{[h \times w]}(t)$ is the difference between consecutive frames of the size $h \times w$ pixels: $\mathbf{d}_{[h \times w]}(t) = \mathbf{f}_{[h \times w]}(t) - \mathbf{f}_{[h \times w]}(t-1)$. The absolute value $|\cdot|$ is taken to remove the gradient directional information and to enhance the movement capture.

For the face tracking, the features used were skin color (albeit very sensitive to illumination variations) (Sigan et al. 2000), and moving face pattern recognition with patterns obtained using a non-negative matrix factorization method (Lee and Seung 1999). Features obtained in this way (details in Rutkowski et al. (2003a) are more localized and correspond to the intuitively perceived parts of faces (face, eyes, nose, and mouth contours).

The extracted facial features, together with the audio synchronized information, permit us to classify a participant in communication into the model of *talking*, *listening*, and *responding* as discussed in section 19.3.3. To enable compatibility with the extracted audio features and for information compression, the two-dimensional digital cosine transformation (DCT) is used, where most of the energy within the processed image is contained in a few uncorrelated DCT coefficients (24 coefficients in our approach: (Rutkowski et al. 2003a, 2004a).

19.3.3 Evaluation of Communicative Interactivity: Higher Level Associations

Communicative interactivity evaluation assesses the behavior of the participants in the communication from the audiovisual channel, and reflects their ability to "properly" interact in the course of conversation. This is quantified by synchronization and interaction measures (Rutkowski et al. 2004a). In Rutkowski et al. (2003a) the *communication efficiency* is defined as *a measure that characterizes the behavioral coordination of communicators*. Figure 19.4 illustrates the model used in our approach, where a measure of the communication efficiency is proposed as a fusion (Gautama et al. 2004; Mandic et al. 2005a,b) of four estimates of mutual information (Shannon and Weaver 1949): (i) two visual (V_i), (ii) two audio(A_i), and (iii) two pairs of audiovisual features (A_i; V_i).

Presence of Communication

Presence of communication is judged based on mutual information(s) for selected regions of interest (ROI) (Hyvarinen et al. 2001), as:

$$\begin{aligned}
I_{A_i V_i} &= H(A_i) + H(V_i) - H(A_i, V_i) \\
&= \frac{1}{2} \log(2\pi e)^n \left| R_{A_i} \right| + \frac{1}{2} \log(2\pi e)^m \left| R_{V_i} \right| - \frac{1}{2} \log(2\pi e)^{n+m} \left| \mathbf{R}_{A_i V_i} \right| \\
&= \frac{1}{2} \log \frac{\left| R_{A_i} \right| \left| R_{V_i} \right|}{\left| R_{A_i V_i} \right|}
\end{aligned} \qquad (19.4)$$

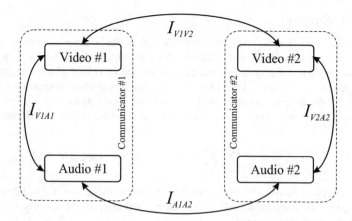

Figure 19.4   Scheme for the communicative interactivity evaluation. Mutual information estimates $I_{A_1 V_1}$ and $I_{A_2 V_2}$ between audio and visual features streams of localized communicators account for the local synchronization. The estimates $I_{A_1 A_2}$ and $I_{V_1 V_2}$ are to detect crosstalks in the same modality

where $i = 1, 2$ and $R_{A_i}, R_{V_i}, R_{A_i V_i}$ stand for empirical estimates of the corresponding covariance matrices of the feature vectors (Rutkowski *et al.* 2003a) (computed recursively).

Simultaneous Activity

Simultaneous activity estimates in the same modes (audio and video respectively) are calculated for the respective video and audio streams, as:

$$I_{V_1 V_2} = \frac{1}{2} \log \frac{|R_{V_1}| \, |R_{V_2}|}{|R_{V_1 V_2}|} \quad \text{and} \quad I_{A_1 A_2} = \frac{1}{2} \log \frac{|R_{A_1}| \, |R_{A_2}|}{|R_{A_1 A_2}|} \tag{19.5}$$

where $R_{A_1 A_2}$ and $R_{V_1 V_2}$ are the empirical estimates of the corresponding covariance matrices for unimodal feature sets representing different communicator activities. Quantities $I_{A_1 V_1}$ and $I_{A_2 V_2}$ evaluate the local synchronicity between the audio (speech) and visual (mostly facial movements) flows and it is expected that the sender should exhibit the higher synchronicity, reflecting the higher activity. Quantities $I_{V_1 V_2}$ and $I_{A_1 A_2}$ are related to the possible crosstalks in same modalities (audio–audio, video–video). The latter is also useful to detect the possible activity overlapping, which can impair the quality of the observed communication. See Figure 19.5 for a track of all four mutual information quantities estimated from ongoing face-to-face conversation. A combined measure of temporal communication efficiency can be calculated as:

$$C(t) = \left(1 - \frac{I_{V_1 V_2}(t) + I_{A_1 A_2}(t)}{2} \right) \left| I_{A_1 V_1}(t) - I_{A_2 V_2}(t) \right|. \tag{19.6}$$

All the above quantities from Equations (19.4), (19.5), and (19.6) are depicted in Figure 19.6 on a frame by frame basis of analyzed face-to-face conversation. A detailed dynamics track of efficiency quantity from Equation (19.6) is illustrated in Figure 19.7.

The Communicator's Role

A role which every communicator takes, that is *sender* or *receiver*, can be estimated by monitoring the behavior of audiovisual features over time. It is intuitively clear that an indication of higher synchronization between the audio and video features characterizes the active member, *the sender*, while the lower

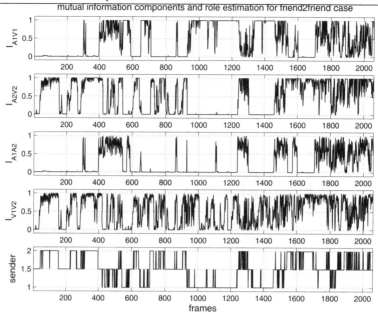

Figure 19.5 Audiovisual activities in the form of four mutual information tracks $I_{A_1V_1}$, $I_{A_2V_2}$, $I_{A_1A_2}$ and $I_{V_1V_2}$. The role in conversation estimation is shown in the bottom panel (value 1 stands for first communicator being recognized as sender and 2 for the second one; 1.5 stands for situations when both communicators are active at a time–less efficient time slots)

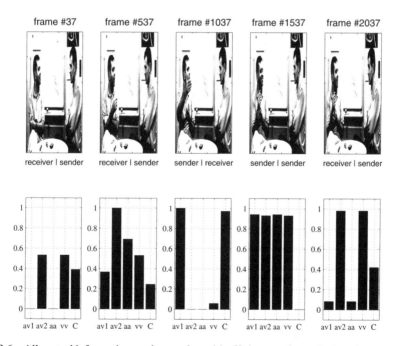

Figure 19.6 All mutual information tracks together with efficiency estimate (av1; av2; aa; vv; C; stand respectively for $I_{A_1V_1}$; $I_{A_2V_2}$; $I_{A_1A_2}$; $I_{V_1V_2}$; $C(t)$)

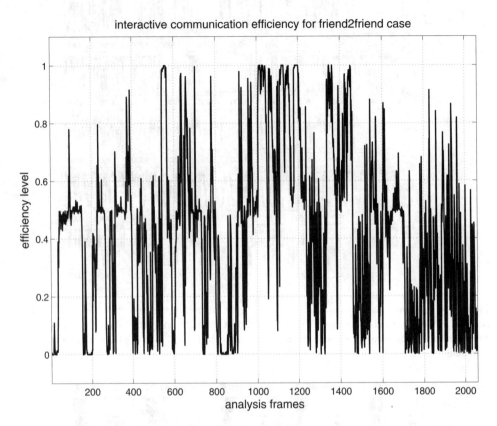

interactive communication efficiency for friend2friend case

Figure 19.7   Efficiency estimate as in Equation (19.6) for a short face-to-face communicative event recored on video and further analyzed. The dynamics of the process are very clearly depicted. The values around 0 stand for inefficient parts of communication when both communicators probably were active equally. Values equal to 1 represent ideal moments, when only one communicator was active fully utilizing his/her sender's role. The original video was taken with speed 30 frames/second

one indicates *the receiver*. This synchronized audiovisual behavior of the sender and the unsynchronized one of the receiver characterizes an efficient communication (Rutkowski *et al.* 2003a, 2004a).

The pair of the mutual information estimates for the local synchronization of the senders and the receivers in Equation (19.4) is used to give clues about concurrent individual activities during the communication event, while the unimodal cross-activities estimates in Equations (19.5) are used to evaluate the interlaced activities for a further classification. Clearly, the efficient sender–receiver interaction involves *action* and *feedback*. The interrelation between the actions of a sender and feedback of a receiver is therefore monitored, whereby the audiovisual synchronicity is used to determine the roles.

In our approach, the interactions between individual participants are modeled within the *data fusion* framework, based on features coming simultaneously from both the audio and video. A multistage and a multisensory classification engine (Hsu and Lin 2002) based on the support vector machine (SVM) approach is used at the *decision making* level of the data fusion framework, where the *one-versus-rest-fashion* approach is used to identify the phases during ongoing communication (based on the mutual information estimates from Equations (19.4) and (19.5)).

The Decision Level

SVMs are particularly suited when *sender–receiver* or *receiver–sender* situations are to be discriminated from the *noncommunicative* or *multisender* cases. In this work, a kernel based on a radial basis function (RBF) is utilized (Cherkassky and Mulier 1998), and is given by:

$$K(\mathbf{x}, \mathbf{x}_i) = e^{-\gamma \|\mathbf{x} - \mathbf{x}_i\|^2}$$

(19.7)

where γ is a kernel-function width parameter.

Using the above concept, an arbitrary multimodal mutual information combination λ can be categorized into four categories:- (i) $f(\lambda) \in (-\infty, -1]$ for the *noncommunicative* case with no interaction (no communication or a single participant); (ii) $f(\lambda) \in (-1, 0]$ for the *sender–receiver* case; (iii) $f(\lambda) \in (0, 1)$ for the *receiver–sender* case; and (iv) $f(\lambda) \in [1, +\infty)$ for the *sender–sender* case. Categories (i) and (iv) are somehow ambiguous due to the lack of clear separation boundaries, and they are treated by separately trained SVM classifiers. Results showing search for decision boundaries for two class classifiers in different mutual information combinations are given in Figure 19.8. A track in time of changing roles in ongoing face-to-face conversation is illustrated in the lower panel of Figure 19.5.

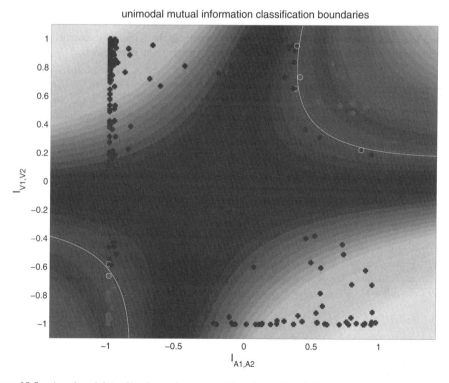

unimodal mutual information classification boundaries

Figure 19.8 A unimodal (audio channels versus video channels only) boundary search: $I_{A1,A2}$ versus $I_{V1,V2}$. Classification results with radial basis function support vector machine (Vapnik 1995)

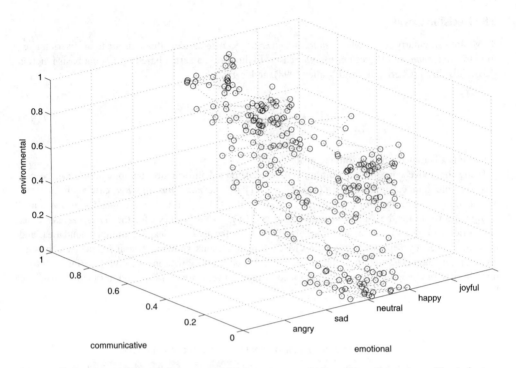

Figure 19.9 The communication atmosphere space spanned along three dimensions. The trajectory represents a face-to-face communication situation in a relatively quite environment

19.4 Experiments

In this section, experiments are reported to illustrate the validity of the proposed communication analysis approach.

Communication atmosphere, as defined in this study, is referred to as a region in the three-dimensional space (Figure 19.9), obtained by independently estimating the environmental impact, the communication efficiency, and the communicators' emotions in the ongoing communication process. This measure allows for the communication process evaluation and reconciliation (e.g., movie authoring).

This communication atmosphere definition can be formalized as follows:

$$A(t) = A(E(t), C(t), M(t)) \tag{19.8}$$

where $A(t)$ represents the communication atmosphere evaluation at a time t, which is a function of the environmental estimate $E(t)$, defined as a sum of multimodal signal-to-noise ratios from Equations (19.1) and (19.2) as follows:

$$E(t) = \frac{A_{SIR}(t) + V_{SIR}(t)}{2}. \tag{19.9}$$

The communicative dimension estimate $C(t)$ (also the communication efficiency measure as in Equation (19.6)), and emotion classification result, $M(t)$, as discussed in section 19.2.3, belongs to the set:

$$M(t) \in \{neutral, sad, angry, happy, joyful\}. \tag{19.10}$$

The estimate $A(t)$ is utilized to characterize communication atmosphere at a given time. Less sensitive measures are the average values for a given time window, given by:

$$A_{avg(t_a,t_b)} = \frac{1}{|t_b - t_a|} \sum_{t=t_a}^{t_b} A(t) \qquad (19.11)$$

where $t_a > t_b$, and short time functions are:

$$A_{t_a,t_b} = \{A(t_a), \ldots, A(t_b)\}. \qquad (19.12)$$

The short time trajectories A_{t_a,t_b} in the three-dimensional space can be subsequently used as inputs for situation classifiers or three-dimensional atmosphere models. These represent a *communicative semantics* modeling problem, in which multiple sequentially structured simultaneous communication processes are in a dialogical relationship. In such models, the particular focus should be put on beats and deixis (Norris 2004), as lower-level actions structure the foregrounding and backgrounding of the higher-level actions that participants are simultaneously engaged in. The problem of classification of *communicative semantics-related* stages is a subject for our further research, which will explore the ultimate relation between the communication atmosphere and communicative situation affordances or communicative situation norms as proposed by Stamper (1996) and modeled by Kryssanov and Kakusho (2005). A plot of short time functions A_{t_a,t_b} of observed real-time face-to-face conversation is presented in Figure 19.9 as a three-dimensional trajectory, and in Figure 19.10 as the three separate bar plots for every dimension showing a vivid independence in time among the chosen dimensions.

19.4.1 The Atmosphere Reconciliation

The communication atmosphere of a recorded meeting can be reconciled according to the user preferences or wishes after the proposed analysis is accomplished. Once the communication atmosphere features are estimated, it is possible to manipulate their values and appropriately post-process the recorded multimedia content to obtain the desired values. It is possible to independently manipulate characteristics of the environmental dimension by increasing or decreasing the auditory or visual presence of the environment. The information separation discussed in previous sections can be used in the opposite way to add or remove the environmental components. In a similar way, it is possible to edit emotional features of the communicators' auditory activities in order to change the emotional component of the overall climate of recorded communication. The communicative dimension can also be reconciled by adjusting the occurrences of communicators' interactions (e.g., by adding or removing silent breaks) in time. An example of the original and reconciled communication atmosphere tracks (only for environmental and emotional dimensions in this case) is presented in Figure 19.11.2

19.4.2 Discussion of Results

The approach presented in this study was tested in two settings. In the first experiment, two sets of cameras and microphones where used to capture ongoing communication events. Two pairs of synchronized video and stereophonic audio streams were recorded. In the second experiment, we utilized a single high-definition digital video camera (HDV) with a stereo microphone. This setup is similar to video recordings broadcasted in television channels. Both setups allowed capture of facial areas with higher resolution, which are highly synchronized with speech waveforms (Rutkowski *et al.* 2003a, 2004a). In both experiments, conducted in laboratory controlled environments, the subjects (communicators) were asked to talk freely in face-to-face situations. We focused on the interlaced communication analysis, so that the subjects were asked to make a conversation with frequent turn-taking (the discussion style).

Such instruction given to subjects had a side effect of increased attention, which had positive impact on our assumption of intentional communication analysis. The experiments were conducted to validate the hypothesis, that the separate analysis of the three dimensions related to communication can be performed, and that this allows for a comprehensive description of the communication process as a whole (feature independence). As for the communicative dimension, estimation of the communication efficiency is based on mutual information in multimedia streams. The track of the integrated communication efficiency value over an ongoing person-to-person communication event is shown in Figure 19.10. The normalized values

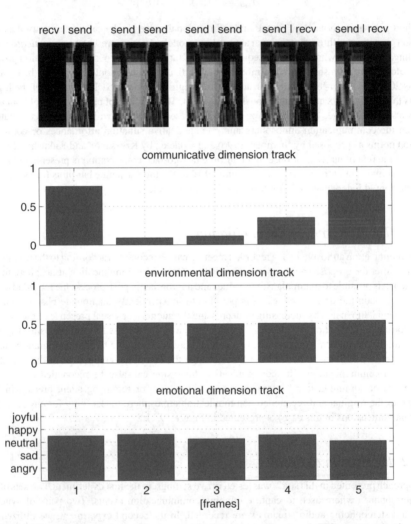

Figure 19.10   The communication track along three dimensions. Top panel: This communication event has the right-side communicator being sender (send | recv), then for two frames both were active (send | send), and finally the left-side communicator became sender (send | recv). There are five sub-scenes presented in which communicative dynamics or differences are hard to capture due to the nature of still images, but the proposed analysis illustrated in three bar plots allows for the tracking and subsequent classification of communicative atmosphere

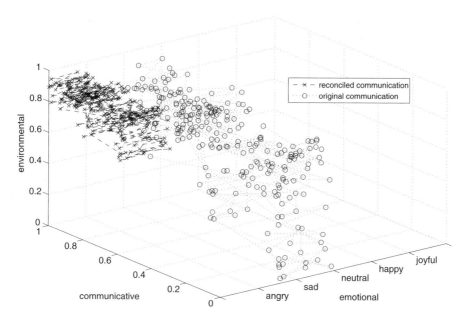

Figure 19.11   Communication atmosphere reconciled: The presented trajectory of a face-to-face communication was reconciled in the environmental (subtraction of environmental noise) and emotional (shifting of the emotional features of the communicators' voices) dimensions

close to unity indicate the moments when the interaction was "proper", cross-talks in audio and visual channels (the local, negative synchronization) did not occur, and there was only a single active person at each time (the lighter area around the person in the top box). The close to zero values correspond to the transitional situations, or when both parties are active at the same time. The communicative situation analysis in this three-dimensional space allows for the tracking of communication events and their sub-sequential classification based on the obtained trajectories (compare the shapes of the trajectories of two independent events with different communicators but with similar discussion topics shown in Figures 19.9 and 19.11). The reconciliation procedure was performed over the recorded communication; in our approach any or all the communication atmosphere dimensions can be considered. An example of manipulated atmosphere trajectories before and after reconciliation is shown in Figure 19.11, where the environmental and mental dimensions were reconciled.

19.5 Conclusions

This study has proposed a three-dimensional communication atmosphere space suitable for the analysis and editing of recorded multimedia communication events. The concept of communication atmosphere analysis and its implementation has been a missing link in contemporary studies dealing with communications situation modeling. The experiments have shown that it is indeed possible to estimate and later modify communicative events by changing the underlaying "affordance" qualitatively, based on behavioral analysis of the communicators. The proposed concept of mutual information evaluation in multimodal sensory data streams makes it possible to identify the participants and to classify them according to their role, and evaluate their emotional states together with environmental interferences. Furthermore, separate adaptation of the proposed atmosphere-related dimensions allows for reconciliation of the event's climate for the use in future virtual reality environments. In the present study, the three dimensions of the communication atmosphere were considered independently. Interdependencies between these dimensions should be addressed in future work, where more background related to communication psychology should be added.

The proposed *information fusion* approach for the evaluation of communicative interaction represents a step toward the modeling of communication situations from the engineering and computer science point of view, as compared to the existing audio and video only approaches (Chen 2003). The experiments have clearly illustrated the possibility to estimate the interactivity level, based on the behavioral analysis of the participants in communication. The mutual information based feature extraction of multimodal audio and video data streams makes it possible to quantify and detect the presence of participants and to classify them according their role. Despite some differences between the conclusions of experts and the outcome of the proposed machine learning framework, our results show strong correlation between the two. In fact, the human judgment is also highly subjective, so further studies will have a larger population of human experts to balance their opinions.

Acknowledgments
The authors would like to thank Professors Toyoaki Nishida, Michihiko Minoh, and Koh Kakusho of Kyoto University for their support and fruitful discussions. This was achieved within the framework of the project "Intelligent Media Technology for Supporting Natural Communication between People", which was partially supported by the Ministry of Education, Science, Sports and Culture in Japan, Grant-in-Aid for Creative Scientific Research, 13GS0003. We are grateful to Professor Victor V. Kryssanov from Ritsumeikan University in Kyoto for his input during the setup stage of this research. We benefited from our discussions with Dr Anil Bharath from Imperial College London. Part of this research was supported by the Royal Society grant RSRG 24543.

References

Adler R.B. and Rodman G. (2003) *Undestanding Human Communication*. Oxford University Press.

Becchetti C. and Ricotti L.P. (1999) *Speech Recognition*. John Wiley.

Birdwhistell R. (1974) The language of the body: the natural environment of words. In *Human Communication: Theoretical explorations*. Lawrence Erlbaum Associates, Hillsdale, NJ, pp. 203–220.

Chen M. (2003) Visualizing the pulse of a classroom. In *Proceedings of the Eleventh ACM International Conference on Multimedia*. ACM Press, pp. 555–561.

Cherkassky V. and Mulier F. (1998) Learning from data. In *Adaptive and Learning Systems for Signal Processing, Communication, and Control*. John Wiley Inc.

Dellaert F., Polzin T. and Waibel A. (1996) Recognizing emotion in speech. In *Proceedings of Fourth International Conference on Spoken Language Processing* (ICSLP'96), vol. 3, pp. 1970–1973.

Dimberg U., Thunberg E. and Elmhed K. (2000) Unconscious facial reactions to emotional expressions. *Psychological Science* **11** (1), 149–182.

Furui S. (2001) *Digital Speech Processing, Synthesis, and Recognition*, 2nd edn. Marcell Dekker, New York.

Gautama T., Mandic D.P. and Hulle M.V. (2004) A novel method for determining the nature of time series. *IEEE Transactions on Biomedical Engineering* **51**, 728–736.

Hsu C.W. and Lin C.J. (2002) A comparison of methods for multi-class support vector machines. *IEEE Transactions on Neural Networks* **13** (2), 415–425.

Hyvarinen A., Karhunen J. and Oja E. (2001) *Independent Component Analysis*. John Wiley & Sons.

Jelfs B., Vayanos P., Chen M., Goh S.L., Boukis C., Gautama T., Rutkowski T., Kuh A. and Mandic D.P. (2006) An online method for detecting nonlinearity within a signal. In *Proceedings of the 10th International Conference on Knowledge-Based & Intelligent Information & Engineering Systems* (KES–06), pp. 1216–1223.

Kryssanov V. and Kakusho K. (2005) From semiotics of hypermedia to physics of semiosis: a view from system theory. *Semiotica*. **54** (1/4), 11–38.

Lee D.D. and Seung H.S. (1999) Learning the parts of objects by non-negative matrix factorization. *Nature* **401**, 788–791.

Mandic D.P. (2004) A general normalised gradient descent algorithm. *IEEE Signal Processing Letters* **11** (2), 115–118.

Mandic D.P. and Chambers J.A. (2001) *Recurrent Neural Networks for Prediction: Architectures, Learning Algorithms and Stability*. John Wiley.

Mandic D.P., Goh S.L. and Aihara K. (2005a) Sequential data fusion via vector spaces: complex modular neural network approach. In *Proceedings of the IEEE Workshop on Machine Learning for Signal Processing*, pp. 147–151.

Mandic D.P., Obradovic D., Kuh A. *et al.* (2005b) Data fusion for modern engineering applications: an overview. In *Proceedings of IEEE International Conference on Artificial Neural Networks* (ICANN'05), pp. 715–721.

Mehrabian M. (1971) *Silent Messages*. Wadsworth, Belmont, CA.

Norris S. (2004) *Analyzing Multimodal Interaction: a methodological framework*. Routledge.

Pantic M. and Rothkrantz L.J.M. (2000) Automatic analysis of facial expressions: the state of the art. *IEEE Transactions on Pattern Analysis and Machine Intelligence* **22**, 1424–1445.

Pease A. and Pease B. (2004) *The Definitive Book of Body Language: How to read others' thoughts by their gestures*. Pease International.

Ralescu A. and Hartani R. (1997) Fuzzy modeling based approach to facial expressions understanding. *Journal of Advanced Computational Intelligence* **1** (1), 45–61.

Rutkowski T.M. and Mandic D. (2005) Communicative interactivity: a multimodal communicative situation classification approach. In W. Duch, J. Kacprzyk, E. Oja and S. Zadrozny (eds), *Artificial Neural Networks: Formal models and their applications*. Springer, Berlin, pp. 741–746.

Rutkowski T.M. and Mandic D. (2007) A multimodal approach to communicative interactivity classification. *Journal of VLSI Signal Processing Systems*. Available from: DOI:10.1007/s11265-007-0081-6. http://dx.doi.org/10.1007/s11265-007-0081-6

Rutkowski T.M., Seki S., Yamakata Y., Kakusho K. and Minoh M. (2003a) Toward the human communication efficiency monitoring from captured audio and video media in real environments. In *Proceedings of 7th International Conference on Knowledge-Based Intelligent Information and Engineering Systems* (KES 2003), pp. 1093–1100.

Rutkowski T.M., Yokoo M., Mandic D., Yagi K., Kameda Y., Kakusho K. and Minoh M. (2003b) Identification and tracking of active speaker's position in noisy environments. In *Proceedings of International Workshop on Acoustic Echo and Noise Control* (IWAENC2003), pp. 283–286.

Rutkowski T.M., Kakusho K., Kryssanov V.V. and Minoh M. (2004a) Evaluation of the communication atmosphere. *Proceedings of 8th International Conference on Knowledge-Based Intelligent Information and Engineering Systems* (KES 2004), pp. 364–370.

Rutkowski T.M., Kryssanov V.V., Kakusho K. and Minoh M. (2004b) Communicating people identification in multimedia streams: an audiovisual fusion approach. In *Proceedings of the IPSJ Forum on Information Technology* (FIT2004), vol. K-007, pp. 405–406.

Rutkowski T.M., Kryssanov V.V., Ralescu A., Kakusho K. and Minoh M. (2004c) From the automatic communication atmosphere analysis to its elements reconciliation in recorded multimedia streams. In *Proceedings of the Fourth National Conference on Multimedia and Network Information Systems and First International Workshop on Intelligent Media Technology for Communicative Intelligence*, vol. II, pp. 125–135.

Rutkowski T.M., Kryssanov V.V., Ralescu A., Kakusho K. and Minoh M. (2005a) An audiovisual information fusion approach to analyze the communication atmosphere. In *Proceedings of Symposium on Conversational Informatics for Supporting Social Intelligence & Interaction* (AISB2005), Workshop on situational and environmental information enforcing involvement in conversation. University of Hertfordshire, England.

Rutkowski T.M., Yamakata Y., Kakusho K. and Minoh M. (2005b) Smart sensor mesh: intelligent sensor clusters configuration and communicative affordances principle. In *Intelligent Media Technology for Communicative Intelligence*. Springer-Verlag, Berlin, pp. 147–157.

Shannon C. and Weaver W. (1949) *The Mathematical Theory of Communication*. University of Illinois Press, Urbana.

Sigan L., Sclaroff S. and Athitsos V. (2000) Estimation and prediction of evolving color distributions for skin segmentation under varying illumination. In *Proceedings of the IEEE Conference on Computer Vision and Pattern Recognition* (CVPR), vol. 2, pp. 152–159.

Stamper R. (1996) Signs, information, norms and systems. In B. Holmqvist, P. Andersen, H. Klein and R. Posner (eds), *Signs of Work*. De Gruyter, Berlin, pp. 349–399.

Vapnik V. (1995) *The Nature of Statistical Learning Theory*. Springer-Verlag.

20

Analysis of Interaction Mechanisms in Online Communities

Naohiro Matsumura

20.1 Introduction

The number of online communities has increased greatly over the last decade. One reason why online communities exist is for individuals to collaborate with others in sharing new information and creating new knowledge about specific topics related to their interests. More specifically, collaborative communication results in creative communication because the synergistic effect of the collaboration provides more creative activities than do the efforts of individuals (Nonaka and Takeuchi 1995). However, to assume that all individuals will recognize and relate to one another because of their exploratory communication is unrealistic. Collaborative communication is based on groups of individuals. Each usually communicates with a limited number of specific others. The formation reflects the hierarchical relationships among individuals in which the positions of individuals play important roles in effective and smooth communication (Krackhardt and Hanson 1993).

It is worth mentioning that the roles are organized spontaneously as individuals recognize their own positions, although no explicit roles exist for individuals in advance. For example, some behave as leaders and some serve as followers, and so forth. The roles for individuals in communication have been a major concern to sociologists. A seminal study was undertaken by Katz and Lazarsfeld (1955), in which the concepts of an opinion leader and a follower were introduced to model the process of communication. Rogers (1995) studied the roles for individuals in communication and classified people into five types (innovator, early adopter, early majority, late majority, and laggard) according to their stage regarding the adoption of new ideas. The models in those studies assumed the flow of influence as one-way: innovator to laggard via adopter and majority.

Regarding online communication, e-mail communication flow has been studied to elucidate human interaction (Shetty and Adibi 2004; Wu et al. 2003). Tyler et al. (2003) analyzed e-mail logs within an organization to identify leaders within informal collaborative communities. These works specifically examined the topological aspect of communication. On the other hand, the contents of communication

Conversational Informatics: An Engineering Approach   Edited by Toyoaki Nishida
© 2007 John Wiley & Sons, Ltd

were analyzed to investigate the roles of individuals in communication (Steyvers *et al.* 2004; McCallum 2005).

This chapter presents further investigation of the behaviors of individuals in communication as a part of Conversational Informatics by analyzing the interaction mechanism of online communities. For this purpose, section 20.2 defines four roles for individuals (leader, coordinator, maven, and follower) in communication based on the influence that individuals receive from others and the influence they impart on others. The influence is measured using the influence diffusion model (IDM), which is presented in section 20.3. Based on these measurements, three types of communication are identified in section 20.4 (leader-led, follower-based, and leader-follower-led communication); each relates to the types of topics based on roles for 119 957 individuals in 3000 message boards on the Internet. Frequent communication patterns of individuals were discerned by mining 53 646 message chains (section 20.5). The conclusion and directions of future work are presented in section 20.6.

20.2 Four Roles for Individuals

In a broad sense, the axis of "leadership" can be an important factor for identifying an individual who initiates, contributes, or makes positive commitments to communication by giving pertinent information or opinions (Hanson 2005). In this sense, such an individual might have the quality of being a leader because that person is capable of leading communication. However, despite giving information or opinions, this person would not be a leader if others did not react to that information. This is often observed when others do not trust an individual. In other words, credibility is a requisite for a leader.

The axis of "coordination" is also an important factor for identifying an individual who has the ability to coordinate and react to others' information or opinions to harmonize with others. Sometimes they pull together or summarize others' ideas. Such behaviors might maintain communication channels and fill the community with a supportive climate that activates communication. Empathy with others is the necessary quality for coordination.

Considering the leadership and coordination mentioned above, the following four types of communication roles are proposed.

- *Leaders*. These are people who take the initiative in communication by giving information and opinions as well as accepting those of others. Such individuals possess high levels of both leadership and coordination.
- *Coordinators*. These are people whose behavior makes the climate of communication better by accepting others' information and opinions rather than giving their own. Their level of coordination is higher than the level of leadership.
- *Mavens*. These people contribute to communication by giving information and opinions rather than accepting those of others. Their level of leadership is higher than their level of coordination.
- *Followers*. These people join communication to seek information or opinions passively. Such individuals have low levels of both leadership and coordination.

Only a small number of active individuals post the majority of messages to message boards on the Internet, and others (often called lurkers) visit the board and read messages but do not post any messages (Barnes 2003). For this reason, lurkers are ignored here; only active individuals are assigned to the four roles described above.

Two indices, $I^{<out>}$ and $I^{<in>}$, identify the roles for individuals. The first, $I^{<out>}$, measures how much influence an individual's messages has on the messages of others, and the other, $I^{<in>}$, measures how much influence an individual's messages receive from those of others. The terms $I^{<out>}$ and $I^{<in>}$ are assumed to correspond respectively to leadership and coordination, and credibility and sympathy

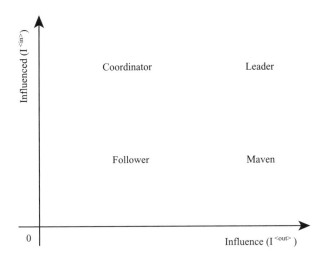

Figure 20.1 Classification of four roles based on $I^{<out>}$ and $I^{<in>}$

are represented as a combination of leadership and coordination. Based on those assumptions, the roles for individuals are roughly determined as represented in Figure 20.1. That is, leaders have high $I^{<out>}$ and $I^{<in>}$, coordinators have low $I^{<out>}$ and high $I^{<in>}$, mavens have high $I^{<out>}$ and low $I^{<in>}$, and followers have low $I^{<out>}$ and $I^{<in>}$. The indices, $I^{<out>}$ and $I^{<in>}$, are described in more detail in the following section.

The conceptualization of those four roles comes from Rogers' five categories: innovator, early adapter, early majority, late majority, and laggard. The main difference between them is the standard; our formalization emphasizes the influence, Rogers centers analyses upon the stage related to the adoption of new ideas.

20.3 Measuring the Influence of Individuals

The idea of measuring influence comes from the influence diffusion model (IDM; Matsumura 2003), in which the influence between a pair of individuals is defined as the sum of propagating terms among them via messages. Our approach simplifies the IDM algorithms to make it more reasonable.

Here, let a message chain be a series of messages that are connected by post-reply relationships; the influence of a message x on a message y (x precedes y) in the same message chain be $i_{x \to y}$. Then, define $i_{x \to y}$ as:

$$i_{x \to y} = |w_x \cap \cdots \cap w_y| \tag{20.1}$$

where w_x and w_y respectively represent the set of terms in x and y. In addition, $|w_x \cap \cdots \cap w_y|$ respectively represent the number of terms propagating from x to y via other messages. Then $i_{x \to y}$ is defined as 0 because the terms in x and y are used in a different context and no influence exists between them if x and y are not in the same message chain.

Based on the influences among messages, the influence of an individual p on an individual q is measured as the total influence of p's messages on the messages of others through q's messages replying to p's messages. Let the set of p's messages be α, the set of q's messages replying to any of α be β, and the message chains starting from a message z be η_z. The influence from p onto q, $j_{p \to q}$, is consequently defined as:

$$j_{p \to q} = \sum_{x \in \alpha} \sum_{x \in \beta} \sum_{y \in \eta_z} i_{x \to y}. \qquad (20.2)$$

The influence of p on q is regarded as q's contribution to the spread of p's messages.

The influence of each individual is also measurable using $j_{p \to q}$. Let the influence of p to others be $I_p^{<out>}$, the influence of others to p be $I_p^{<in>}$, and the set of all individuals except p be γ. Then, $I_p^{<out>}$ and $I_p^{<in>}$ are defined as:

$$I_p^{<out>} = \sum_{q \in \gamma} j_{p \to q} \qquad (20.3)$$

$$I_p^{<in>} = \sum_{q \in \gamma} j_{q \to p}. \qquad (20.4)$$

As an example of measuring the influence, the simple message chain in Figure 20.2 is used, where Anne posted message 1, Bobby posted message 2 as a reply to message 1, and Cathy posted messages 3 and 4 as respective replies to message 2 and message 1. In that figure, solid arrows show the replies to previous messages, and dotted arrows show the flows of influence. Here, the influence between a pair of individuals is as follows.

- The influence of Anne on Bobby is 3 (i.e., $j_{Anne \to Bobby} = 3$), because two terms ("Environment" and "Convenience") were propagated from Anne to Bobby, and one term ("Environment") was propagated from Anne to Cathy via Bobby.
- The influence of Anne on Cathy is 1 (i.e., $j_{Anne \to Cathy} = 1$), because one term ("Price") was propagated from Anne to Cathy.

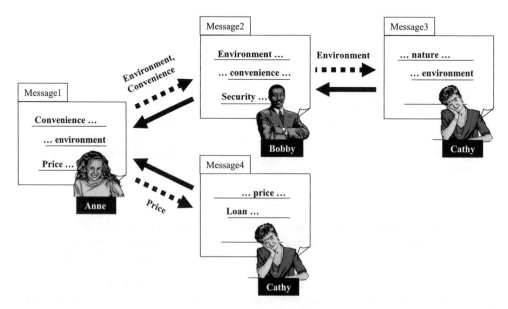

Figure 20.2 Message chain of four messages sent by three individuals. (With kind permission from Springer Science and Business Media)

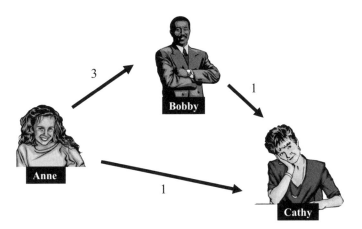

Figure 20.3 A social network showing the influence among three individuals from Figure 20.2. (With kind permission from Springer Science and Business Media)

- The influence of Bobby on Cathy is 1 (i.e., $j_{Bobby \rightarrow Cathy} = 1$), because one term ("Environment") was propagated from Bobby to Cathy.
- The influence of Bobby on Anne and of Cathy on Anne is 0 (i.e., $j_{Bobby \rightarrow Anne} = 0$ and $j_{Cathy \rightarrow Anne} = 0$), because no term was propagated to Anne from either Bobby or Cathy.

Note that the influence lines between individuals only show the direct mutual influence. Therefore, the indirect influence of Anne on Cathy via Bobby is the contribution of Bobby, added to the influence of Anne on Bobby.

By mapping the influence among individuals, a social network showing influence can be obtained as shown in Figure 20.3 where their relationships are shown as directional links and their mutual influence. The $I_p^{<out>}$ and $I_p^{<in>}$ of each individual is measured as follows:

$$I_{Anne}^{<out>} = 4, \qquad I_{Anne}^{<in>} = 0 \qquad (20.5)$$

$$I_{Bobby}^{<out>} = 1, \qquad I_{Bobby}^{<in>} = 3 \qquad (20.6)$$

$$I_{Cathy}^{<out>} = 0, \qquad I_{Cathy}^{<in>} = 2 \qquad (20.7)$$

According to the influence described above, the influential relationships among individuals can be inferred in addition to their roles in communication. For example, Anne would be a *maven* because she provides the most influence on others without accepting any influence from others. Bobby would be a *coordinator* because he was a recipient of influence and transmitted some of it to Cathy. Cathy would be a *follower* or *coordinator* because she only received influence from others. In this case, no *leader* was identified because no person had both high $I^{<out>}$ and high $I^{<in>}$.

Diffusion of influence serves an important role in distilling and purifying their messages, where the formation of individuals has a strong effect on the diffusion process. In the following sections, results of statistical analyses of numerous communication logs are used to show features of communications.

20.4 Three Types of Communication

Various communication types might exist according to the individuals or topics discussed. For example, results presented in our earlier work showed that a discussion type and a chitchat type of communication are affected by the tendency of using specific expressions in anonymous online communities (Matsumura

Table 20.1 Distribution of four roles measured from 3000 message boards from 15 categories in Yahoo!Japan. The values are normalized to 1 for each category

Yahoo!Japan categories	Leader	Maven	Coordinator	Follower
Family & Home	0.5263	0.0289	0.0426	0.4022
Health & Wellness	0.5137	0.0453	0.0605	0.3804
Arts	0.5331	0.0312	0.0438	0.3920
Science	0.5337	0.0372	0.0505	0.3785
Cultures & Community	0.5752	0.0415	0.0428	0.3405
Romance & Relationships	0.5277	0.0450	0.0533	0.3741
Hobbies & Crafts	0.5158	0.0293	0.0496	0.4054
Regional	0.5088	0.0407	0.0595	0.3910
Entertainment	0.4844	0.0275	0.0540	0.4341
Government & Politics	0.4291	0.0224	0.0558	0.4927
Business & Finance	0.4506	0.0365	0.0596	0.4534
Schools & Education	0.4212	0.0348	0.0591	0.4849
Recreation & Sports	0.4101	0.0420	0.0626	0.4852
Computers & Internet	0.3611	0.0289	0.0691	0.5409
Current Events	0.2931	0.0303	0.1010	0.5756

et al. 2005a). Results presented in another paper classified social networks in online communities into three types based on the structural features (Matsumura *et al.* 2005b). Those studies analyzed the group dynamics in online communication, although this section specifically addresses the communication types from the roles for the individual's point of view. Next, 3000 message boards were selected, each having more than 300 messages, from among 15 categories of Yahoo!Japan message boards. The first 300 messages from each message board were downloaded to equilibrate the number of messages. Thereby, 3000 message boards were prepared, each having 300 messages. As in other online communities (Wallace 1999), only ten % of individuals provided 69% of all messages on the boards.

From those messages, $I^{<out>}$ and $I^{<in>}$ were measured using the approach described in section 20.3, and the role of each was identified by applying the classification in Table 20.1. In this study, the levels of $I^{<out>}$ and $I^{<in>}$ (e.g., "High" or "Low") are defined experimentally as follows:

- The level of $I^{<out>}$ (or $I^{<in>}$) is set to "High" if the value is more than double the mean value.
- The level of $I^{<out>}$ (or $I^{<in>}$) is set to "Low" if the value is no more than double the mean value.

By this definition, I identified the roles for 119 957 individuals in all, where 17% of individuals are assigned as "High" in $I^{<out>}$ or $I^{<in>}$ and they provide 62% of all messages. Note that the "Middle" level is ignored because the range of this level is too narrow to determine.

The normalized distribution of posted messages by each role for each category is shown in Table 20.1. Here, to investigate the relationships between four roles and categories, hierarchical cluster analysis was applied to the data. This analysis merges clusters based on the mean Euclidean distance between the elements of each cluster (Rogers 1995). Figure 20.4 shows that a tree-like diagram, called a dendrogram, is then constructed. Three major clusters are readily identifiable, each corresponding to a type of communication. The clusters were named as follows.

- *Leader-led communication*. This cluster includes eight categories ("Romance & Relationships", "Science", "Regional", "Health & Wellness", "Arts", "Family & Home", "Hobby & Crafts", and

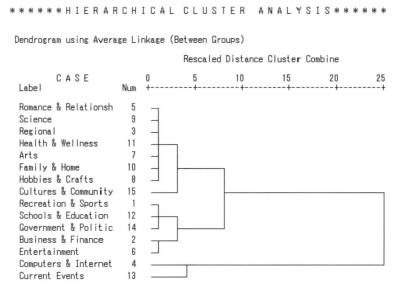

Figure 20.4   Dendrogram produced through hierarchical cluster analysis

"Cultures & Community"). In this cluster, leaders posted the most messages. The topics in these categories are common and familiar to many individuals who share these interests.
- *Follower-based communication.* This cluster includes two categories ("Computers & Internet" and "Current Events"), where the messages by followers are more prominent than those in other clusters. The topics in this cluster are transitory information.
- *Leader-follower-led communication.* This cluster includes five categories ("Recreation & Sports", "Schools & Education", "Government & Politics", "Business & Finance", and "Entertainment"), where leaders and followers post messages at the same level. In this cluster, the topic specificity requires high-level knowledge for giving information or opinions. For some reason, the communication in this cluster proceeds with harmonic dialogues between leaders and followers.

20.5 Frequent Communication Patterns

It was assumed from the results described in section 20.4 that the formation of individuals is related to the topics and that its context as shared with individuals in the community. Frequent communication patterns of four roles were extracted from message chains to reveal the patterns of the formation.

A communication pattern is defined as an ordered sub-tree in which each node corresponds to a role of the sender and the order shows post-reply relationships between them. For example, in Figure 20.5 where message chains comprise 36 messages posted by six participants' forms, Alex replies to Charley six times and Bonnie replies to Alex three times. Such specific post-reply relationships signify frequent communication patterns of participants.

We can now describe the frequent communication patterns of participants from the point of their roles, by recasting each message into the sender's role (leader, maven, coordinator, or follower). Then, a frequent communication pattern corresponds to an ordered sub-tree occurring more than N times in a set of ordered trees, where N is a user-given threshold called "minimum support". For example, if a sub-tree

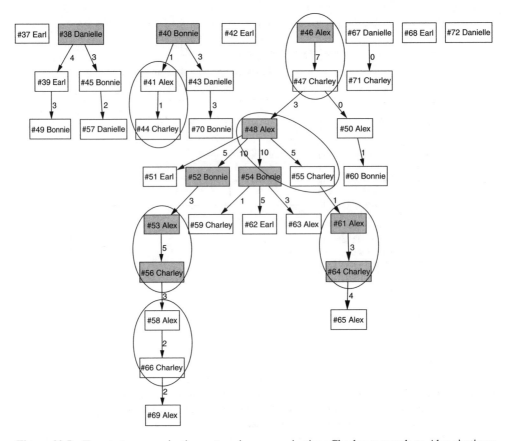

Figure 20.5 Frequent communication pattern in communication: Charley responds to Alex six times (shown as circles)

occurs more than N times in a set of ordered trees, the sub-tree is extracted as a frequent communication pattern.

For this study, 53 646 ordered trees extracted from the same message boards in section 20.4 were classified into three communication types (31 110 ordered trees for leader-led communication, 5967 ordered trees for follower-based communication, and 16 569 ordered trees for leader-follower-led communication) from which frequent communication patterns were extracted under the condition of 5% minimum support using the FREQT algorithm (Abe *et al.* 2002).[1] All extracted patterns represented as S-expressions (McCarthy 1960) are shown in Table 20.2.

No dominant differences of frequent communication patterns between three categories are recognized. Instead, some typical frequent communication patterns were identified as follows:

- *Leader-chain pattern* (represented as (L(L(L)···)))). Leaders repeatedly communicate with other leaders.

---

[1] FREQT free software is available at http://chasen.org/~taku/software/freqt/

Table 20.2 The top 10 most frequent formation patterns of individuals' roles for each communication type. The patterns are extracted from 53 646 message chains in all. L, M, C, and F respectively stand for leader, maven, coordinator, and follower. The patterns are represented as S-expressions, and the appearance percentages (the number of trees including sub-tree pattern divided by the number of all trees) are shown as support values

Leader-led communication		Follower-led communication		Leader-follower-led communication	
S-expression	Support	S-expression	Support	S-expression	Support
(L(F))	0.52	(L(F))	0.42	(L(F))	0.50
(F(L))	0.49	(F(F))	0.40	(F(L))	0.43
(L(L))	0.45	(F(L))	0.39	(L(L))	0.40
(L(F(L)))	0.35	(L(L))	0.36	(L(F(L)))	0.30
(L(L(L)))	0.29	(L(F(L)))	0.25	(F(F))	0.30
(F(F))	0.26	(L(L(L)))	0.24	(L(L(L)))	0.26
(F(L(F)))	0.26	(F(L(F)))	0.21	(F(L(F)))	0.21
(L(L(L(L))))	0.19	(F(F(F)))	0.20	(L(L(L(L))))	0.17
(L(F(L(F))))	0.19	(L(L(L(L))))	0.17	(L(F(L(F))))	0.16
(F(L(F(L))))	0.18	(L(L)(F))	0.15	(L(L(F)))	0.15

- *Follower-chain pattern* (represented as (F(F(F)···))). Followers repeatedly communicate with other followers.
- *Leader-follower-repetition pattern* (represented as (L(F(L(F)···))))). Leaders and followers alternately communicate with each other.

In these cases, the relationships between leaders and followers are symmetric; the same is apparent between leaders and coordinators in the top 25 ranking. On the other hand, asymmetric relationships are apparent in the top 30 ranking:

- *Maven-leader pattern* (represented as (M(L))). A leader frequently replies to a maven.
- *Follower-coordinator pattern* (represented as (F(C))). A coordinator replies frequently to a follower.

It is worth noting that the symmetric and asymmetric communication patterns were extracted automatically. These patterns might show the flexibility and rigidity of communication among individuals in online communication, and contribute to providing mid-level design principles to motivate the contributions of individuals, as Beenen *et al.* (2004) found. Nevertheless, no evidence from this study supports that. In any case, these patterns are certainly basic constituent factors of our interaction mechanism.

20.6 Conclusions

This chapter has described an approach for identifying the roles of individuals in communication. Results of this investigation revealed three communication types in addition to frequent communication patterns. These results show one aspect of the norms that govern our communication in online communities irrespective of conscious or unconscious intention. Understanding and exploiting these norms would promote smooth communication and achieve fertile communication.

At the present, useful applications are considered as the following.

- *Community monitoring.* Identifying the roles of individuals and communication types can be helpful to monitor the state of online communities.
- *Community management.* Followed by the monitoring described above, undesirable states of communication can be managed if they occur.
- *Community design.* Based on analyses of various online communities, design principles and guidelines for online communities can be obtained.

Future work will include further research into other online communities to clarify the variety of interaction mechanisms. Integration of other research on Conversational Informatics that are explained in other chapters of this book might also suggest exciting investigations.

References

Abe K., Kawasoe S., Asai T., Arimura H. and Arikawa S. (2002) Optimized substructure discovery for semi-structured data. In *Proc. PKDD-2002*, pp. 1–14.

Barnes S.B. (2003) *Computer-Mediated Communication*. Allyn & Bacon.

Beenen G., Ling K., Wang X., Chang K., Frankowski D., Resnick P. and Kraut R.E. (2004) Using social psychology to motivate contributions to online communities. *Proc. ACM CSCW'04*, pp. 212–221.

Hanson M.P. (2005) *Clues To Achieving Consensus: a leader's guide to navigating collaborative problem-solving*. ScarecrowEducation.

Katz E. and Lazarsfeld P. (1955) *Personal Influence*. Free Press, New York.

Krackhardt D. and Hanson J.R. (1993) Informal networks: the company behind the chart. *Harvard Business Review*, July-August, pp. 104–111.

Matsumura N. (2003). Topic diffusion in a community. In Y. Ohsawa and P. McBurney (eds), *Chance Discovery*. Springer Verlag, pp. 84–97.

Matsumura N., Miura A., Shibanai Y., Ohsawa Y. and Nishida T. (2005a) The dynamism of Nichannel. *Journal of AI & Society* **19** (1), 84–92.

Matsumura N., Goldberg D.E. and Llorà X. (2005b) Mining social networks in message boards. *Proc. WWW2005*, pp. 1092–1093.

McCallum A., Corrada-Emmanuel A. and Wang, X. (2005) Topic and role discovery in social networks. In *Proc. IJCAI-05*, pp. 786–791.

McCarthy J.L. (1960) Recursive functions of symbolic expressions and their computation by machine I. *CACM* **3** (4), pp. 184–195.

Nonaka I. and Takeuchi H. (1995) *The Knowledge-Creating Company*. Oxford University Press.

Rogers E.M. (1995) *Diffusion of Innovations* 4th edn. Free Press, New York.

Shetty J. and Adibi J. (2004) The Enron email dataset database schema and brief statistical report (technical report). Information Sciences Institute.

Steyvers M., Smyth P., Rosen-Zvi M. and Griffiths T. (2004) Probabilistic author-topic models for information discovery. *Proc. ACM KDD2004*, pp. 306–315.

Tyler J., Wilkinson D. and Huberman B.A. (2003) E-mail as spectroscopy: automated discovery of community structure within organizations, Available at http://www.lanl.gov/arXiv:cond-mat/0303264

Wallace P. (1999) *The Psychology of the Internet*. Cambridge University Press.

Wu F., Huberman B.A., Adamic L.A. and Tyler J.R. (2003) Information flow in social groups. Available at http://arXiv.org/abs/cond-mat/0305305

21

Mutual Adaptation: A New Criterion for Designing and Evaluating Human–Computer Interaction

Kazuhiro Ueda and Takanori Komatsu

21.1 Introduction

Researchers in artificial intelligence (AI), which is a field of information science and technology, have tried to understand what human intelligence is and how it can be realized mechanically, especially by using a computer. Many of their efforts have been aimed at developing automatic theorem provers, computer chess programs, computer-aided diagnostic systems, and so on. A supercomputer named Deep Blue, which beat chess champion Gary Kasparov in 1997, was especially representative of such efforts. At that time, the matches it played with the champion received the full attention of the public and mass media; it could calculate 200 million moves in a second and analyze all of the possible moves 14 moves ahead in 3 minutes, which is considered to be far beyond a normal human's ability. On the other hand, Deep Blue did not have other abilities; for example, it lacked the ability to talk with people, or even to play other board games such as Go and Shogi (Japanese chess). Therefore, while most people appreciated that Deep Blue had a certain kind of greatness, they did not think it had a human level of intelligence.

While AI researchers have pursued computational realizations of specialized abilities or expertise that human experts possess, researchers in cognitive science, which is a research field closely related to AI, are concerned with clarifying what that expertise is. Moreover, considering that human intelligence is not limited to such expertise, cognitive science researchers have also begun to focus on "mundane intelligence", which is the intelligence that we all exercise when we perform everyday activities that we do not think of as important or significant; for example, our communication skills and abilities to collaborate with others and our ability to understand and memorize things by making use of the external world or environment are all aspects of mundane intelligence.

Conversational Informatics: An Engineering Approach   Edited by Toyoaki Nishida
© 2007 John Wiley & Sons, Ltd

What are the essentials of mundane intelligence? Although researchers do not have a unified view on this question, we can consider the following abilities as candidates:

- the ability to communicate
- the ability to adapt to a new situation that one has never encountered.

The cognitive science community has treated these abilities separately; for example, the ability to communicate has been addressed in studies on pragmatics, conversational analysis, and the use of gestures, while the ability to adapt to new situations has been examined in relation to analogical reasoning and learning by transfer. Despite this separation, the two abilities are closely related to each other because the knowledge necessary for adaptation, or the way of adaptation itself, can be transmitted or transferred through communication. That is, human intelligence seems to be based on social learning, meaning that humans are social beings who learn by interacting with others and their environment.

This above phenomenon would be observed not only in human–human relations but also in human–animal relations, especially human–pet relations. So this means that most pet animals also have "mundane intelligence". Imagine the relationship between a dog and its owner; the dog and the owner who spend years together would form a nested adaptation between them (called "mutual adaptation" here). This is especially evident when an owner succeeds in teaching his or her dog verbal commands like "stay" or "sit" even though the dog cannot understand the literal meanings of the owners' verbal commands. In this case, the dog learns the meaning of the owner's command by means of rewards such as beef jerky, petting, etc., and the owner also learns the appropriate way to train the dog by watching its behaviors. That is, the dog will adapt its behaviors to its owner's speech and the owner will adapt his or her speech to the dog's behaviors. We assume that most readers accept the above empirical perspectives about the relationship between a dog and its owner; however there are no scientific studies to reveal the mechanism of such mutual adaptation process between humans and animals, while this is scientifically observed in all pairs that communicate smoothly between children and parents (Snow 1977a,b; Fernald et al. 1989; Fernald and McRoberts 1996).

What are the requirements for enabling pets to achieve mutual adaptation with people? If we could answer this question, the basic mechanism of mutual adaptation and a part of mundane intelligence would be clarified, and moreover, the pet robot, which could create intimate communication with humans, would be realized and be our companion. The above considerations about interaction between people and pet animals suggest that the following two competences would be then enumerated:

- One is the ability to recognize the reward information in the paralanguage information of human expression; for example, a smiling face is a "positive reward". Here, the reason why not linguistic but "paralinguistic" information is focused is that pet animals are considered to learn the meanings of human commands, triggered by human facial expression, gaze and prosody when the commands are given, because they are assumed not to directly understand and use the symbolic aspects of human language system.
- The other is the possession of a learning system about humans that can use two different types of rewards, the usual and direct one – a successful performance of a specific task is a "positive reward" – and the other indirect one, from human paralanguage information, as mentioned previously.

The purpose of this chapter is to clarify, based on the combined use of cognitive analysis and constructive approach, what the actual mechanism of the learning system is, which is triggered by para- or non-language information and is based on the above two different types of rewards. These have yet to be revealed, and a way of finding them could be that of observing and analyzing actual relationships between pet animals and their owners. However, it is quite difficult to include an actual pet animal in an actual controlled experimental setting; we guess that this is the most significant reason that there are no scientific studies about this issue up to now. Therefore, an experiment to reveal the specific contents of the rewards and the actual mechanism of the learning system was conducted by observing how two people could create smooth communication through a mutual adaptation process like that between pets and their owners (see section 21.2).

Especially in this study, prosodic information, one kind of paralanguage information in speech sound was the focus. Prosodic information cannot be written as texts or characters, but rather is expressed as stress or inflection and is believed to reflect more directly a speaker's internal state, a speaker's attitude and emotions, than does phonemic information. In actual relationships between pet animals and owners who have spent some years together, the animals can behave in the way their owners command them, even though they cannot understand the literal meanings of the owners' verbal commands. This indicates speech communication without phoneme information is possible, and that prosodic information has some role in it.

If the communication experiment reveals the specific contents of rewards and the actual learning mechanism for such communication, we can make the learning or meaning-acquisition model of a learner (i.e., a pet animal), which can be said to reflect a part of the underlying "mundane intelligence". In fact, such a meaning-acquisition model was proposed and actually constructed (section 21.3). The constructed model can be applied to autonomous artifacts, especially pet robots such as Sony's AIBO[1] and NEC's PaPeRo.[2] And it is also possible to apply this to an adaptive speech interface system that is required for communicating smoothly with users by learning the meanings of their verbal commands. Therefore, the pet robots and the adaptive speech interfaces will behave as the users wish, and robot and owners (or interface and users) can form a mutual adaptation process. For this purpose, the pet robots will be required to handle several communication modalities to achieve the mutual adaptation process (e.g., visual, speech, and tactile information), whereas the adaptive speech interface system will only require "speech" modality. Therefore in this study, as a first step to revealing requirements to enable people and pet robots to achieve a mutual adaptation, the focus is on a meaning-acquisition model that will provide the basic structure of an adaptive speech interface. This meaning-acquisition model should be designed to learn the meanings of verbal commands given by users so that the interface will be able to behave as users intend. This chapter also reports on the tests used to determine whether the meaning-acquisition model had sufficient competence to communicate with the participants in this experiment in a way that would be effective for users of pet robots or other machines (section 21.4).

Up to now, many speech interface researchers have agreed that all speech interfaces virtually require that a user engage in some learning and adaptation processes. However, most traditional interface studies have not paid much attention to the user learning process or its competences, and have not conducted psychological experiments to observe and analyze actual human learning or adaptation processes. Some have studied an adaptive (learning) interface system that provides a smooth operating environment for users by learning and adapting to the users' pattern of operations (e.g., Sears and Shneiderman 1994; Masui 1999). These studies only focus on the interface's learning mechanism and on unilateral adaptation of an interface to a user.

A lot of studies in psychoacoustics have already focused on the role of prosodic information in speech communication, for example, rising intonation being interpreted as an interrogative or disagreement (Ladd *et al.* 1985; Scherer *et al.* 1991; Johnstone and Sherer 2000) or a turn-taking signal (Pirrehumbert and Hirschberg 1990; Hobbs 1990; Koiso *et al.* 1998). However, few interface researches have focused yet on the previously mentioned aspects of prosodic information, especially for the learning mechanism of a speech interface system, such as rising intonation disagreement.

Recently, due to increasing complexity of information equipment, designing an interface that humans can use has become a major concern (e.g., Carroll 2000; Reeves and Nass 1996; Shneiderman 1997). The study described in this chapter would complement the traditional interface studies and be a key to create

---

[1] See http://www.sony.net/products/aibo/
[2] See http://www.incx.nec.co.jp/robot/

a mutual adaptation process between machines and users. It is expected that the acquired result would contribute to a principal methodology for realizing an auto customizing or personalizing function for a human–computer interface that would be practical for any user and one that would provide insights for continued research into pet robots and human–agent interaction (HAI) (Yamada and Kakusho, 2002).

Many studies on artificial agents and autonomous robotics have tried to achieve smooth interactions between humans and computers or robots. However, these studies assume that the robot or agent would learn autonomously; they did not consider that the user's learning process would affect the robot or agent (e.g., Breazeal 2003; Kubota and Hisajima 2003; Nicolescu and Mataric 2002). In particular, the learning agents of these studies (autonomous agent or robot) would carry out various tasks, from learning to decision making, as an independent subject, which means that the agent was separated from its environment and other agents. But do humans, who are far superior to agents in cognitive ability, usually behave in such a way? Not at all. A human seldom acts solely to accomplish a cognitive task such as problem solving and decision making; he or she is helped or given appropriate instruction by and/or collaborates with others. Namely, an instructor or collaborator often provides a learner with social meaning and a code, and the learner makes use of the meaning and code. Why not have learning agents engage in social learning as humans do?

This question was the starting point of our research. This study therefore places importance on social learning based on instruction, as described above, which consists of human mundane intelligence. In this point, this study can be said to be original. This study is also original in that not only the learner (agent or robot) but also the human instructor are considered as the constituents of the learning system. In addition, the authors think the learning ability of the human instructor should be made most of because humans are more adaptive than agents are.

As mentioned above, the proposed model would contribute to a principal methodology for realizing an auto customizing or personalizing function for a human–computer interface that would be practical for any user. Its application is not in constructing an effective system or artifact, which engineers in general pursue, but in building a unique system, or "the one and only artifact", for a user, as his/her family or pet actually is. This is explained in section 21.5.

21.2 Communication Experiment

21.2.1 Purpose and Settings

A construct of a desired meaning-acquisition model, revealing the specific contents of the rewards expressed and recognized in human paralanguage, and the actual mechanism of a learning system about the model's human partner are needed. To meet these needs, a communication experiment was conducted to observe how human participants created smooth communication by acquiring meaning from utterances in languages they did not understand (Komatsu et al. 2002).

In this experiment, pairs of participants, one teacher and one operator in each pair, participated. The teachers were placed in room A and the operators in room B (Figure 21.1). The goal of the participants in each pair was to work together to get the highest possible score in the video game "Pong" (Figure 21.2). Ten points were awarded to the participants each time they hit the falling ball with a paddle, and ten points were deducted each time they missed it. The total score was displayed on the right side of the game window. This scoring system was mainly used for operators to understand whether their current action was successful or not. The teachers' role was to give instructions to the operators, and the operators' role was to move the paddle to hit the ball to obtain as high a total score as possible. The operators' display, however, did not show the ball, which was their target, so that they needed to understand the intentions and meanings of the teachers' instructions in order to operate the paddle. The teachers' instructions were made linguistically incomprehensible. Therefore the scoring system had an important role in telling the operators whether their current action was successful or not.

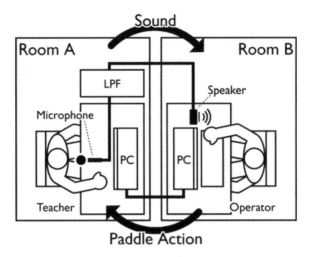

Figure 21.1   Pong video game environment

21.2.2 Participants

There were two groups of participants. In both groups, the operators could not linguistically understand the instruction of the teacher. The experimenter told the teachers that they could use any verbal instructions, whatever words or sentences they wanted.

Group 1 consisted of 11 pairs of 22 Japanese (22–28 years old; 18 men and 4 women). All 22 participants had sufficient experience to use a Windows PC. Eleven participants were graduate students, five were undergrads, three had university degrees, and three had junior college degrees.

The members of each pair of participants spoke the same native language (in this case, only Japanese was used). To make the teachers' instructions linguistically incomprehensible to the operators, the teachers' instructions were transmitted through a low-pass filter (LPF). The LPF masked the phonemes in teachers' speech, but this did not affect the prosodic features, such as the number of syllables. A cut-off frequency of about 150 Hz was set for the LPF for the voices of male teachers and 250 Hz for voices of female teachers.

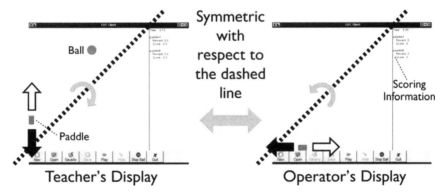

Figure 21.2   Different game display setting for teachers and operators in participating pairs

Even though the instructions were masked by the LPF, an operator still might be able to guess the instructions given, based on the number of syllables. For instance, suppose that both participants see the same display setting as that shown on the right side in Figure 21.2. In this case, the operator would intuitively speculate that the teacher would use the instructions "right" and "left" ("migi" and "hidari" in Japanese). If the teacher used those instructions, the operator could easily recognize which instruction is "migi" or "hidari" based on the number of syllables, regardless of the effects of the LPF. To avoid this operator bias, the displays of both participants were made symmetrical with respect to the dashed line in Figure 21.2, so that the teacher would use instructions, "ue" ("up" in English) and "shita" ("down" in English). The two words have the same number of syllables, and there is no difference between the utterance duration of "ue" and "shita" in Japanese.

Group 2 consisted of six pairs (23–26 years old; 10 men and 2 women). All 12 participants had sufficient experience to use a Windows PC. Eight participants were graduates, and four had university degrees. The participant pairs were an Indonesian (teacher) and an American (operator), a Spanish speaker and a Filipino, a Korean and a Chinese, and three pairs of Chinese and Japanese.

Individual members of these pairs did not speak the same native language. Moreover, before this experiment, it was confirmed that the operators did not have any experience of studying the teachers' native language. The teachers were asked to speak only their native language. Therefore, in this group, the operators could not linguistically understand what the teachers were saying even though no LPF was used. Therefore, the teachers' instructions were transmitted directly to the operators. The display setting was the same as used by group 1.

21.2.3 Procedures

The experimenter first explained to each participant pair that the purpose of this experiment was to score as high a total score as possible by working together. Then, in case of group 1, the experimenter let the pair hear a vocal sample, with and without LPF masking, to demonstrate the LPF's effects. At this point, the experimenter did not mention that different display settings would be used. After this explanation, the test was started. Each pair played two consecutive ten-minute games, with three minutes of rest between them. The roles of teacher and the operator were fixed, and the participants had no opportunity to chat during the experiment.

21.2.4 Results

To evaluate the pairs' performance (i.e., whether they could score effectively in this Pong game), two values were assigned to each of the operators' actions of moving the paddle and hitting the ball. For each action, if an operator moved the paddle in the direction the teacher intended, 1 point was awarded as the correct direction value (CDV); if they moved it in a different direction, the CDV score was 0. When the operator moved the paddle multiple times, just one CDV score (0 or 1 point) was assigned only for the last action just before the ball reached the bottom of the display or hit the paddle. If the operators hit the ball, 1 point was awarded, as the hit value (HV), for each action; if they missed it, the HV score was 0.

A statistical testing hypothesis was used to categorize the participants at the end of the experiment. The testing hypothesis, using a binominal distribution, was formed by means of only ten of the participants' CDV and HV scores. Specifically, the hypothesis stated that when operators did not understand the teachers' intended direction, and when the instructions that meant up or down remained unknown, the probability they would move the paddle toward the teachers' intended direction was 0.5. In this case, the probability P_{CDV} that the average CDV would become more than 0.8 during the previous 10 actions – that is, the percentage of moving in the correct direction would be 80% – was calculated as $P_{CDV} < 0.0547$ by means of a testing hypothesis formed by a binominal distribution. Thus, when the average CDV was

Table 21.1   Average correct direction value (CDV) and hit value (HV) at the end of experiment for group 1 per pair

Category	(Average CDV, Average HV)
Category 1 (2 pairs)	(0.5, 0.5), (0.3, 0.2)
	Ave (0.40, 0.35), S.D. (0.14, 0.21)
Category 2 (5 pairs)	(0.9, 0.3), (1.0, 0.2), (1.0, 0.5), (0.8, 0.6), (1.0, 0.6)
	Ave (0.94, 0.44), S.D. (0.09, 0.18)
Category 3 (4 pairs)	(1.0, 0.9), (1.0, 0.7), (1.0, 0.7), (0.9, 0.8)
	Ave (0.98, 0.76), S.D. (0.05, 0.09)

greater than 0.8, it was concluded that the operator had succeeded in understanding the teacher's instruction about the direction. In addition, when the operator understood the teacher's intended direction, the probability of hitting the ball was 0.34, as calculated both with the velocity of the paddle and the ball and with the length of the paddle. In this case, the probability P_{HV} that the average HV became greater than 0.7 during the previous 10 actions – that is, the percentage of hitting the ball was 70% – was calculated as $P_{HV} < 0.023$. Thus, when the average HV became more than 0.7, it was concluded that the operator could hit the ball effectively. Then, the pairs of participants were divided into three categories:

- *Category 1.* The average CDV at the end of the experiment was less than 0.8. It is considered that the participants assigned to this category failed to understand any instruction.
- *Category 2.* The average CDV was more than 0.8, and the average HV at the end of the experiment was less than 0.7. It is considered that the participants assigned to this category succeeded in moving the paddle in the direction the teachers intended, but could not hit the ball well.
- *Category 3.* The average CDV was more than 0.8, and the average HV was more than 0.7. It is recognized that these participants could move in the teachers' intended position and hit the ball well.

Tables 21.1 and 21.2 show the average CDV and HV values during the last ten actions at the end of experiment for the three categories. In group 1, out of 11 pairs, 2 operators were assigned to category 1. Among the 9 remaining pairs, 5 operators were in category 2. The 4 remaining ones were in category 3. In group 2, 2 out of 6 pairs were in category 1, 2 in category 2, and the remaining 2 were in category 3. Here, when the operators scored above a 0.8 average CDV (i.e., pairs in category 2 and 3), it was concluded that the operators in these pairs somehow succeeded in understanding the teachers' instructions and communicated efficiently. How could the pairs in category 2 and 3 successfully communicate even in such a poor communication environment? The processes of the participants' behaviors were investigated

Table 21.2   Average correct direction value (CDV) and hit value (HV) at the end of experiment for group 2 per pair

Category	(Average CDV, Average HV, mother tongue of teacher-operator)
Category 1 (2 pairs)	(0.5, 0.5, Chinese-Japanese), (0.4, 0.4, Chinese-Japanese)
	Ave (0.45, 0.45), S.D. (0.07, 0.07)
Category 2 (2 pairs)	(1.0, 0.6, Indonesian-American), (0.8, 0.6, Chinese-Japanese)
	Ave (0.90, 0.60), S.D. (0.14, 0.00)
Category 3 (2 pairs)	(1.0, 0.9, Spanish-Filipino), (1.0, 0.7, Korean-Chinese)
	Ave (1.00, 0.80), S.D. (0.00, 0.14)

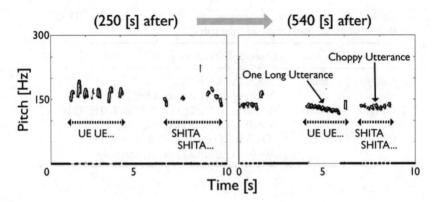

Figure 21.3   Typical transition in pitch curves of teacher's instruction. Left: 250 [s], Right: 540 [s]

in terms of the prosodic elements of the instructions, the movement of the paddle, and the variety of instructions the teachers used.

Participants' Behaviors in Group 1

The operators in 9 of the 11 pairs in group 1 succeeded in somehow understanding the teacher's instructions. A particular teaching–learning process was observed in most of these 9 pairs. At the beginning of the experiment, the teacher gave a wide variety of instructions to the operator, for example, "move to the center of the display"; "move a bit up"; "do not move!" etc. The pitch curve of this initial instruction type is depicted on the left side of Figure 21.3. At this point, it would have been difficult for the operators to discriminate between "ue, ue, ue. . . (up, up, up. . .)" instructions to move to the right and "shita, shita. . . (down, down. . .)" instructions to move to the left. These instructions were cluttered and undifferentiated for the operators so that they could not recognize the intentions and meanings of the teachers' instructions. Thus, the operators' behaviors (paddle actions) were hesitant, and they sometimes looked confused.

As they continued to give instructions to the operators, the successful teachers gradually recognized that the operators did not properly interpret the instructions being given them. The successful teachers tried to simplify their instruction patterns to make them easier for the operators to understand. Specifically, these teachers started to gradually decrease the variety of instructions and typically converged on only two, such as "ue" and "shita". In contrast, the teachers in unsuccessful pairs (category 1) did not simplify their instructions and continued to use a wide variety of instructions (Figure 21.4). A value named "variety of instruction" was defined as the total of the types of instructions used during ten teaching opportunities (during the 10 times the ball was falling).

These results show that the teachers apparently made it easier for the operators to understand their instructions by decreasing their variety and differentiating them. In conjunction with this gradual decrease in the variety of instructions, differences in the pitch curves of the two instructions became more distinct. Most teachers started to repeat the instructions; e.g., "ue, ue, ue. . ." and "shita, shita. . .". The former instruction was audible as one long, continuous voice, whereas the latter sounded like many choppy utterances (see the right side of Figure 21.3). In Figure 21.3, the results show that the teacher's instructional pattern at this time was audibly simplified and differentiated.

At this point, to show that they recognized the intentions and meanings of instructions, the operators were required to make their paddle actions correspond to these differentiated instructions. Two methods were used to do so. One was that the operators recognized that a given instruction indicated that their last

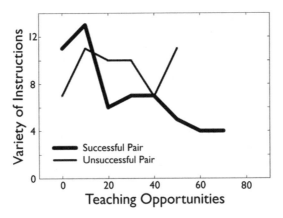

Figure 21.4 Typical transition in varieties of the teacher's instruction in a successful and an unsuccessful pair: This figure shows that the varieties of instructions gradually decreased in the successful pair but did not in the unsuccessful pair

action succeeded in hitting the ball even though this was done by chance during trial-and-error attempts. The other one was that a given instruction did not indicate that their last action was successful when the teacher expressed their frustration by high-pitched vocalizations that sounded as if the teacher "got angry".

Actually, when the pitch of teachers' voices at a certain point in a series of instructions became about 20 Hz higher than the pitch at the onset of the series, and when this higher pitch continued for at least 500 ms (depicted in the circle in Figure 21.5), some operators intuitively recognized this sound feature as

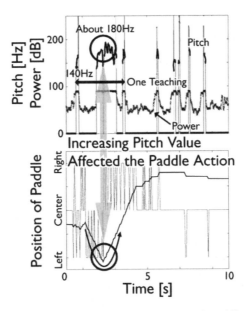

Figure 21.5 High-pitch element in teaching voice affected operator's paddle action. Specifically, when an operator heard a high-pitched voice, he or she immediately changed moving direction (see the circles)

a warning about their last action. Specifically, when operators heard such high-pitched utterances, they recognized that their last action was wrong and modified it immediately. This prosodic sound feature was named "Attention Prosody (AP)." Thus, hitting the ball served as a positive reinforcement for the operator's last action, while the AP vocalization by the teachers served as a negative reinforcement or reward.

The operators thus learned to understand the teacher's instructions by exploiting these two kinds of reward signals, a positive reward (hitting the ball) and a negative reward (hearing an AP sound). At first glance, missing the ball and having ten points deducted from the total score seemed to work as a negative reward for their learning processes. However, at least three cases occurred in which the operator missed the ball: (a) The operator moved the paddle in a different direction from that the teacher intended; (b) the operator moved in the correct direction but passed the position where the ball would fall; (c) the operator moved in the correct direction but failed to reach the ball's falling position. Therefore, interpreting why they failed to hit the ball was too complicated for the operators and they could not use this event for their learning process. Therefore, it was observed that the participants did not use this "falling ball" event as a negative reward in their learning processes.

Learning to recognize the instructions was due not only to teachers' efforts, such as decreasing the variety of instruction and using APs, but was also due to the efforts of operators. As already mentioned, even in successful pairs, the operators' actions were initially hesitant as they tried to concentrate on inferring the teachers' intentions and meanings from the given instructions. The operators in successful pairs started reacting actively after they inferred the teachers' intentions and meanings. Their actions thus indicated their comprehension of given instructions. In this case, the teachers could recognize the level of the operators' comprehension of instructions from their actions, so the teachers could improve their ways of teachings to fit the operators' learning modes. On the other hand, the operators' behaviors in the unsuccessful pairs were also hesitant at the beginning of the experiment, but over time they started moving continuously even when no instructions were being given. These operators seemed to disregard the teachers' instructions. In this case, an operator's actions did not indicate any comprehension of the instructions, and the teacher could not figure out the operator's learning mode.

Participants' Behaviors in Group 2

The operators in 4 of 6 pairs succeeded in understanding the teacher's instruction. In fact, there was only one difference between the successful pairs in groups 1 and 2. The operators in group 2 could recognize the types of instruction according to the differentiated sound patterns from the beginning of the experiment, even though they did not know anything about the teachers' spoken language. Therefore, the time consumed in this meaning-acquisition process was shorter than for group 1. Except for this point, most of the processes that the pairs in group 2 demonstrated were nearly the same as those in group 1.

Comments

A general meaning-acquisition process, such as a listener trying to figure out what a speaker says, seems to be achieved only by the listener's efforts to adapt unilaterally to the speaker. However, from the result of this communication experiment, it was found that the teachers (speakers) also adapted to the operators' behaviors and contributed to the meaning-acquisition process by improving their instructions according to the operators' (listeners') learning modes. This was ascertained from the operators' paddle actions in terms of their reaction to the teachers' instructions. Thus, it can be said that mutual adaptation processes were observed. Not only did the operators try to learn the intentions and meanings of the teachers' instructions, but also the teachers simultaneously revised their manner of giving instructions to fit the operators' learning modes.

In successful pairs, the following specific behaviors were observed:

1. Teachers
    - decreased the types of instructions to fit the operators' learning modes while the experiment was proceeding
    - made the operators focus on their actions by pronouncing APs (see Figure 21.5).
2. Operators
    - moved the paddle to indicate their comprehension of a given instruction whenever an instruction was given
    - moved the paddle differently according to different types of instruction
    - corrected their paddle actions in response to APs.

The experimental results revealed that most operators could recognize the intentions and meanings of teachers' verbal commands by inducing the teachers' adaptations and by using these adaptations in a meaning-acquisition processes. In addition, it was also observed that AP sound features were universally interpreted, by the operators, as "a caution about their current action" and had a significant role as negative reward information in the meaning-acquisition process.

21.3 Proposal for a Meaning-acquisition Model

21.3.1 Requirements

Based on the results of the communication experiment, the observed operators' behaviors were used to achieve a meaning-acquisition model that could recognize the meanings of verbal commands given. To do so, the model needed to satisfactorily demonstrate these abilities:

1. Recognize that a given verbal command indicates a certain paddle action.
2. Find critical sound features in speech to distinguish different types of instructions.
3. Acquire the meanings of instructions without having a-priori prepared knowledge about instructions (e.g., a dictionary-like database).
4. Extract AP sound features from the given verbal commands to use them in a meaning-acquisition process.

To meet these abilities, a meaning-acquisition model was proposed in consideration of the following assumptions.

- The model assumes that an incoming instruction sound indicates a certain paddle action and that recognizing the meaning of an instruction is equivalent to creating an appropriate mapping between the incoming instruction and the appropriate paddle action.
- When a paddle action is correct (when the model succeeds in hitting the ball), the model should recognize that the meaning of the given instruction identifies the last action as successful.
- Conversely, when an action is incorrect (when the model receives an AP sound from the teacher), the model should recognize that the meaning of the given instruction does not identify the last action as successful.
- Each instruction sound is represented by an eight-dimensional vector consisting of the average values of the eight different sound features (such as pitch, zero-cross number, etc.; see Figure 21.6). In addition, each action is represented by a scholar value (its absolute value indicates its velocity, and plus or minus indicates right or left, respectively). The model assumes that this combined nine-dimensional sound-action vector (instance) is explained by a normal distribution in a mixture of normal distributions. Each distribution in the mixture expresses each intention or meaning of a given instruction.

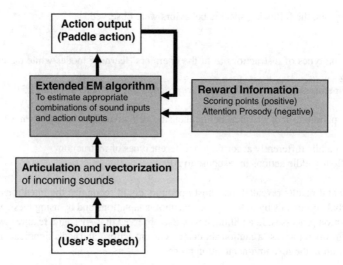

Figure 21.6   Outline of procedure of the proposed meaning-acquisition model

This model must acquire the appropriate parameter sets (average μ and standard deviation σ in each dimension) of each normal distribution in a mixture of normal distributions to explain the incoming instances in order to recognize the meanings of given instructions.

21.3.2   Methodology

As a basic methodology for acquiring the appropriate parameter sets, the EM algorithm (Dempster *et al*. 1977; Mitchell 1997) was used. This algorithm can be used even for variables whose values have never been directly observed, provided the general form of the probability distribution governing these variables is known. To estimate each distribution's parameter, the EM algorithm can use positive instances that are acquired when the positive reward is given; however, this algorithm cannot use the negative instances that are acquired when a negative reward is given (specifically the model receives an AP sound from the teacher), because this speech sound does not indicate this user's paddle action. Therefore, we developed an extended EM algorithm so that it could use negative instances for estimating the parameters. The proposed meaning-acquisition model is schematically depicted in Figure 21.6. For a more detailed description of this learning model, refer to (Komatsu *et al*. 2005).

21.3.3   Testing Experiment

To evaluate the basic competence of this meaning-acquisition model, a testing experiment was conducted to find whether this model could learn to recognize the meanings of instructions through interaction with a human instructor. As a testing environment, this meaning-acquisition model was incorporated into the paddle component of the "Pong" game software. As already described in section 21.3.1, this model does not use phoneme information and does not have a-priori knowledge about the instructions given for recognizing the meanings of instructions. Therefore, this model was required to learn to recognize the meanings of instructions through prosodic sound features so as to distinguish the different instruction types, regardless of the actual language being spoken. Moreover, it can do so if enough of these prosodic sound features are given.

Table 21.3 Average CDV and HV at the end of experiment, number of instructions, consumed time resulting from testing experiment

Instruction type (upward/downward)	(Average CDV, Average HV, number of instructions, consumed time [s])
1. High/low	(1.0, 0.7, 81, 469)
2. Long/choppy	(1.0, 0.5, 92, 420)
3. Ue/shita (Japanese)	(1.0, 0.6, 87, 445)
4. Up/down (English)	(1.0, 0.6, 87, 435)
5. Low/high	(1.0, 0.8, 37, 193)

In order to test the learning performance of this model, an instructor gave different instructions to this model. It was then observed whether this model could learn to recognize the meanings of the following five types of instructions:

1. high-pitched utterances for upward and low-pitched ones for downward while saying "ahhh"
2. long utterances for up and choppy ones for down while saying "ahhh"
3. "ue," Japanese for up, and "shita," Japanese for down
4. "up" and "down" in English
5. the inversion of 1.

The instructor in this testing experiment was a teaching expert who already had enough training to give ideal instructions and knew well the basic structure of this model. In each type of instruction, this instructor played the game Pong, giving verbal instruction until the model's most recent 10 actions scored an average CDV that exceeded 0.8 and an average HV that exceeded 0.7. These criteria are the same as those used in the communication experiment to determine whether a human operator succeeded in recognizing instructions.

Each final average CDV and HV for each instruction, the number of instructions, and the time consumed are summarized in Table 21.3. In these results, it was observed that this model learned to recognize the meanings of all five types of the given instructions during the first 2–5 minutes of the game. This suggests that the model had sufficient ability to recognize the meanings of given instructions through interaction with an actual human instructor, without any a-priori knowledge about those instructions, when the instructor used the salient prosodic features.

21.4 Interaction between the Proposed Model and Users

21.4.1 Purpose and Settings

The instructor used for the previously described study was an ideal one because he already knew and understood the meaning-acquisition model well. Therefore, to apply this model to interface systems that would work in actual everyday situations in the near future, confirmation was required of whether this model could recognize the instructions of users with no specific knowledge of the model.

Another testing experiment was conducted with participants who would use the model in the same way as ultimate end users without knowledge of the model (Komatsu *et al.* 2003). The goal of each participant was to teach verbal commands to the constructed meaning-acquisition model that was driving the "Pong" paddle and to make the paddle move as desired.

Recently, Harada (2002) reported that most people hesitate to talk naturally to agents without an actual physical entity such as a life-like agent in a computer. However, finding the conditions that will induce humans to speak to computers naturally is worthwhile. In this experiment, the focus was on the effect of experimenters' instructions on such condition. For instance, an experimenter gave the instruction to the participant, "please talk to this model as if talking to a person". It was assumed that this kind of instruction would be a catchphrase helping users intuitively understand how to use and interact with this model without reading thick instruction manuals or other documents.

21.4.2 Participants and Procedures

Two groups of participants took part in this study. The experimenter gave different instructions to each group. In addition, the experimenter told the participants that they could use any verbal instructions – whatever words or sentences they wished.

Group A had 10 participants (9 Japanese and 1 Filipino; 21–29 years old; 6 men and 4 women). All participants were graduates in computer science and cognitive sciences. In addition, they all had sufficient experience not only in using computers but also in computer programming. The experimenter gave them these instructions: "The game paddle is operated by a learning computer and your task is to teach this computer to make the paddle move as you want by using verbal instructions. So please start teaching this computer." The aim of this instruction was to notify the participants that the teaching target was a computer agent in the display.

Group B also had 10 participants (10 Japanese; 22–39 years old; 8 men and 2 women). Again, all participants were graduates in computer science and cognitive sciences and had experience not only in using computers but also in computer programming. The experimenter gave the participants these instructions: "So please start teaching as if talking to a person." Unlike the instructions for group A, the aim of this different instruction was to notify the participants that the teaching target was something other than a simple computer.

21.4.3 Results

To judge whether the model succeeded in recognizing the intentions and meanings of the participants' instructions, the same CDV value was used as in the communication experiment described earlier. That is, when the average CDV exceeded 0.8, it was considered that the model had succeeded in at least recognizing a participant's instructions. In addition, an AP ratio was used to distinguish whether participants in this experiment talked to the model operating the game paddle as if talking to human participants in the same way as did those participating in the previous communication experiment. This AP ratio was defined as the value calculated by dividing the number of APs observed by the total number of instructions.

In the previous communication experiment (section 21.2), an AP ratio of about 5% was observed for most successful pairs of participants. Actually, 13 participants' average AP ratio (9 in group 1 and 4 in group 2) was 4.93 % (SD = 2.32). Therefore, it was assumed that an AP ratio of around 5% would mean that the participant talked to the model as if talking to humans, like the previous communication experiment with two human participants.

The participants played the game for about 30 minutes, and if they scored a CDV exceeding an average of 0.8 in their most recent 10 actions (the model could move the paddle in the direction the participants intended in 80% of 10 attempts), the experiment was terminated. Afterward, the interviews were conducted to collect the participants' impressions of this experiment and to ask them how you felt about their partners in this experiment (the model operating the game paddle). Tables 21.4 and 21.5 show the consumed time, AP ratio, types of final instructions, maximum and final varieties of instructions

Table 21.4 Consumed time, AP ration, final instruction used, varieties of instruction, and average CDV and HV results for participants in group A

Participants	Time (sec)	AP ratio (%)	Final instruction used (upward/downward, and [others])	Max-final varieties of instruction	(CDV, HV)
A	1425	0.8	Taas/Baba [*Tagalog]	2-2	(1.0, 0.6)
B	1690	1.8	UEUE/SHITASHITA	2-2	(0.9, 0.7)
C	485	0.0	Sony/Aiwa	2-2	(1.0, 0.8)
D	770	0.0	UEUE/SHITASHITA	2-2	(1.0, 0.7)
E	1237	0.0	UE/SHITA	3-2	(0.6, 0.2)
F	1401	2.5	UEUE/SHITASHITA	4-2	(0.9, 0.5)
G	1109	2.7	UEUE/SHITA	3-2	(0.6, 0.1)
H	1955	0.0	[I, E, O, KA, A, KE, SE]	6-7	(0.4, 0.2)
1	1891	0.6	UE/SHITA	2-2	(0.4, 0.1)
J	729	1.2	AAAAA/Uhhh	4-2	(0.8, 0.6)
Average	1269.2	0.96		3.0–2.5	(0.75, 0.45)
SD	500.9	1.05		1.3–1.6	(0.24, 0.27)

Table 21.5 Consumed time, AP ration, final instruction used, varieties of instruction, and average CDV and HV results for participants in group B

Participants	Time (sec)	AP ratio (%)	Finally instruction used (upward/downward, and [others])	Max-min varieties of instruction	(CDV, HV)
K	1351	3.5	UEUE/SHITASHITA	10-2	(1.0, 0.7)
L	1574	0.3	UE, UE-E (toward)/SHITA, SHITA-E [MOU-CHOTTO (a bit more)]	8-5	(0.6, 0.4)
M	1670	3.2	UE/SHITAAAA	3-2	(1.0, 0.7)
N	373	6.1	UE/SHITA	5-2	(0.9, 0.5)
O	1569	2.3	UEUE/SHITASHITA [SUKOSHI (a bit), IKISUGI (passed)]	9-4	(0.4, 0.4)
P	1166	4.1	UEUE/SHITA	4-2	(1.0, 0.5)
Q	1347	4.0	UE, UE-NI-UGOITE (move upward)/SHITA, SHITA-NI (toward) [YOSHI (good)]	8-5	(0.6, 0.4)
R	1208	0.0	[A, I, U, E, O]	2-5	(0.6, 0.4)
S	1565	12.4	UE, UE-DESU(*formal usage)/SHITA [OSOIDESU (slow)]	4-4	(0.8, 0.3)
T	1798	8.7	AGATTE (move upward)/SAGARU, SAGATTE, SHITA, SHITA-NI-UGOKU	8-5	(0.3, 0.1)
Average	1362.1	4.46		6.1-3.6	(0.72, 0.44)
S.D.	402.1	3.78		2.8-1.4	(0.26, 0.18)

during the experiment, and average CDV at the end of the experiment and HV values for participants in groups A and B, respectively.

Results for Group A

Six out of 10 participants succeeded in teaching verbal instructions to the model and in making the paddle move as desired. Interestingly, all participants (regardless of performance, whether they succeeded or failed in teaching) showed nearly common behaviors. They did not change the types and varieties of instructions they gave. In concrete terms, they consistently used only two types of instructions (corresponding to "up" and "down"). Therefore, the model had no opportunities to induce the participants' adaptations (to induce that the participants decreased the types of instructions given, for example) and to use such adaptations in meaning-acquisition processes. Thus, the data suggest that the relations between these pairs were different from the ones between the participants in the communication experiment.

In addition, they did not use AP sounds very frequently to teach the meaning-acquisition model (Table 21.4 shows an 0.96% average AP ratio for this group.). This means that they consistently gave unemotional instruction to the model. According to the interviews taken after the experiment, all participants in group A reported that they felt great stress while giving instructions to this model. They reported that they assumed that computers in general would not understand emotional expression. Therefore, they suppressed use of emotional expressions and tried to remain calm even if the paddle moved in the direction opposite to the one intended. Additional analysis is required to investigate the relations among unemotional speeches, user's mental stresses, and the existence of mutual adaptation.

In this group A, nearly no behavioral differences existed between the participants who succeeded and those who failed. It was then assumed that whether the participants failed or succeeded in teaching the game paddle may have been decided by whether the model moving the paddle hit the ball effectively at the very beginning of the experiment, even by chance. In any case, although this model succeeded in responding correctly to the participant's instructions, the relations between the model and participants were completely different from those observed in the communication experiments with human partners.

Results for Group B

Five out of 10 participants succeeded in teaching the verbal instructions to the model and these participants showed nearly common behavior. They decreased the varieties of instructions provided according to the learning modes of the meaning-acquisition model and used AP sound features to make the model focus on its last action. Moreover, they achieved an AP ratio of about 3%, and no participants reported that they felt any stress during the experiment. The successful participants in group B used more natural instructions compared with those used in group A. These participants' behaviors were quite similar to those observed in successful pairs in the communication experiment. Interestingly, participants M and N who succeeded (and participant R who actually failed) reported that they did not try to teach the instructions to the model; instead, they tried to discover how the model preferred to receive instructions. Such behaviors were completely different from those participants in group A, who consistently used the same instructions. Therefore, these participants' behaviors might be evidence that these teachers actually adapted to the model by using its behaviors to improve their teaching strategy.

The data for this group suggest that the model succeeded in recognizing the participants' verbal commands by using the participants' adaptations (AP utterances and decreasing the types of instructions) for meaning-acquisition processes. Therefore, it can be said that a partial mutual adaptation process existed between the participants and the model. The model recognized the meanings of the user's verbal commands by exploiting the user's adaptation to it. The differences between a true mutual adaptation in the communication experiment and the partial mutual adaptation observed here resulted from the participants' different behaviors. In a true mutual adaptation, the teachers induced the operators' adaptations and used them. In a partial mutual adaptation, the users (teachers) just used the model's responses without inducing the model's adapted responses. Therefore, if the participants successfully induce the model's adapted response, one may assume that a true mutual adaptation between them was achieved.

Out of 10 participants in group B, 5 failed to teach the instructions to the model. Their behaviors were nearly the same as those who failed in the communication experiment (pairs in category 1). They used a wide variety of instructions and did not decrease them. In addition, they did not use instructions with AP sound features or they used them too frequently.

Comments

The results of this experiment suggest that the proposed meaning-acquisition model had sufficient competences to recognize the intentions and meanings of users' instructions. The model could recognize the given verbal instructions by exploiting the uses' adaptations for its meaning-acquisition process, with the scoring paddle action being recognized as a positive reward and the received AP sound features being recognized as a negative reward. This was possible when the experimenter gave the participants appropriate instructions, such as those given to the participants in group B. In turn, the participants' adapted behaviors, such as AP utterance and decreasing the types of instructions, were induced not by the meaning-acquisition model itself but rather by the instruction provided to the participants in group B. Therefore, these instructions would be available as catchphrases for a speech interface system based on this meaning-acquisition model, and they may help users understand intuitively how to use and interact with such an interface.

At first glance, the experimenter's instruction "please start teaching as if talking to a person" appeared to force the participants to act as if they were talking to person. However, as already mentioned previously, no participants in group B reported that they felt any stress during the experiment, so it is said that the experimenter's instructions actually induced the participants' natural behaviors. Up to the present, no scientific proof has been reported about the relation between such instructions and the participants' behaviors. Therefore, additional studies are required to investigate the effectiveness of this kind of instruction. Especially, the attitude of participants about whether they felt any mental stress from being given this instruction or from other conditions (such as agent's appearance, physical entity, etc.) needs to be studied.

21.5 Discussion

21.5.1 Applicability of this Meaning-acquisition Model

The testing experiment described in the previous section revealed that the proposed meaning-acquisition model could learn the meanings of verbal instructions and could also behave in the way its users intended when appropriate instructions were given to the participants. In addition, the model and the participants could also form a partial mutual adaptation process when the experimenter gave them appropriate instructions. Based on that result, it is expected that this meaning-acquisition model can be applied to hands-free speech interface systems for personal computers or car navigation systems.

In related technologies, some studies on interfaces have applied the prosodic features to a sound input devise for personal computers to carry out tasks such as scrolling a display upward when speech with rising intonation is continued (Goto *et al.* 2001; Tsukahara and Ward, 2001). For example, Igarashi and Hughes (2001) developed a speech interface for use with personal computers that allows users to give verbal commands such as "move up ahhh". This interface starts to scroll the screen of a computer upwards after recognizing the meaning of a part of the command as "move up". The interface continues to scroll upward as long as the user pronounces "ahhh" and changes the scrolling velocity according to the user's inflection. One of the advantages of this method is that this interface can start the action immediately after recognizing a minimum meaningful sentence such as "move up" so that the calculation time of the sound recognizer is less than that of a traditional one. Another is that users can interactively control the scrolling distance and velocity with the "ahhh" pronunciation while watching the actual screen motion. Therefore, the user does not need to rigidly define the scrolling distance and velocity before giving the verbal commands. In addition, prosodic information has a strong tendency to reflect a user's intention or attitude; therefore, one can assume that the user can control the scrolling intuitively.

In this method, however, mapping between particular speech elements and a desired resulting action is required to be defined and given in advance, just as in a traditional speech interface. That is, obeying the prepared mapping database is strongly required even if a user wants to use "roll up" instead of "move up". The model proposed in this study would complement the study done by Igarashi and Hughes (2001) in terms of the functions of personalization or auto-customization. This is because one of the advantages of the proposed meaning-acquisition model is to learn the meanings of given speech commands through the interaction with users via their emotional expressions (Attention Prosody is a "negative reward".). This could provide the flexibility to help users personalize or customize the meanings of verbal commands or scrolling direction through actual everyday use.

For example, in daily life some users unconsciously complain verbally about their computers' unexpected reactions in an angry voice. In such a case, this learning model could detect this user's anger and reflect this voice feature in its learning process as a negative reward.

21.5.2 Towards Realizing the One and Only Artifact for a User

If a user is supposed to carry out a task, such as scrolling a display upward or downward, through a hands-free speech interface, the number of commands which the user gives to the interface is limited; for example, "Roll up or down" and "Move up or down". Moreover, it provides not a light workload for the user to make the interface learn mappings between the verbal commands he/she uses and the interface's behaviors (e.g., scrolling up or down). Someone therefore may think that the method proposed by Igarashi and Hughes (2001) is superior to that proposed here, for such a simple task. The authors agree with this opinion, considering the effectiveness of interface operation.

What on earth is the advantage of building an artifact, based on the proposed model, that achieves mutual adaptation through communication with humans? To answer this question, it is useful to refer to the work of Karatani (1989). Karatani proposed that there were two criteria for evaluating "intelligence": one "general – specific/peculiar" and the other "common – special/unique". Let us take "dog" for example. The former criterion shows the place that an individual dog (e.g., named "Snoopy") takes in the distribution of some feature (e.g., size; big or small) which dogs have as a species, whereas the latter shows to what extent the individual dog (named "Snoopy") is special for an individual person (e.g., the owner of "Snoopy"). It can be said that AI researchers or cognitive scientists have tried to evaluate intelligence based on the former criterion. That is, because expertise is peculiar to a certain person, as described in section 21.1, cognitive scientists have appreciated it and AI researchers have tried to embody it in computers. By contrast, we would ask "Why not appreciate (or why not seek to embody) intelligence based on the second criterion?"; that is, the creation of a unique system ("the one and only artifact") for a user, as in the case of a family or pet.

As you can see with considering the explanations provided in section 21.1, we aim to develop "the one and only artifact" for its user because an artifact implemented with a mutual adaptation ability can form an intimate mutually adaptive relationship with its user. In other words, we aim to construct an artifact that learns through interaction to follow only its user's commands, or we aim to create an interactive media, such as pet robots, that place importance on interactions with their users, instead of building an artifact that is intended to be useful to many people as traditional AI has sought to.

Pet robots, such as Sony's AIBO and NEC's PaPeRo, are becoming increasingly popular with people of different generations and genders. In fact, pet robots have actually become a part of domestic life. Although most people enjoy interacting with these robots, they do not seem to regard them as actual pet animals and eventually become bored with them. Their owners do not consider pet robots to be "the only ones". What is required to improve this situation? Some people may think the answer to be in making a robot with a cuddly appearance and others may think it to be in interactions such as eye contact and unintentional movements where one can feel animacy. It is very difficult to answer this question, but we believe that, to overcome the above problem, a pet robot should have a learning ability that enables it and its owner to adapt to each other continuously. That is, we believe it is necessary to realize an aspect of "mundane intelligence".

As you can see from the discussion so far, the final purpose of this study is to construct an artifact that can communicate smoothly and intimately with a particular individual; that is, the final goal is to build a conversational artifact that realizes an aspect of "mundane intelligence". To do so, how constrained human–human communication could be established was analyzed, in terms of the use of paralinguistic information. Namely, in this study, designing the conversational artifact was based on the results of scientific analysis for a kind of actual conversation. So this study adopts an integrating method of scientific and engineering means for investigating and realizing conversation, which is one of the main scopes of conversational informatics.

21.5.3 Limitations of the Proposed Model

Although the testing experiment showed that this model could learn the meanings of user's verbal commands through the interaction with users, the result of this experiment suggested that this model left at least two issues unresolved. One issue was the model's unnatural reactions, and the other was its scalability for a multifunctional interface system.

Most participants in both groups (A and B), regardless of success in teaching, felt that the model's actions were unnatural and unpredictable in relation to their behaviors. Specifically, these situations can be divided into the following two cases:

- *Case 1: Could not react to a sudden strategy changes.* At the beginning of the experiment, the typical teacher used a type of instruction, such as "ue" for upward and "shita" for downward. Then, if the operator's performance was impaired in the first 2–3 minutes, some teachers changed their instruction types; for example, "ueeee" and "shitaaa", in order to further differentiate the sounds of these two instructions. In this case, human operators in the communication experiment could immediately understand that the teaching pattern was somehow changed and then could start trying to improve their paddle actions, while the constructed model could not.
- *Case 2: Moved in only one direction.* Sometimes the model continued to move only in one direction, even though the participants used various types of instructions, such as "down", "up", "opposite!" or whatever.

These cases sometimes inhibited participants' ability to induce the model's adapted response and to show natural behaviors toward it, and this might be the reason that half of the participants in the testing experiment failed to teach their instructions to the model. It was assumed that the characteristics of the statistical learning algorithm, implemented in this model as an EM algorithm, was the principal cause of these two cases. In case 1, it was assumed that this issue resulted from one of the characteristics of statistical learning; that is, learning consequences depend on the number of accumulated instances. For example, if a user changed teaching strategy and started using different instructions not previously experienced by the model, the model did not have any information about these newly arriving instructions at that point. Therefore, this model could not handle these experienced instructions appropriately, so the model's output from the learning was outside the scope of the user's expectation. In case 2, this issue resulted from another characteristic of statistical learning; that is, a local minimum problem. For example, the paddle moves only in one direction consistently, even though the participant uses various types of instructions. In this case, only one distribution in a mixture of normal distributions was selected to explain all incoming instances. This local minimum problem, which is intrinsic to most methods of statistical learning, is nearly inevitable because its principal cause is the initial parameter values randomly assigned at the beginning of the learning process.

Thus, these characteristics of statistical learning can be said to eventually cause a model to behave unnaturally. These characteristics disabled the participants to induce the model's adapted responses, to behave naturally toward this model and, in some cases, to teach instructions.

This model's competence was only tested in the simple "Pong" game environment, so it is not sure whether this model could be applied to a multifunctional everyday interface. A user would need to use a wider variety of instructions for a multifunctional interface than can be used with the current proposed model. Therefore, the following bottlenecks are foreseen.

1. *The model cannot recognize an adverbial or evaluation instruction.* As already described in section 21.3.1, this model was constructed on the basis of an assumption that "an instruction indicates a certain action", so that the model can learn only the meaning of action instructions such as "move up", "move down" and so on. Therefore, when a user used adverbial or evaluation instructions such as "a bit" or "good", this model tried to assign certain action meanings to these non-action instructions. This meaningless learning process would interfere with the model's meaning-acquisition.
2. *The limited number of instructions that the model could recognize.* The number of instructions that the model could recognize is the same as the number of distributions in a mixture of normal distributions in EM algorithms. Therefore, when a teacher used a wider variety of instructions than the number of distributions the model actually had, the model could not recognize the meanings of all instructions the teacher used. Conversely, if the model was equipped with tens of distributions to enable it to recognize a wide variety of instructions, the learning speed would be decreased.

Thus, the current meaning-acquisition model does not have sufficient ability to be applied to an actual multipurpose interface. To resolve the issue of scalability, the following two abilities will be required. One is that a function capable of automatically customizing the number of distributions in a mixture of normal distributions would need to be constructed according to a user's instructional patterns; the other is that an ability to handle adverbial or evaluation instructions should be implemented. To accomplish this, referring to Suzuki's learning method (Suzuki *et al*. 2002) would be worthwhile. Suzuki's method enables an agent to learn the meanings of action and evaluation instructions simultaneously by using a reinforcement learning framework. If this method had been incorporated into the current learning model, the improved model could have acquired meanings of adverbial and evaluation instructions. Moreover, this method has the potential of enabling the model to acquire new reward information; for example, the verbal instruction "good" could became a positive reward, based on the two existing primitive rewards (hitting the ball or receiving an AP).

21.6 Conclusions

When interactive agents such as pet robots or an adaptive speech interface system are required to form a mutual adaptation process with users, these agents should be equipped with the following two competences: the ability to recognize the reward information expressed in human paralanguage information, and the possession of a system for learning about humans by using reward information. The purpose of this study is to clarify the specific contents of the rewards and the actual mechanism of the learning system by observing how two persons could create a smooth communication such as that between a dog and its owner.

Communication experiments to observe the process of how humans could establish communication were conducted, and based on the results of those experiments a meaning-acquisition model was constructed. It was confirmed that this model could recognize the intentions and meanings of users' verbal commands by exploiting the users' adaptations in a meaning-acquisition process. However, this successful learning was observed only when an experimenter gave the participants appropriate instructions equivalent to catchphrases (e.g., "Please start teaching this model as if talking to a person.") that helped users learn how to employ and interact with the model intuitively. As a result, the participants' adapted behaviors such as AP utterance and decreasing the types of instructions were induced not only by the meaning-acquisition model itself but also by the experimenter's instructions. Thus, this result indicated the need for a subsequent study to discover how to induce the participants' adaptations or natural behaviors without giving these kinds of instructions. A method of doing so is required for an effective learning of this meaning-acquisition model.

Progress to date has derived some unresolved issues from this learning mechanism, which include the model's unnatural action and its scalability for future use. If these issues are settled, it will contribute to development of basic technology for achieving auto-customization or personalization of a speech interface, or to development of an interface for a pet robot that can create an intimate mutually adaptive relationship such as that between a real dog and its owner. The authors believe that this will lead to realization of a part of "mundane intelligence" and eventually to creation of "the one and only artifact" for its user, which research direction is quite different from the one that AI has thus far established.

References

Breazeal C. (2003) Towards sociable robots. *Robotics and Autonomous Systems* **42** (3-4), 167–175.

Carroll J. (2000) *Making Use: Scenario-based design of human–computer interactions.* MIT Press.

Dempster A., Laird N. and Rubin D. (1977) Maximum likelihood from incomplete data via the EM Algorithm. *Journal of Royal Statistical Society B* **39**, 1–28.

Ferrnald A., Taeschner T., Dunn J., Papousek M., de Boysson-Bardies B. and Fukui I. (1989) A cross-language study of prosodic modification in mother's and father's speech to preverbal infants. *Journal of Child Language* **16**, 477–501.

Fernald A. and McRoberts G.W. (1996) Prosodic bootstrapping: a critical analysis of the argument and the evidence, In J.L. Morgan and L. Denuth (eds), *Signal to Syntax: Bootstrapping from speech to syntax in early acquisition.* Lawrence Erlbaum Associates, pp. 365–388.

Goto M., Ito H., Akiba T. and Hayamizu S. (2001) Speech completion: new speech interface with on-demand completion assistance. *Proceedings of HCI International 2001.*

Harada T.E. (2002) Effects of agency and social contexts faced to verbal interface system. In *Proceedings of the 19th Japanese Cognitive Science Society Annual Meeting* (in Japanese), pp. 14–15.

Hobbs J.R. (1990) The Pierrehumbert–Hirschberg theory of intonational meaning made simple: comments on Pierrehumbelt and Hirschberg. In P.R. Cohen and M.E. Pollack (eds), *Intentions in Communication.* MIT Press.

Igarashi T. and Hughes J.F. (2001) Voice as sound: using nonverbal voice input for interactive control. In *Proceedings of the 14th Annual Symposium on User Interface Software Technology*, pp. 155–156.

Johnstone T. and Scherer K.R. (2000) Vocal communication of emotion. In M. Lewis and J.M. Haviland-Jones (eds), *Handbook of Emotions*, pp. 220–235.

Karatani K. (1989) *Tankyuu II* (in Japanese). Kodansha.

Katcher A.H. and Beck A.B. (1983) *New Perspectives on Our Lives with Companion Animals.* University of Pennsylvania Press.

Koiso H., Horiuchi Y., Tutiya S., Ichikawa A. and Den Y. (1998) An analysis of turn-taking and backchannels based on prosodic and syntactic features in Japanese Map Task dialogs. *Language and Speech* **41**, 295–321.

Komatsu T., Suzuki K., Ueda K., Hiraki K. and Oka N. (2002) Mutual adaptive meaning acquisition by paralaguage information: experimental analysis of communication establishing process. In *Proceedings of the 24th Annual Conference of the Cognitive Science Society*, pp. 548–553.

Komatsu T., Utsunomiya A., Suzuki K., Ueda K., Hiraki K. and Oka N. (2003) Toward a mutual adaptive interface: an interface induces a user's adaptation and utilizes this induced adaptation and vice versa. In *Proceedings of the 25th Annual Conference of the Cognitive Science Society*, pp. 687–692.

Komatsu T., Ohtsuka S., Ueda K., Komeda T. and Oka N. (2004) A method for estimating whether a user is in a smooth communication with an interactive agent in human–agent interaction. *Lecture Notes in Artificial Intelligence*, vol. 3213, pp. 371–377.

Komatsu T., Utsunomiya A., Suzuki K., Ueda K., Hiraki K. and Oka N. (2005) Experiments toward a mutual adaptive speech interface that adopts the cognitive features humans use for communication and induces and exploits users' adaptations. *International Journal of Human–Computer Interaction* **18** (3), 243–268.

Kubota N. and Hisajima D. (2003) Structured learning of a partner robot based on perceiving-acting cycle. In *Proceedings of the IEEE CIRA2003 Computational Intelligence in Robotics and Automation* (CD-ROM).

Ladd D.R., Silverman K., Tolkmitt K., Bergman G. and Scherer K.R. (1985) Evidence for the independent function of intonation contour type, vocal quality and F0 range in signaling speaker affect. *Journal of the Acoustic Society of America* **78**, 453–444.

Masui T. (1999) POBox: an efficient text input method for handheld and ubiquitous computers. In *Proceedings of the International Symposium on Handheld and Ubiquitous Computing (HUC'99)*, pp. 289–300.

Mitchell T. (1997) *Machine Learning*. McGraw-Hill College.

Nicolescu M. and Mataric M.J. (2002) A hierarchical architecture for behavior-based robots. In *Proceedings of the First International Joint Conference on Autonomous Agents and Multi-Agent Systems*, pp. 227–233.

Pirrehumbert J.B. and Hirschberg J. (1990) The meaning of intonational contours in the interpretation of discourse. In P.R. Cohen and M.E. Pollack (eds), *Intentions in Communication*. MIT Press.

Reeves B. and Nass C. (1996) *The Media Equation*. Cambridge University Press.

Scherer K.R., Bense R., Wallbott H.G. and Goldbeck T. (1991) Vocal cues in emotion encoding and decoding. *Motivation and Emotion* **15**, 123–148.

Sears A. and Shneiderman B. (1994) Split menus: effectively using selection frequency to organize menus. *ACM Transactions on Computer–Human Interaction* **1** (1), 27–51.

Shneiderman B. (1997) *Designing the User Interface*. Pearson Addison Wesley.

Snow C.E. (1977a) Mother's speech research: from input to interaction. In C.E. Snow and C.A. Ferguson (eds), *Talking to Children: Language input and acquisition*. Cambridge University Press.

Snow C.E. (1977b) The development of conversations between mothers and babies. *Journal of Child Language* **4**, 1–22.

Suzuki K., Ueda K. and Hiraki K. (2002) A computational model of an instruction learner: how to learn "good" or "bad" through action. *Cognitive Studies* (in Japanese) **9** (2), 200–212.

Tsukahara W. and Ward N. (2001) Responding to subtle, fleeting changes in user's internal state. In *Proceedings of Computer Human Interaction 2001* (CHI2001), pp. 77–84.

Yamada S. and Kakusho K. (2002) Human–agent interaction as adaptation. *Journal of the Japanese Society for Artificial Intelligence* (in Japanese) **17** (6), 658–664.

Index